TRILOGY OF CONNECTORS

TRILOGY OF CONNECTORS

Basic principles, and connector design explanations

IMPRINT

PUBLISHER
Würth Elektronik eiSos GmbH & Co. KG
Max-Eyth-Str. 1 · 74638 Waldenburg · Germany
Tel. +49 (0) 79 42 945-0
eiSos@we-online.de · www.we-online.com

REALISATION
Andreas Aigner

AUTHORS
Dr. Robert S. Mroczkowski // Romain Jugy // Alexander Gerfer // Thomas Robok

ILLUSTRATIONS // Würth Elektronik
GRAPHIC DESIGN // DIE NECKARPRINZEN · 74074 Heilbronn · Germany
TYPE SETTING // Ideen im Kopf · 75050 Gemmingen · Germany
PRINT // Schweikert Druck · 74182 Obersulm · Germany
PUBLISHING HOUSE // Swiridoff Verlag · 74653 Künzelsau · Germany

EDITION
1st edition November 2010
2nd edition November 2012
3st edition November 2015
4th edition November 2024

ISBN 978-3-89929-201-5

CONTENT

PREFACE

Dear readers,
It's been a while since the „Trilogy of Connectors" received a major update and I am thankful that we were able to realize a complete rework - not only to restructure the chapters, but to add valuable new content to give you more knowledge at hand.

The fourth edition of the "Trilogy of Connectors" still contains the principles about materials and contact physics that are still the foundation on which application-specific knowledge is based, because a connector on close consideration is more complex than at first glance. New topics like Connectors for shielded cables and their EMC consideration and temperature rise and the impact of inrush current are added and SKEDD and press-fit technology are introduced to complete this chapter.

Especially the product-based chapter 2 has grown significantly. Circular connectors are added and explained and by adding SPE (single pair Ethernet) we started adding information about the latest industrial network technology. On many other product types, you will find helpful information about their use and how they work.

Of course, these are just a few highlights and the applications chapter is still here to help you benefit from experiences. Many more topics have been added and existing topics were reworked to fit to the latest state of knowledge.

To create all this content, meet your expectations and hopefully add some surprise for you, many experts from Product Management, our Field Application Engineers, partners and above all our customers were involved and supported us with knowledge, application examples and great ideas.

Thank you for sharing your experience and for taking the time for this project!

I look forward to your comments and hope you enjoy reading the book!

Sincerely,

Thomas Robok
Würth Elektronik eiSos GmbH & Co. KG

THANK YOU!

Special thanks for their constructive assistance go to:
Jean Patrick Penlou, Würth Elektronik, France
Cyril Messageot, Würth Elektronik, France
Cyril Hernandez, Würth Elektronik, France
Julien Canal, Würth Elektronik, France (Packaging Note)
Guillaume Grenier, EKTRO, France (THR & Crimping Note)
Wladimir Jelisarow, Würth Elektronik, Germany
Alain Lafuente, Würth Elektronik, France (Ethernet Note)
Adrien Reynaud, Würth Elektronik, France
Sylvain Grattard, DB Products (Dek Franchise), France
Stéphane Chauvin, DB Products (Dek Franchise), France
Quentin Laidebeur, Würth Elektronik, France
Cabelvar, France
Geraldine Morisson, Würth Elektronik, France
Nicolas Prou, Würth Elektronik, France
Benjamin Bulot, Würth Elektronik, France
Andreas Aigner, Würth Elektronik, Germany
Günther Klenner, K&K Prime Engineering, Germany
Dr. Mathias Bachmayer, Aevum Mechatronik GmbH, Germany
Dr. Anton Gilg, Anton Gilg Elektronik, Germany
Gerhard Stelzer, Würth Elektronik, Germany
Dr.-Ing. Heinz Zenkner, Würth Elektronik, Germany
Stefan Hellwig, Würth Elektronik, Germany

Suggestions and criticisms on this book gratefully received by:
Würth Elektronik eiSos GmbH & Co. KG
Thomas Robok
Email: thomas.robok@we-online.de
Tel. +49 7942 945 0

THE AUTHORS

Thomas Robok, born in 1977, worked for several years as a specialist for audio and media live solutions and installed sound with customer specific solutions. He is with Würth Elektronik since 2010, where he started in sales and developing to technical training and support as FAE. Since the beginning of 2016, he is working for the technical Academy of Würth Elektronik, organizing and developing international training for electromechanical parts.

Dr. Robert S. Mroczkowski, founded connNtext associates in 1998, a firm providing consulting services in connector applications to the electronics industry. Dr. Mroczkowski has more than 30 years experience in the electronics industry. He joined AMP Inc. in 1971. While at AMP, his responsibilities included consulting on connector design, materials, and reliability concerns, and he provided an interface to AMP customers on these issues. In 1990 he joined the AMP Advanced Development Laboratories, where he developed microstrip cable connectors and a new micro coaxial connector for medical ultrasound diagnostic equipment. Dr. Mroczkowski retired in 1998 as an AMP principal. He wrote more than 20 technical papers. He holds seven patents. In 1997, Dr. Mroczkowski received the Lifetime Achievement Award of the International Institute of Connector and Interconnection Technology. He holds a bachelor's, master's, and doctorate of science degrees in physical metallurgy from the Massachusetts Institute of Technology (MIT).

Romain Jugy has worked in the field of connectors for almost 25 years. He started at Radiall where he was coaxial connector product manager. He was then a successful international product manager at HARTING. Since 2005, he has been the person in charge of internationally developing the Würth Elektronik connector range and he is now General Manager of Wurth Elektronik France and International Electromechanic Business Unit Director for the Wurth Elektronik eiSos group worldwide. He has an engineering diploma in Chemistry and Material (ENSCM Montpellier), as well as a Master in Marketing and Company Administration (IAE Montpellier).

Alexander Gerfer, born 1965, worked in the field of research and development for precision measuring instruments following his training as a radio and television technician. This was followed by a degree on electrical engineering at the Technical University of Cologne. While studying, Alexander Gerfer published numerous application circuits and construction guidelines from the field of consumer electronics. After his degree, he worked in electronic component distribution and from 1997 on, he was head of R&D department at Würth Elektronik. Today Alexander Gerfer is Managing Director and Chief Technology Officer of Würth Elektronik eiSos GmbH & Co. KG.

Trilogy of Connectors

Basic Principles

Part 1: Basic Principles of Connectors

1. Basic Principles of Connectors

A list of electrical or electronic devices using connectors would be long and varied both in function and construction. At the beginning of the list, both in function and construction, is an incandescent lamp. At the complex end of the list, both in function and construction, is a collection of computer servers interconnected to provide the immense functionality of a search engine.

Let's take the easiest example, the function of an incandescent lamp is to provide light by passing an electric current through a filament, a technology developed in the late 19th century. While the functionality and technology of the various servers linked together in a search engine is beyond the scope of this handbook, it must be noted that its construction includes a power source and a multitude of electronic devices that must be connected together.

The common link between the two examples is interconnections, interconnections provided, in some manner, by connectors. Every electrical or electronic device includes at least one connector, the interconnection between the power source and a functional device. The discussion will begin with the incandescent lamp system which contains all the basic connector construction and functionality issues in a search engine, but at a more accessible level. The search engine will be discussed later in terms of the various levels of interconnection in electronic devices and equipment.

The example incandescent lamp system shown in Figure 1.1 includes a power source, the house wiring, and a wall receptacle/plug, arguably the most familiar connector, which provides the interconnection between the power source and the lamp. There are two types of connections shown in Figure 1.1.

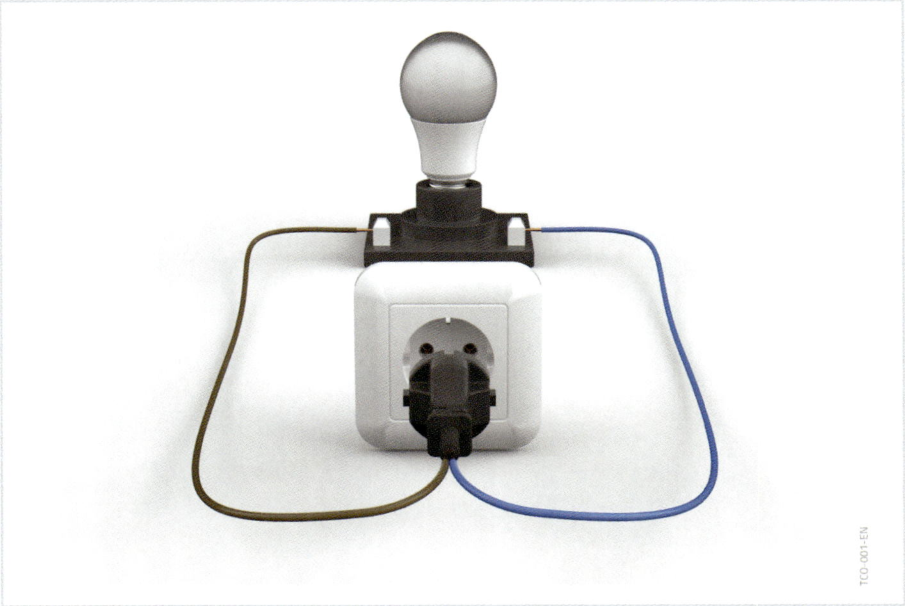

Fig. 1.1: Incandescent light

The power connection consists of two sections the wall receptacle and the wall plug. This connection is a **separable connection**, the wall plug is separable from the wall receptacle to enable the lamp to be moved to different locations. The wall plug has two metal pins that are inserted into the wall receptacle and the wall receptacle contains two metal spring systems that are deflected as the pins are inserted. Two pins/receptacle springs are necessary to provide both input and output power connections. The force created by the deflection of the receptacle spring when a pin and the receptacle spring come in contact creates and maintains the metal-to-metal interface that enables the power, voltage and current, from the power source to be transmitted to the lamp. The properties of this interface will be discussed in detail later in this chapter, but it is worth noting at this point **that creating and maintaining a metal-to-metal interface is the prime goal of connector design**.

In addition to this separable connection, there are two **permanent connections**. The wall receptacle is connected to the house wiring, usually by a mechanical connection. Similarly, the wall plug is connected to the lamp cord wiring, also through a mechanical connection. **These two types of connections, separable and permanent, are characteristics of connectors.**

1.1 Connector Function

With this basic introduction to connectors in hand, consider now a more detailed introduction to connectors and the components of a connector. A functional definition of a connector is:

A connector is an **electromechanical** system which provides a **separable connection** between two subsystems of an electronic system without an **unacceptable effect** on the performance of the device.

The bolded terms in this definition are the most significant characteristics of a connector.

As previously mentioned, deflection of mechanical springs creates a force between the two halves of the connector on mating which creates areas of metal-to-metal contact at the mating interface. The metal-to-metal interface provides the connection where the current flow takes place.

A separable connection, separability, is the primary reason for using a connector. Separability may be required or desirable for a variety of reasons. Manufacturability considerations include allowing for independent manufacturing of subassemblies or subsystems with final assembly taking place at a central location. As the complexity and functionality of electronic systems continues to increase this manufacturing flexibility is increasingly important. One example is the ability to "custom build" a personal computer to your own specifications. A second example is to allow for maintenance or upgrading of components or subsystems as increased functionality becomes necessary. Finally, portability and the ability to support an increasing range of peripherals in a laptop computer require separable connections for multiple use of an input port or usage at multiple locations. The number of **mating cycles**, mate and unmate, needed depends on the reason separability is required. Manufacturability reasons generally require only a few mating cycles while portability and multiple port usage may require several hundred mating cycles in a laptop computer.

While these separability capabilities are virtues, separability, by definition, introduces an additional interface into the system. This interface must not introduce any unacceptable electrical effects, on the system performance. The most significant potentially negative effect relates to the resistance across the mating interface: both the magnitude and stability of the resistance. Resistance increases are of particular importance in power transmission for two reasons. First, the Joule, or I^2R, heating as current, I, flows increase as the electrical resistance, R, increases. Joule heating increases the temperature of the contacts which, in turn increases the degradation rate of the interface, up to and including, melting or open circuiting of the interface. Second, large increases in resistance can impact the signal transmission characteristics in RF or digital applications. Thus, the magnitude and stability of the resistance across a mating interface is a major consideration in connector performance and reliability. Interface resistance will be discussed in detail later in this chapter.

1.2 Connector Structure

Consider now a structural definition of a connector. There are four basic components in a connector. They are:

A. the connector housing
B. the contact springs
C. the contact finish
D. the contact interface

This chapter will provide a basic introduction to connector housings, contact springs and contact finishes with details to be provided in following chapters. The contact interface will be discussed in detail in this chapter.

A few general comments are appropriate at this point. All connectors consist of two mating halves, a plug half which is inserted into the receptacle half. This plug/receptacle structure exists in both the housings and the contact springs. With respect to the contact springs, as mentioned earlier, in most connectors the plug contact deflects a spring system in the receptacle contact. This spring deflection produces the contact force which creates and maintains the desired metal-to-metal contact interface.

Consider these components as realized in a wall plug/receptacle connector as shown in Figure 1.2. The housing of the plug connector is overmolded onto the straight copper alloy plug contacts which are crimped onto the lamp cord wire. The housing of the receptacle connector is assembled around the copper alloy receptacle contacts which are pressure connected to the house wiring. When the connector halves are mated, the receptacle contact spring is deflected by the plug contact creating the contact interface. In this example, there is no contact finish, no plating or coating, on the contacts. The functions of a contact finish will be discussed in a following section.

Fig. 1.2: Components of a wall plug/receptacle connector

A cross section of a more typical connector is shown in Figure 1.3 with all components illustrated. Each of the connector components will be discussed briefly in this chapter and in detail in the following chapters.

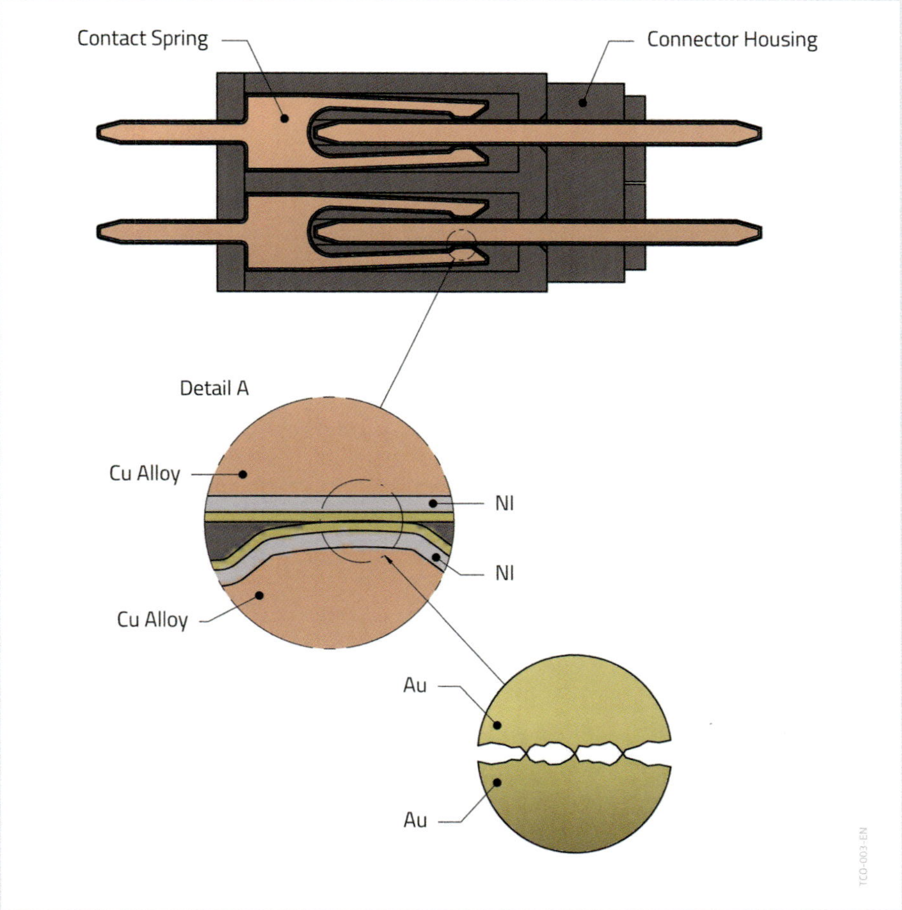

Fig. 1.3: Schematic connector cross section

1.3 Polymer Materials and Processing

The majority of connector housings are manufactured from thermoplastic polymers due to their combination of electrical and mechanical properties and the manufacturing efficiencies of injection molding. This section will provide an overview of polymer materials and properties and a brief introduction to injection molding.

1.3.1 Polymer Materials

The word "polymer" describes the structure of polymers in that its meaning is "many mers" and the structure of a polymer is the repetition of a characteristic unit, the mer, many times, thousands of times. The properties of the polymer depend on the type of mer and the number of repetitions. The mers commonly used in engineering polymers are generally compounds among carbon and hydrogen, nitrogen, sulphur, chlorine and fluor.

The simplest carbon-based compound is methane, CH_4, a gas. The structure of methane is tetrahedral as shown in Figure 1.4 In contrast to metals where the outer electrons can be considered as free , (more about in Chapter 1.5.2.1), the four outer electrons of carbon form shared directional bonds with four hydrogen atoms, each of which have one electron, to form methane gas. The electrons are shared between the carbon and hydrogen and are called covalent bonds. Covalent bonds are quite stable and methane is a stable molecule.

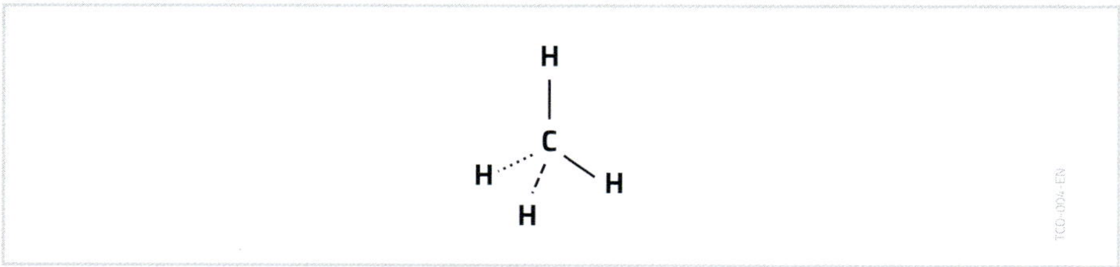

Fig. 1.4: Methane tetrahedral structure and bonds

There are two variants of a carbon-hydrogen compound containing two carbons, ethane and ethylene. Ethane bonds use two electrons, the same as methane (Figure 1.5a). Ethylene used two electrons between carbon and hydrogen, but four electrons between the carbons, a so-called double bond (Figure 1.5b).

Ethylene is the simplest mer that can be polymerized and is the source of a family of polymers, polyethylene. The polymerization reaction uses heat, pressure and catalysis to break the double bonds so that each carbon atom has a "dangling bond" (Figure 1.5c), which can, in turn combine with a "dangling bond" of an adjacent ethylene mer producing a chain of two ethylene molecules (Figure 1.5d). This process is called addition polymerization, mers are simply added on at each end of a polymer chain. A second polymerization process, condensation polymerization, will be discussed in a following section.

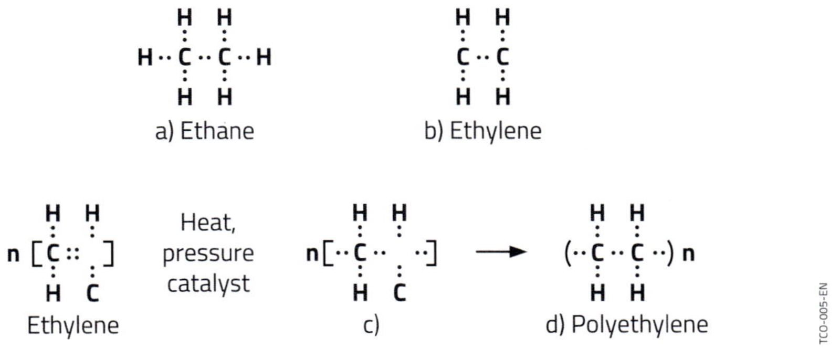

Fig. 1.5: Polymerization

As noted, the properties of the polymer vary with the number of mers. The number of mers is generally characterized in terms of the molecular weight of the polymer. For example, the molecular weight of the ethylene mer is

$$2\,(C_{12})+4\,(H_1)=28 \qquad \text{(EQ 1)}$$

C_{12}: Carbon – Molecular weight 12,000 g.mol^{-1}
H_1: Hydrogen – Molecular weight 1,008 g.mol^{-1}

As will be discussed later, polyethylene goes from a gas to a liquid to a grease to a wax, to a microcrystalline wax to a plastic as the molecular weight, the number of mers in the polyethylene chain increases. High Density High Molecular Weight (HDHMW) polyethylene has a molecular weight in the millions of ethylene mers. The structure and properties of a polymer chain depend on factors in addition to molecular weight.

Polymer Microstructure and Types

Consider now some other aspects of polymer bonding and how it affects the structure and properties of polymer chains. Two of the major aspects are bond angles and side groups. Figure 1.6 shows a schematic polyethylene chain. The carbon atoms are shown as black circles and are the backbone of the polymer chain. The hydrogen atoms, in grey, are the side group in polyethylene. There is a characteristic covalent bond angle, 110 degrees, between the carbon atoms in the backbone of a polyethylene chain. That angle depends on the bond energy between carbon and the side group and the size of the side group. For example, other side groups in an ethylene type structure are methane, chlorine and fluorine which result in the polymers polypropylene, polyvinylchloride (PVC) and polytetrafluoroethylene (PTFE) respectively. Figure 1.7 shows the mers of these variants. Naturally the bond strengths, therefore the bond lengths, and bond angles of these polymers would be different from one another. Thus, the interactions of the chains in the bulk polymer would be different resulting in the different properties.

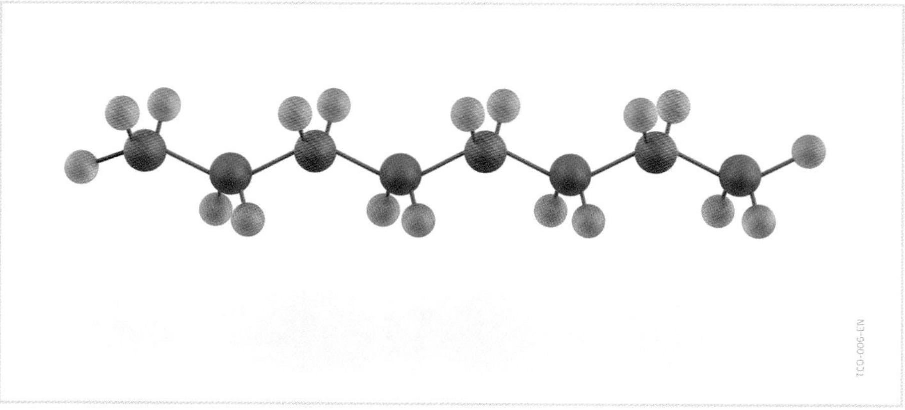

Fig. 1.6: Polyethylene chain

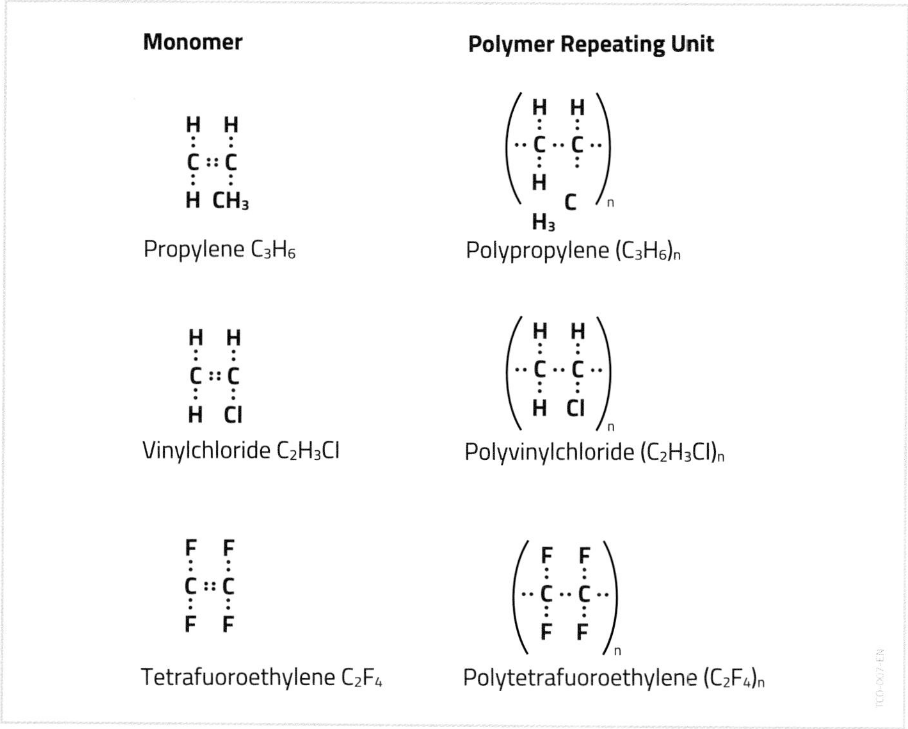

Fig. 1.7: Mer variants

As the length of the polymer chains increases, the directionality of the bonds tends to cause the chain to twist. The degree and form of twisting depends on the side groups on the chain and how they attach to the chain, randomly or at preferred sites, which would, in turn affect how the chains pack together. The reactivity of the side groups can also cause branching, an offshoot chain, or cross linking between chains resulting in three chain configurations, linear, branched and crosslinked as schematically illustrated in Figure 1.8.

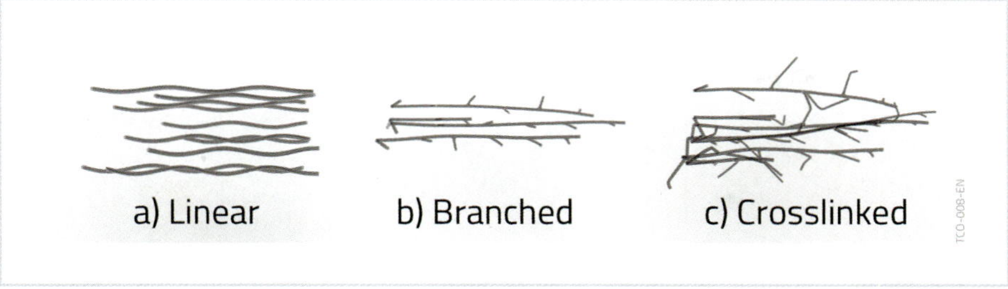

Fig. 1.8: Chain configurations

These differences in chain interactions result in two fundamentally different micro-structures, amorphous and crystalline polymers and two fundamentally different thermal characteristics, thermosetting and thermoplastic polymers. The former significantly affects the mechanical properties of a polymer and the latter the processing characteristics. Most of the engineering polymers used in connectors are crystalline or semi-crystalline thermoplastic materials: crystalline for the properties and thermoplastic for processing efficiency.

Polymer Structure and Mechanical Properties

The following discussion will be rather qualitative than quantitative to highlight the mechanisms of polymer deformation and how they vary with the polymer molecular weight and crystallinity. As noted previously, as the molecular weight of a polymer increases the length of the individual polymer chains increases which leads to two significant mechanisms of polymer strength, one physical and one chemical. First, as the length of the chains increases, they are more likely to become entangled with one another. A common analogy being based on spaghetti; cut spaghetti reacts as individual strands while full length spaghetti reacts as a mass. Second, the increased chain length causes an increase in electrostatic bonding between the chains. Suffice it to say that there are several variations of dipolar bonding that can be active between polymer chains. Though the individual bond energies are small, the number of bonds between chains can be very large. In addition the electrostatic bonding tends to align the chains enhancing the opportunity for further bonding. This tendency for alignment is the source of crystallinity in polymers. Figure 1.9a illustrates the random entanglements of an amorphous polymer, Figure 1.9b an intermediate stage of crystallinity and Figure 1.9c a highly crystalline polymer. The crystallinity of a polymer can be controlled by composition and processing. Note in Figure 1.9b that there are chains that are included in more than one crystallite cluster. These so called "tie molecules" in combination with the crystallites are an additional "entanglement" mechanism that is more stable than the physical entanglement previously described.

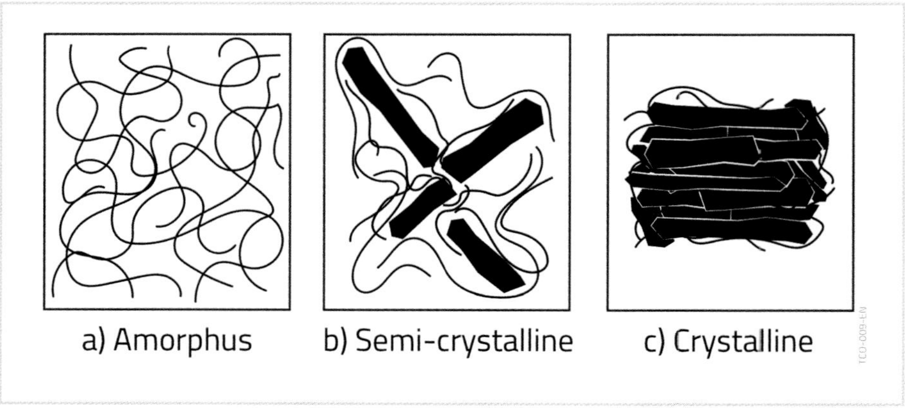

Fig. 1.9: Amorphous and crystalline polymer structures

At low molecular weights, very short chains, the chains are free to slide over each other under an applied load. In the liquid and grease stages the chains flow freely and disentangle from each other. Eventually the increase in chain length promotes interchain bonding so that not all entanglements are disrupted, and the polymer becomes a wax. Increasing dipole bonding promotes increasing crystallinity leading to microcrystalline waxes and, eventually to rigid polymers, plastics. It should be noted that the dipole bond strength and frequency on the chain depend on the side groups in the chain so the tendency towards crystallinity is dependent on the side groups, the mers of the polymer. Crystalline polymers used in connectors include, polyamides (nylon), polyesters, and polyphenylenesulfide. Comments on these polymers will be provided later.

There is another class of polymers important to connector technology, **Liquid Crystal Polymers (LCP)**. As the name indicates, LCP are crystalline in the liquid state consisting of rod like structures that readily self-align into crystallites. At the temperatures where the LCP is liquid, thermal agitation limits the degree of crystallinity that is realized, but as the temperature decreases during the molding process a high degree of crystallinity is achieved.

The mechanical properties of many engineering polymers are further improved by the addition of glass fibres. The strengthening effect of glass fibres arise from chain bonding to the fibres. For example, addition of **glass fibres** to the polyester PBT (polybutylene terephthalate) can double the tensile strength and increase the flexural modulus (analogue to elastic modulus in bending) by 250 %.

A comparison of the mechanical properties of metals and polymers may be useful at this point. Metal strength data is generally provided via tensile stress. Polymer strength data may be obtained from either tensile or bending stress tests. The table in Figure 1.38 contains tensile data for copper30zinc brass in a temper commonly used in connectors and bending data for a commonly used PBT polyester polymer. The polymer data includes neat (unfilled) and 15 and 30 % glass reinforced grades. Elastic modulus corresponds to flexural modulus and yield strength corresponds to flexural strength in the table.

Material	Elastic Modulus Flexural -Modulus GPa (ISO)/Mpsi (ASTM)	Yield Strength (0.2 % offset) Flexural Strength MPa (ISO)/Kpsl (ASTM)
Brass (Cu70/Zn30)	110/16	495/70
PBT	2.5/0.37	80/12.5
PBT – 15 % Glass	5.2/0.70	150/21.0
PBT – 30 % Glass	9.7/1.2	210/28.0

Tab. 1.1: Tensile date for Cu30Zn Brass

There are several significant differences between the mechanical properties of metal and polymers in Table. 1.1: First, metals are much stiffer and stronger than polymers. For reference, the elastic moduli of copper alloys used in connectors fall in the range of 110 to 140 GPa (16 to 20 Mpsi) while for typical connector engineering polymers the corresponding range is 1 to 19 GPa (0.3 to 2.8 Mpsi); the corresponding comparison for tensile (flexural) strength are 230 to 760 MPa (35 to 110 Kpsi) for metals and 40 to 185 MPa (6 to 27 Kpsi) for polymers. Second, the effect of reinforcing glass fibres on polymer mechanical strength is significant. In the example polymer the flexural modulus and flexural strength are doubled with 15 % glass, and nearly tripled with 30 % glass. Similar improvements occur with other polymers. Third, it merits noting that the higher values of flexural modulus and strength are for Liquid Crystal Polymers and that LCP do not benefit from reinforcing fibres. A final, and significant difference between metals and polymers is their temperature and strain rate sensitivity. The mechanical properties of metals are essentially constant with respect to temperature and strain rate. Polymers, in contrast are sensitive to both. The mechanical properties of polymers are reduced as a function of increasing temperature and decreasing strain rate. A simplistic, but basically correct, explanation of these sensitivities revolves around the movement of polymer chains as a function of temperature and load. Increasing temperature increases the amount of thermal motion of the chains and facilitates their untangling and sliding over each other, thus reducing the strength of the material. Slow strain rates have a similar effect, the chains have more time to untangle and slide as the load slowly increases, viscoelasticity. Glass reinforced polymers are less sensitive to these temperature and strain rate effects due to the polymer bonding to the glass fibres restricting the chain motion. Metals are not immune to thermal effects as noted earlier with respect to stress relaxation and its effect on contact force. Polymers, however, show much larger effects, material flow, creep, rather than stress relaxation which is a localized atomic level rearrangement.

1.3.2 Polymer Processing

There are two processes used to convert polymer resin pellets to a connector housing, compression/transfer molding and injection molding. Before discussing those processes, however, another method of polymerizing, the starting resin must be mentioned for completeness, condensation polymerization.

Recall that addition polymerization occurs by the addition of individual mers to existing polymer chains under the influence of temperature, pressure, and catalysis. Some typical addition polymers include polyethylene, polypropylene, polyvinyl chloride (PVC) and polytetrafluoroethylene (PTFE).

Condensation polymers are so called because the bond breaking process that leads to polymerization occurs through a chemical reaction between different mers, an organic acid and an alcohol, esterification, with the production of by-products, condensates. The most common condensates are water and methyl alcohol, CH4O. Polymers produced by the esterification reaction are called polyesters. Nylons, polyamides, are also produced by condensation polymerization.

Compression/Transfer Molding

Compression and transfer molding processes are used primarily for molding of cross linking, thermosetting, polymers. The processes are similar, differing in how the polymer resin is presented to the mold. In compression molding a fixed amount of material is placed into the mold which is then pressurized and heated to the molding temperature. In transfer molding the material is transferred from a transfer pot into the mold at controlled temperature and pressure. Molding temperatures are high and time in the mold is long in order to optimize the cross linking of the polymers with a significant cost impact. Any molding scrap, from sprues and runners in the mold to underfilled parts is scrap.

Injection Molding

Injection molding is similar to transfer molding in that the resin pellets are held in a hopper and transferred, after heating/mixing/plasticating, by injection under pressure into a mold. While some thermosets are injection molded, the vast majority of connector housings are thermoplastic polymers.

There are two major components of an injection molding system, the injection system, and the clamping system, shown in Figure 1.10. The injection system includes:

Hopper: The hopper holds the polymer resin pellets prior to transfer into the plastication section. The hopper can be refilled as needed because the pellets are not held under controlled temperature.

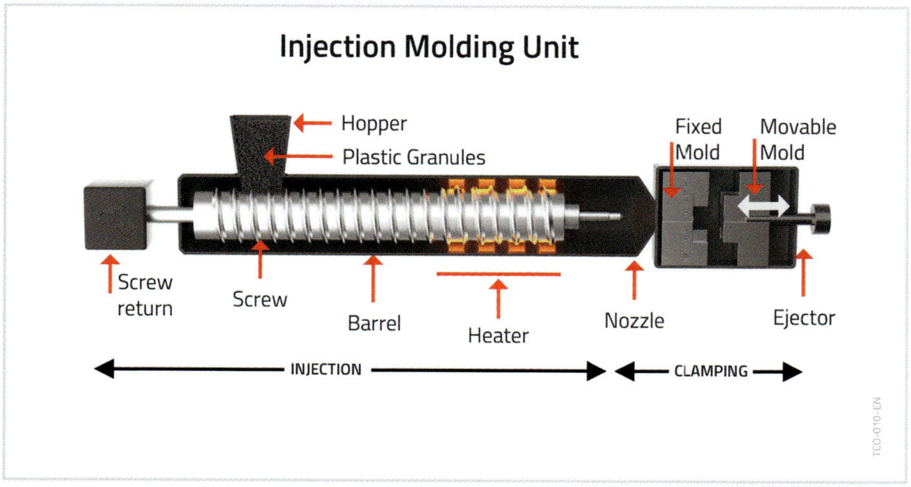

Fig. 1.10: Schematic injection molding system

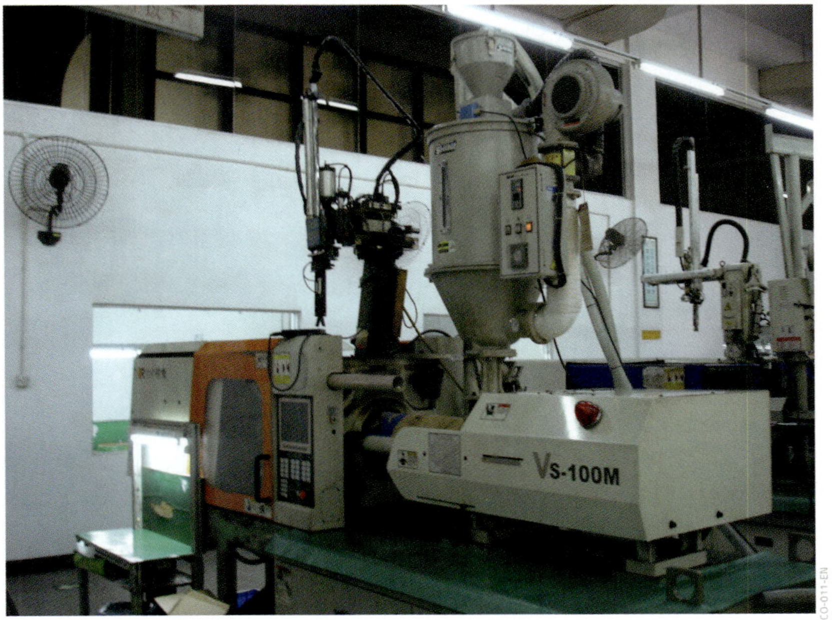

Fig. 1.11: Molding machine

Plastication: Plastication includes two steps, heating of the resin and mixing of the heated pellets into a homogenous polymer melt at a controlled temperature. Heat is supplied by heater bands around the injection barrel. A reciprocating screw within the injection barrel mixes the heated polymer until it is ready for injection into the mold. The reciprocating screw then moves forward injecting the polymer under controlled pressure through a nozzle into the gating system of the mold. The pressure is maintained during the cooling cycle of the process. The nozzle then retracts and the hopper fills the plastication section with polymer for the next molding shot.

The molding system includes

Clamping System: The clamping system contains and supports the mold. It consists of a fixed front plate which engages the nozzle during injection of the polymer and a movable back platen which holds the mold closed under a clamping pressure dependent on the size of the mold and polymer being molded and moves backward on completion of the cooling time to open the mold and eject the molded parts.

Mold: The mold may be a single or multi-cavity mold depending on the economics of the part being molded. A multi-cavity mold will contain sprues and runners to interconnect the individual cavities. Mold design of multi-cavity molds can be complex in order to balance the polymer flow to the individual cavities. For thermoplastic polymers, the sprue and runner parts may be salvaged and reused depending on the requirements on the part. An example of a multi-cavity mold is provided in Figure 1.12.

Fig. 1.12: Plastic injection mold

As noted with the processing of metals from casting to strip, the principles of injection molding are straight forward, but many design challenges must be met to ensure consistent high volume manufactured parts.

1.3.3 Polymer Families

The number of individual engineering polymer products has exploded in the past years due to specialized requirements on the polymer being addressed by a variety of additives. Additives and the function they provide include:

Flame Retardants: Most connector housings are required to meet flammability requirements, burning rates and combustion products, defined by various product safety organizations. Some polymers are intrinsically flame retardant, but the majority require additives to provide that function (most of them are halogen).

Reinforcing Agents: Reinforcing agents are intended to enhance the mechanical properties of polymers. As noted previously, glass fibres are the most common reinforcing agents. The enhancement is dependent on the percentage of fibre, the length of the fibres, the bonding between the polymer and the fibres and the processing.

Stabilizing Agents: Stabilizing agents are intended to reduce the sensitivity of a polymer to moisture, oxidation, Ultra-violet radiation etc.

Colorants/Plasticizers: Colorants, of course, are used to add colour to the polymer. Plasticizers are additives intended to improve polymer processibility.

Given the variety of products available in these families the discussion will be qualitative. Refer to the supplier literature for details on specific products in each family.

Polyamides: Polyamides are generally known as "nylon". There are several varieties of nylon use in connectors depending on the component mers used for polymerization. The generic designations are 66 and 46 with 66 being the oldest and most commonly used. Nylons are semicrystalline polymers and have a number of desirable characteristics and one unfortunate negative.

66 nylon once dominated the connector market as a housing material due to its combination of strength, toughness, abrasion resistance and chemical stability. The negative characteristic of nylons is high moisture absorption, with 66 nylon being the most susceptible. High moisture absorption results in growth in dimensions of the part, and subsequent shrinkage if the part later dries out. This compromises the dimensional stability and limits the size of housings that can be made with 66 nylon.

Moisture absorption also compromises the mechanical and electrical properties of nylons. For 66 nylon the flexural modulus and flexural strength both decrease by more than 50% from the Dry as Molded to samples conditioned at 100% Relative Humidity. The volume resistance also degrades from 10^{15} to 10^{9} Ω/cm, a significant decline, but still a very high resistance. Other nylons are less susceptible to this problem. Nylons are still widely used in automotive, commercial, white goods and industrial markets.

Polyesters: Polyesters, too, are semicrystalline polymers with a good balance of electrical and chemical properties and good mechanical properties with glass reinforcement. There are three polyesters used in connectors, PET (Polyethylene Terephthalate), PBT (Polybutylene Terephthalate) and PCT (PolyCyclohexylenedimethylene Terephthalate). The different mers used for each polyester affect the polymer chain stiffness which affect the mechanical properties directly, but also affect the crystallization of the polymers which influences processing parameters.

Apart from Surface Mount Technology (SMT) soldering applications where the temperature capability of PBT is not adequate, it is the most widely used of the polyesters. PBT has excellent chemical resistance, very good electrical properties and moderate mechanical properties. It is also sensitive to moisture during processing.

PET is similar in properties to PBT but has better mechanical properties and temperature capabilities and could be used in some SMT applications (before lead free issue). It is also sensitive to moisture during processing.

PCT has better short-term temperature capabilities than PBT and PET and was suitable for SMT applications before lead free but not so much used any more.

Polyphenylenesulfide (PPS): PPS is a crystalline polymer and has high temperature capabilities for SMT and continuous use applications. The crystallinity of PPS, and, therefore, the properties, are process sensitive. Most PPS polymers are glass reinforced for both mechanical properties and to reduce brittleness and flash for enhanced molding performance. PPS is inherently flame retardant and has outstanding chemical resistance. The properties of PPS are stable over a large temperature range.

Liquid Crystal Polymers (LCP): LCP, as noted previously, is crystalline even in the molten state due to its rod like structure. This impacts several processing and performance characteristics of LCP. The high crystallinity of LCP results in a good balance of flexural modulus and strength over a range of temperatures. The high temperature stability of LCP makes it suitable for SMT and high continuous use applications. However, the intrinsic crystallinity of LCP during mold flow can produce highly anisotropic properties due to the rod alignment in the flow direction. The anisotropy in crystallinity creates a tendency towards part warpage and results in weak "weld lines", the interface where two fronts of injected polymer meet in the mold. This tendency is countered by the use of glass reinforcement and other fillers to reduce the anisotropy. LCP is inherently flame retardant and has outstanding chemical resistance. Furthermore, LCP has outstanding mold flow allowing the molding of thin sections and excellent dimensional stability an important characteristic for molding of miniaturized connector housings.

1.4 Connector Housing

The connector housing, A in Figure 1.3, is the most obvious connector component and has four major functions:

Electrical insulation: All of the individual contacts within the connector housing must be electrically insulated from one another in order to perform their designated function. Connector pin counts of more than 1000 have been realized. In some cases, power distribution for example, multiple contacts may be interconnected, usually on the power and/or ground plane of a PCB. As noted in chapter 1.3 "Polymer Materials and Processing" the polymer materials used for connector housings have very high volume and surface resistivities. Volume resistivity is an intrinsic property and typical engineering polymer values are in the range of 10^{16} Ωcm. Surface resistivities are dependent on surface contamination and degradation. Two other electrical parameters related to these resistivities merit mention, the dielectric strength and Dielectric Withstanding Voltage (DWV). The dielectric strength quantifies the ability of a polymer to withstand an applied voltage without electrical breakdown through the thickness of the polymer, in volts/cm and correlates with volume resistivity. DWV specifies the voltage that can be applied across the surface of the polymer without creating surface conduction paths over the polymer and is a measure of surface degradation. The stability of these resistances with respect to time, temperature and environmental degradation must be considered during material selection. The electrical insulation characteristics of a connector housing are dependent on both the design and material of the housing as will be discussed.

Dimensional stability: The two halves of the housing must mate together and, therefore, their mating features must be of corresponding shapes and on the same centerlines. The dimensional control required, obviously, depends on the size and the centerline spacing, of the housing. Centerline spacings can vary from 2.54 mm (0.100 in) to 0.4 mm (0.016 in.) with corresponding decreases in dimensional tolerance requirements. Connector housings greater than 30 cm, (12 inches) have been molded successfully. The ability to realize these levels of dimensional control, both during molding of the housings and in field applications is an important consideration in housing design and material selection.

Mechanical support: In many connector designs the contacts are "latched" into the housing for mechanical support as well as dimensional control. Because spring deflections are responsible for creating the contact force, the mechanical stability of the support provided by the connector housing can have a significant impact on the spring characteristics of such connectors. For example, polymer creep at a retention groove in a housing may result in a change in the contact spring deflection, with an accompanying decrease in the contact normal force. Once again, this requirement impacts on the design and materials choices for the housing.

Environmental shielding: The effectiveness of the environmental shielding provided by a connector depends entirely on the design of the housing. The basic guideline is to keep the openings in the housing to a minimum to restrict the access of the environment to the separable connection, in all cases, and to the permanent connections depending on the design. There is one exception to this general guideline. In the case of power connectors, air flow channels through the housing may be beneficial in reducing the temperature rise of the connector due to Joule, I^2R, heating.

Most contact springs are made from copper alloys which are susceptible to corrosion reactions with oxygen, sulphur and chlorine, all of which are present in typical connector applications. Details of these processes will be provided in chapter 1.7 "Contact Finishes". At this point, it suffices to say that the design of the connector housing can have a significant impact on the environmental stability of a connector by shielding the contacts and contact interfaces from the environment.

Figure 1.13 illustrates this shielding effect. The coupon shown is a copper plate, it was exposed to a sulphur bearing environment while mated with the ZIF connector shown. Note the severe corrosion of the exposed portion of the coupon as compared to the shiny surface of the portion contained within the housing. Note also the wiping marks made by the receptacle contact springs as the coupon is inserted into the connector. This illustrates the cleaning effect that occurs during mating to wipe away surface films and contaminants.

Fig. 1.13: Copper coupon

More important, however, is the data shown in Figure 1.14. This data will be discussed in detail in chapter 1.7 "Contact Finishes". At this point it is sufficient to note that the data showing significant increases in contact resistance was from connectors which were exposed to the test environment **unmated**. That is, the plug and receptacle halves were exposed separately, while the stable resistance data was taken from **mated** connectors. The effectiveness of housing environmental shielding is clear. The shielding effect of connector housings provides an important contribution to the ability of connectors to remain functional in harsh environmental applications.

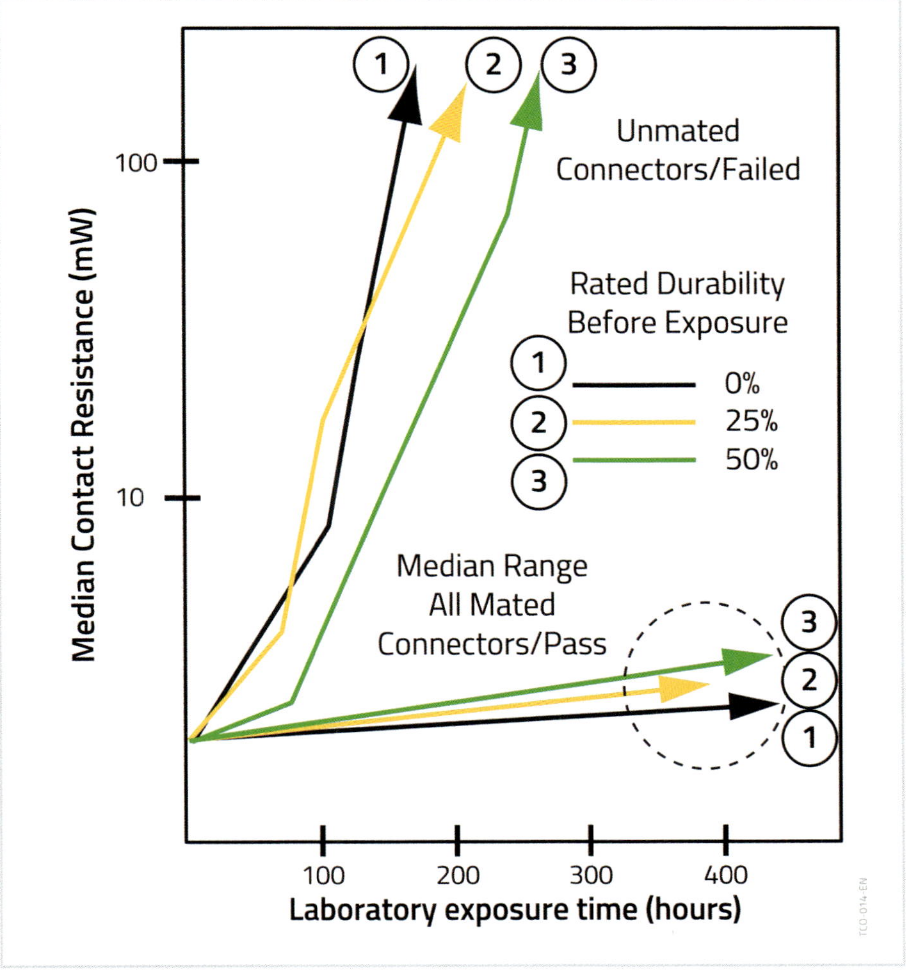

Fig. 1.14: Resistance data

Connector housings vary in pin count from one to over a thousand, and in complexity from simple headers, Figure 1.15, to housings incorporating channels for air flow, Figure 1.16.

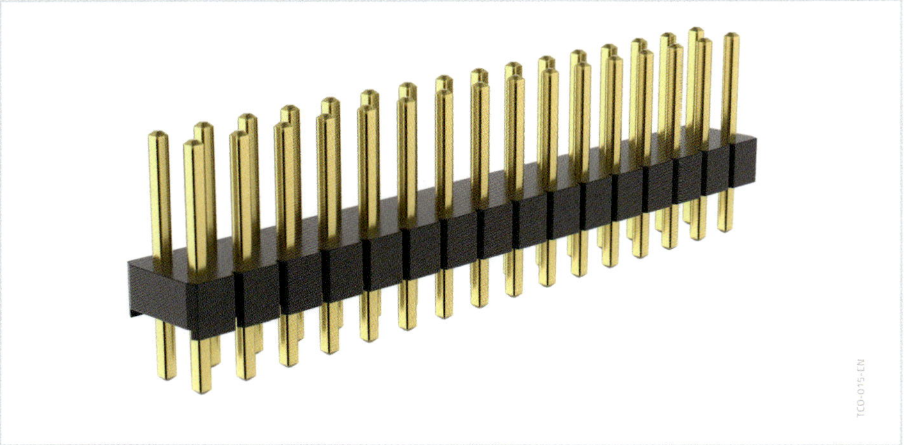

Fig. 1.15: Pin Header

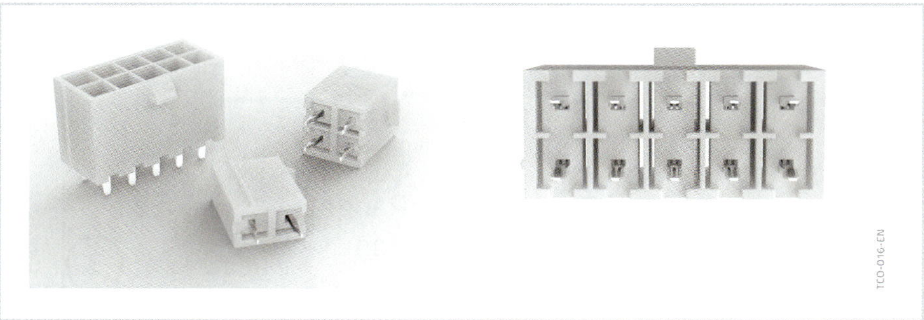

Fig. 1.16: Mini Power Connector top view (gap for air flow area)

1.4.1 Material Selection

In some cases, the selection of an appropriate polymer for a connector housing is determined by a particular advantage or limitation of a polymer. For example, Liquid Crystal Polymers (LCP) have high temperature capabilities that make them suitable for SMT applications. Similarly, the dimensional growth of nylon 66 housings due to moisture absorption limits the dimensions of nylon connector housings. More generally, there are three considerations in the evaluation of whether a particular polymer is a suitable choice for a given connector: the connector function (signal or power), housing manufacturing (connector dimensions and tolerances) and application requirements (SMT).

Signal/Power: There are two categories of signal applications, low current, less than an amp to a few amps, and high speed or high data rate applications, where the impedance of a connector becomes a concern. For the first category, polymer issues are primarily in molding miniaturized connector housings. These concerns will be addressed in chapter 1.4.2 "Housing Design". For high-speed applications the dielectric constant and dissipation factors of the polymer and their stability with respect to temperature and environmental degradation merit consideration. It is the differences in stability that distinguishes among polymers rather than their intrinsic properties, which are generally similar. It should be noted that the impedance of a connector is strongly dependent on connector dimensions as well.

In power connectors, surface resistivities can be important due to the potential for voltage breakdown and surface leakage currents. Temperature stability may also be a consideration due to coupling of the Joule heating of the contacts into the housing and the associated increase in housing temperature. As noted in chapter 1.3 "Polymer Materials and Processing" the mechanical properties of polymers are temperature sensitive.

Manufacturing: Both property variations with molding and dimensional tolerances are of interest. As just noted, polymer variances in dielectric constant and dissipation factor may affect high frequency performance and the molding process is a significant factor in controlling such variances. Mold design also influences the knit line configuration in the molded housing and weakness in knit lines can be a source of weakness and "brittleness" in the housing. Knit line strength varies with the polymer with LCP being a particular concern because the rod like structure of LCP polymers does not cross the interface between the flow fronts as they meet in the mold.

Dimensional tolerances come into play at both ends of connector housing size. Large connector dimensions, particularly long and narrow housings are susceptible to warp or bow due to polymer anisotropies created by mold fill. Crystalline and glass reinforced polymers tend to orient along direction of polymer flow during molding. Thicker sections, connector outside walls, tend to be more crystalline than the thinner section walls separating connector cavities and shrink more than the thin sections creating the potential for warping and bowing.

Miniaturized connectors, with small centerlines and thin walls, require good polymer flow during the molding process. High flow polymers, however, are more susceptible to flash, extrusion of polymer through seams in the mold, which is undesirable.

It must be noted that control of the injection molding process is critical to minimizing these polymer and design issues. Melt and mold temperatures, injection and holding pressures, hold-time and cooling profiles, among others, affect crystallization rate and percentage, shrinkage, bow and warp of connector housings.

Application Considerations include component assembly application thermal and chemical environments and application current and voltage requirements. The thermal profiles required for SMT soldering profiles are arguably the most significant assembly issue. The requirements have become even more demanding with the drive towards lead free solders which require higher soldering temperatures. LCP dominate the SMT market. The potential for degradation of polymer surface electrical properties is the major concern with respect to the chemical environment of the application. Moisture must be considered among the chemicals of concern for some polymers, e.g. nylons. The Continuous Use Temperature (CUT) of an application is also significant because, as noted, the mechanical properties of a polymer are sensitive to temperature as are the rates of chemical reaction. When the contact force system of a connector is supported by the polymer, either directly by a pre-load system, or indirectly, by a retention lance to maintain contact position, the temperature dependence of polymer creep can become a significant concern. With respect to application voltage and current requirements, the concerns are different in the two cases. Voltage issues are primarily safety related and impact on connector design as will be discussed in the next section. Current issues are performance related, in particular the effect of Joule heating on the connector temperature and the effect of temperature rise on polymer properties as noted previously. The effects of Joule heating will be discussed in chapter 2.1 "Level of Interconnection".

1.4.2 Housing Design

Housing design considerations are different for signal, low current and voltage, and power, high current and voltage, connectors. For signal connectors, as noted previously, the major issue is connector miniaturization and its impact on mold design and molding practices. For power connectors there are safety and performance related housing design issues.

There is an additional signal connector issue to be addressed with respect to IDC connectors, strain relief features. As noted in chapter 1.9 "Contact Springs". IDC connectors can benefit from strain relief features to diminish the potential for stresses parallel to the axis of the IDC slot disturbing the IDC connection. Some of these features are incorporated into the housing directly while the use of external covers is also an option. Figure 1.17 shows a strain relief cover for an IDC connector.

Fig. 1.17: IDC connector strain relief

There are two safety related housing design features included in power connectors. Creepage and clearance distances are defined in standards issued by organizations such as VDE and UL. Figure 1.188 illustrates these distances. Clearance is the distance between two contacts in air and creepage is the distance from contact to contact along the surface of the housing. Creepage distance is directed to ensuring that dielectric breakdown of air, arcing, does not occur between contacts. Creepage addresses the potential for electrical degradation of the housing surface resistance of the polymer. A second approach towards the same end is to put the contacts in recesses or silos in the housing. An example of contacts in a recesses and silos is provided in Figure 1.19. One purpose of these features is to "finger proof" the contacts, that is to prevent a finger from shorting across a live contact pair resulting in an electric shock. Contacts within a recess or silo do not have a clearance issue, but creepage distance is still a requirement.

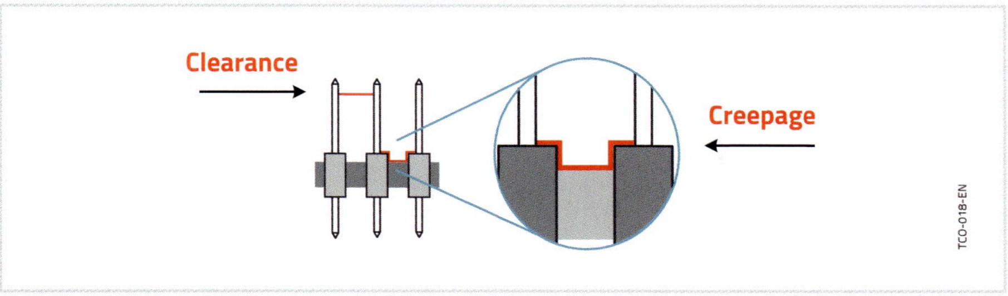

Fig. 1.18: Creepage and clearance distances

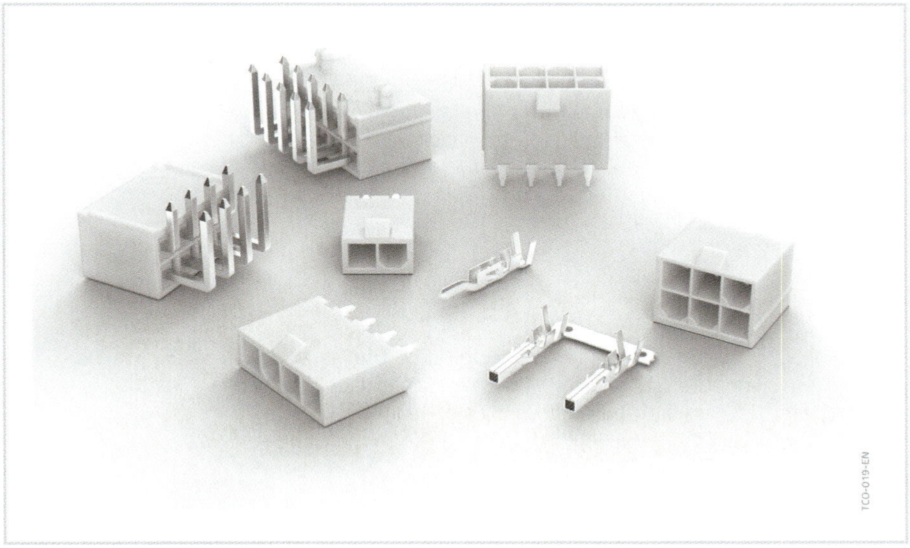

Fig. 1.19: Silos and recess in WR-MPC4 family

An additional power connector housing design feature is to provide passages for air flow through the housing for cooling of the contacts to minimize the effects of Joule heating on the contact temperature. An example of such features is shown in Figure 1.20.

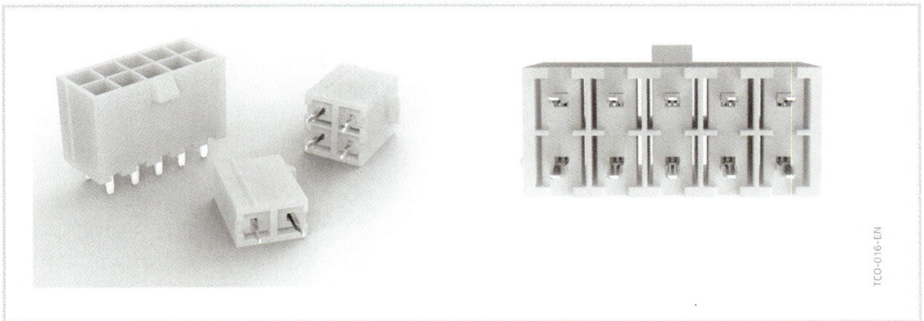

Fig. 1.20: Air passage into a WR-MPC4 connector

1.5 Connector Materials and Processes

This section will provide a basic discussion of connector materials and manufacturing processes. It will give insights in contact finishes with electroplating as well as copper alloy metallurgy.

1.5.1 Contact Finish Electroplating

The majority of connector contact finishes are applied by continuous electroplating of reels of stamped and formed contacts with tin, tin-lead, silver and gold over nickel plating's in thickness ranges from 0.1 to 5.0 μm (4 to 200 micro-inches). Three plating practices will be described, overall, selective and duplex. In **overall plating** the entire contact surface is plated with a single material; tin finishes are often overall plated.

In **selective plating** the plating is applied only to defined functional areas of the contact; gold plating is generally done selectively, with the gold applied only at the portion of the plug and receptacle contacts that define the mating interface. In **duplex plating**, different plating's are applied at the separable connection and termination portions of the contact. The most common duplex plating is tin at the termination end, for soldering, and gold over nickel at the separable connection end. Electroplating and selective/duplex plating practices are highly proprietary and only a general description will be provided in this discussion.

1.5.1.1 Electroplating Basics

Electroplating, or electrodeposition, in connector technology refers to the plating of a material, selected for dedicated properties which will be explained deeper in section 1.7 and apply the needed contact finishes onto a contact spring material.

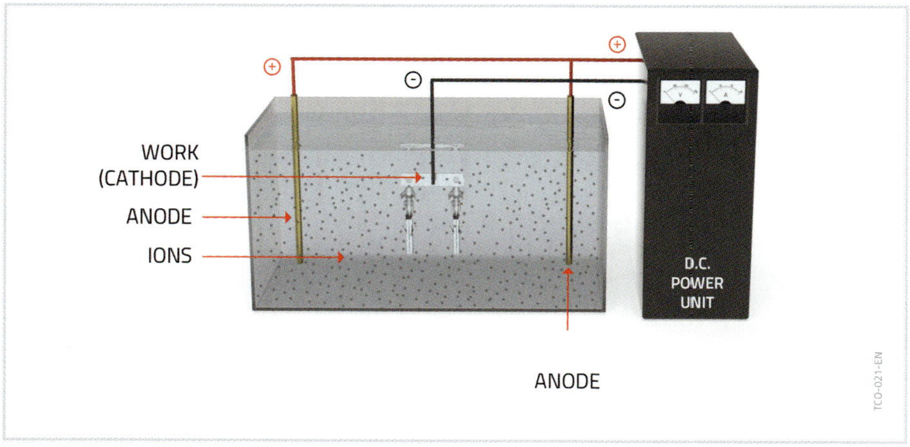

Fig. 1.21: Plating principle

Figure 1.21 schematically illustrates a simple electrochemical plating system. The "electro" part of the system includes the voltage/current source and the electrodes, anode and cathode, immersed in the "chemical" part of the system, the electrolyte or plating bath, with the circuit being completed by the flow of ions from the plating bath to the electrodes. The metal to be deposited may be the anode and be ionized and go into solution in the electrolyte, or come from the composition of the plating bath. Copper, tin, silver and nickel metal usually comes from anodes, while gold salts are usually added to the plating bath in a controlled process to maintain the composition of the bath. The plating bath generally contains other ions to facilitate current flow between the electrodes. The deposition of metal takes place at the cathode.

The overall plating process occurs in the following sequence:

1. Power supply pumps electrons into the cathode.
2. An electron from the cathode transfers to a positively charged metal ion in the solution and the reduced metal plates onto the cathode.
3. Ionic conduction through the plating bath completes the circuit to the anode.
4. At the anode two different processes take place depending on whether the anode material is soluble, the source of the metal to be plated, or insoluble, inert. If the anode material is soluble, a metal atom gives up an electron and goes into the solution as a positively charged metal ion replenishing the metal content of the plating bath. If the anode is inert a negatively charged ion from the plating bath gives up an electron to the anode.
5. The electron flows from the anode to the power supply completing the circuit.

Note that the deposition of metal at the cathode requires an electron so the rate of deposition depends on the flow of electrons, that is, the current flowing from the rectifier. The thickness of the deposit, therefore, depends on the current and the length of time the current is applied. This relationship is a result of Faraday's law which relates the weight of a substance produced by an anodic or cathodic electrode reaction during electrolysis as being directly proportional to the quantity of electricity passed through the cell.

While Faraday's law fixes the amount of metal deposited, the distribution of that metal is dependent on the distribution of the current. This fact allows for selective deposition of the metal, a process commonly used in connector contact electroplating. Figure 1.22 schematically illustrates the relationship between current distribution and deposited metal thickness. The same amount of current flows in the two geometries so the total metal deposited is the same, but the distribution is very different. The anode and cathode geometries and spacings in connector electroplating practice are designed to shape the current flow between the anode and cathode so as to control the deposition of the finish material, both thickness and distribution, onto the contact. This capability is critical to the successful implementation of selective and duplex plating processes. The sizes, shapes and distance between electrodes are among the proprietary processes of connector electroplating. Plating bath compositions and applied current practices may also be proprietary.

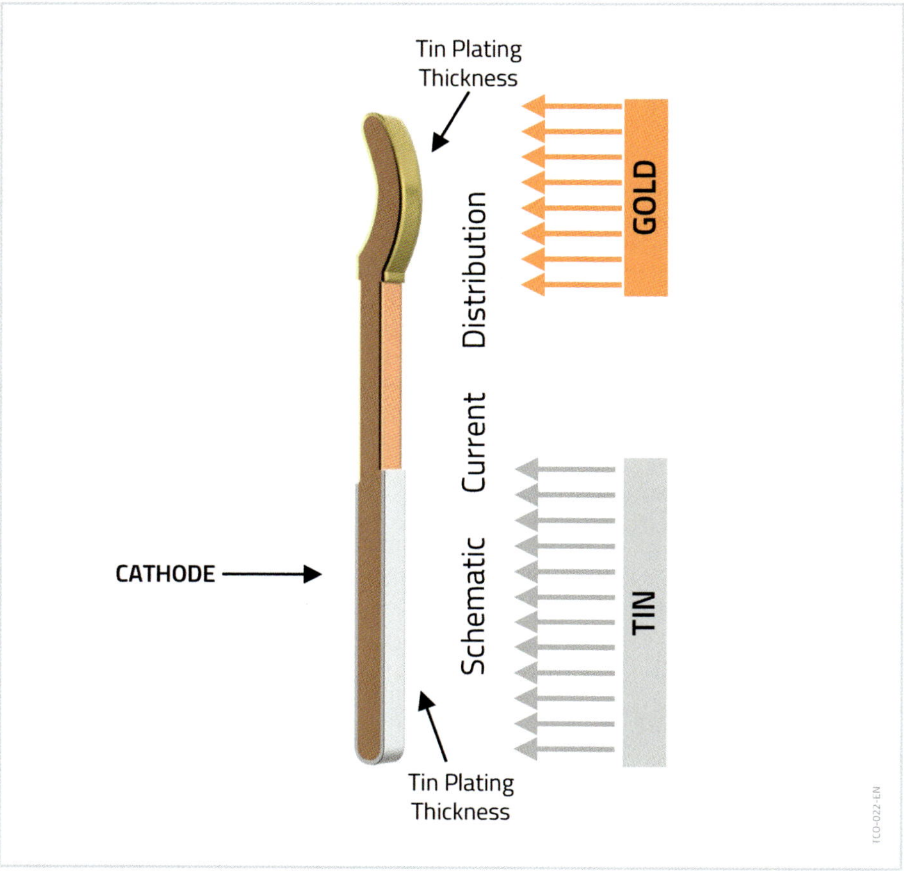

Fig. 1.22: Current distribution – metal thickness

1.5.1.2 Reel-to-Reel Strip Plating

The following discussion provides a basic overview of the stages of electrodeposition during a reel-to-reel strip plating process. The function of each stage will be described, but the details of accomplishing the function are not included because, as noted previously, electroplating practices are highly proprietary.

Stage 0 – Incoming inspection:

The quality of the electroplated coating is directly related to the quality of the incoming roll. Successful electroplating requires a "clean" metal surface: the material to be electroplated must not have any residues of punch oil, scratches or punch damage, otherwise a smooth, continuous and defect-free coating will not be produced.

Stage 1 – Precleaning:

Given the previous comment on surface cleanliness it is not surprising that the first stage of the electroplating process is to clean the contacts. Many different cleaning processes may be included in this stage, ultrasonic organic cleaning to remove residual stamping oils or contaminants, acid, electrocleaning/electropolishing etc. A rinsing station may be included between each of the cleaning stages as well. Cleanliness is critical to electroplating success.

Stage 2 – Plating:

The selected metal is now electroplated onto the cleaned contacts. While the need for a nickel underplate with gold contact finishes has been emphasized, there are other reasons for using an underplate with other finishes. For example a thin layer of copper may be applied as a "strike" to facilitate adhesion onto some copper alloys, or a thicker layer may be used to smooth the surface, a levelling copper. The same electroplating practices will be used for these supplementary plating as for the finish plating. Plating bath pH , composition and temperature need to be carefully monitored during the electroplating process.

Stage 3 – Rinse/Dry:

On completion of electroplating the contacts are rinsed and dried before being re-reeled and sent on to the next stage of the manufacturing process. Rinsing is important because residual plating solution and salts can cause corrosion of the contacts if not removed.

Multiple "Stage 2" stations can be used for multiple layer plating's, gold over nickel, and, if so, it is necessary to ensure that the plating solutions do not transfer from one station to the next so a rinse station is generally included between all plating stations. In the case of gold over nickel plating's, it is essential that the contacts do not dry out between the nickel and the gold stations because nickel exposed to air "passivates", forms a thin oxide layer, inhibiting the effective plating of the following gold plate.

A few comments on selective and duplex plating are in order at this point. Figure 1.23 illustrates a contact which is both selectively and duplex plated. The first plating applied is an overall nickel underplate, that is, the entire contact surface is plated with nickel. The second plating applied is a selective gold at the mating interface, a. The third plating is a selective tin, b, at the contact tail to provide a solderable surface.

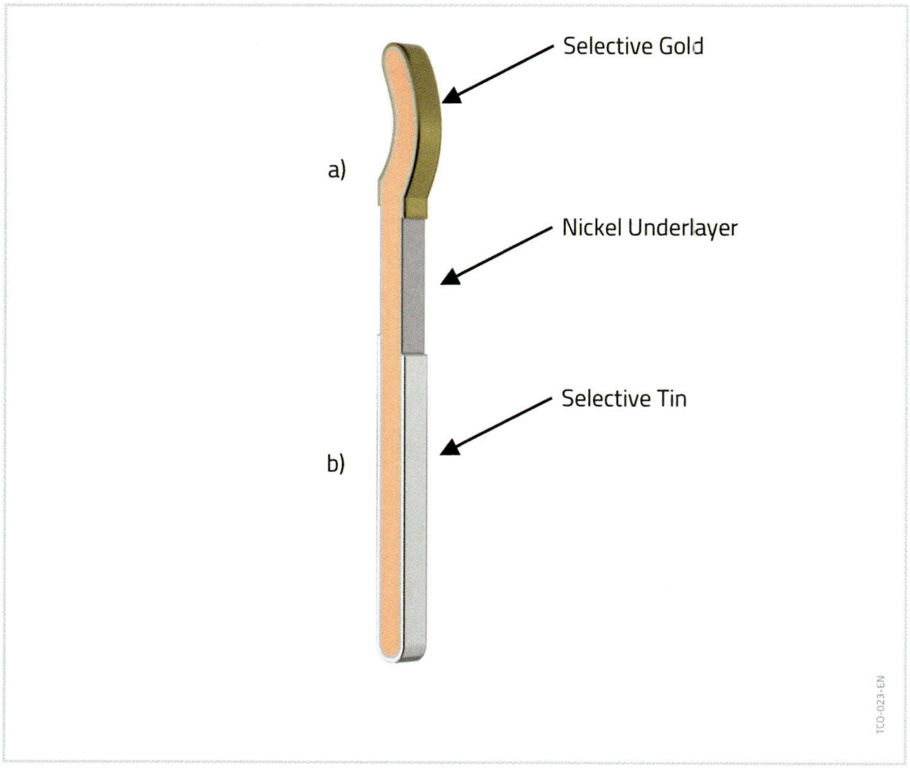

Fig. 1.23: Selective and duplex plating

The selective plating is obvious; the tin and gold platings are applied only where needed for functionality. The contact is duplex plated because two separate areas have been plated. In this example the two plating's are different, gold over nickel and tin over nickel. In some cases, a gold flash is used to preserve the solderability of the contact tail section. This combination would also be considered duplex plating. If the gold flash was used, the contacts could be plated using two separate gold plating stages in the same configuration in the plating bath. The anode/cathode location and geometry would be positioned appropriately in the plating bath to plate desired areas and thicknesses. In the case of duplex tin, however, it is common to apply the gold plating and then invert the contacts in a transition stage before immersing only the contact tails into a supplementary bath to plate the tail section with tin.

The photos of Figure 1.24 show different plating bathes (a) Ni, b) Tin and c) where the bandolier pins are rotated to perform selective plating.

a) Nickel bath b) Tin bath c) Selective plating

Fig. 1.24: Plating lines

1.5.1.3 Miscellaneous

Not all connector contacts are reel-to-reel plated. There are two batch plating processes and two non-electrolytic practices worthy of a brief discussion.

Batch Plating

The two batch plating processes differ in how the contacts are presented to the plating bath. In one case, commonly called **"barrel plating"** the contacts are simply placed in the barrel and incidental contact between individual contacts is relied on to retain continuity of the cathode. The contacts are generally tumbled to promote greater uniformity of plating thickness. In the second case, the contacts are attached in some manner to a frame or rack which is made the cathode during the plating process. Batch plating is used primarily for tin plated contacts intended for industrial and white goods applications.

Electroless Plating

The electrochemical reactions that give rise to electroless plating can be quite complicated. For our purposes it is sufficient to say that electroless plating consists in one metal replacing another due to a difference in electrochemical potentials without any external power, hence, electroless. Electroless plating is sometimes called **immersion plating** because it occurs on immersion without external power. Nickel, gold, silver and copper can be electroless plated. Benefits of electroless plating are that it plates uniformly over a surface, thus can plate into recesses and blind holes not amenable to electroplating. Limitations of electroless plating are speed and limited thickness capability, fractions of a micron. The process essentially stops when the entire surface being replaced has been covered over.

Hot Air Reflowed Tin (HART)/Hot Air Solder Levelling (HASL)

These processes are similar in that a tin or solder coating is applied, and a hot air "knife" is used to level the coating. HASL is typically used in printed circuit board applications.

1.5.2 Copper Alloy Metallurgy and Processing

The majority of connector contacts are manufactured from copper alloys due to the combination of high electrical conductivity and good mechanical strength and formability characteristics. This section will provide an overview of copper alloy metallurgy to explain how those characteristics are obtained and used in the manufacture of connector contacts.

1.5.2.1 Copper Alloy Metallurgy

With the exception of silver, copper has the highest electrical conductivity of any metal. Mechanically, copper has moderate strength and good formability. Formability refers to the ability to take a metal and bend or form it into various contours, such as the connector receptacle contacts shown in Figure 1.25, without cracking. By adding other elements into the copper matrix, alloying, the mechanical characteristics of copper can be improved significantly without unacceptable loss in formability. The changes in mechanical performance characteristics depend on the alloying element, e.g. tin, zinc or beryllium, and the amount of the alloying element added to the copper. Most copper alloys used in connectors are lightly alloyed with alloy contents of less than 10 %. Cartridge brass, arguably the most common contact material due to heavy use in white goods and industrial applications, is the major alloy with a higher alloy content, being nominally 70 % copper and 30 % zinc.

TCO-025-EN

Fig. 1.25: Contact bending and forming

The improvements in mechanical performance, however, are accompanied by a significant decrease in electrical conductivity. A common benchmark for the electrical conductivity of copper alloys uses the **International Annealed Copper Standard, IACS,** as a reference. Against this standard, **copper has a conductivity of 102 % IACS**[1] while phosphor bronze, copper/5 % tin has an IACS conductivity of 20 %, a significant decrease. Even at these conductivity levels, however, copper alloys are still highly conductive. For comparison the conductivities of aluminium and nickel, in IACS values, are 61 % and 16 % respectively.

Consider now some of the details of alloying of copper with respect to electrical conductivity and mechanical performance. The section begins with the effects on electrical conductivity.

Alloying and Electrical Conductivity

A simplistic, but basically accurate explanation for the high conductivity of metals is, that they contain free electrons, electrons that are not tied up in the inner shells of the atomic structure. When an electric field is applied to a metal, these free electrons are accelerated by the field and result in current flow. While these free electrons are free from individual atoms, they are surrounded by other atom shells which can impede the flow of the free electrons, that is, scatter the free electrons and introduce resistance into the system. In pure copper, all the atoms are on fixed positions in a close packed lattice, as schematically illustrated in Figure 1.26a, and this fixed structure introduces a characteristic resistance, the resistance of copper. If other elements are added to the copper, they will have different characteristics, size and electronic structure, so they will distort the lattice, as shown in Figure 1.26b, increasing the scattering of the free electrons and, thus, the electrical resistance. The degree of increased scattering depends on the alloying elements and the amount of alloying additions. Figure 1.27 shows the effect of small additions of zinc, tin and nickel to copper.

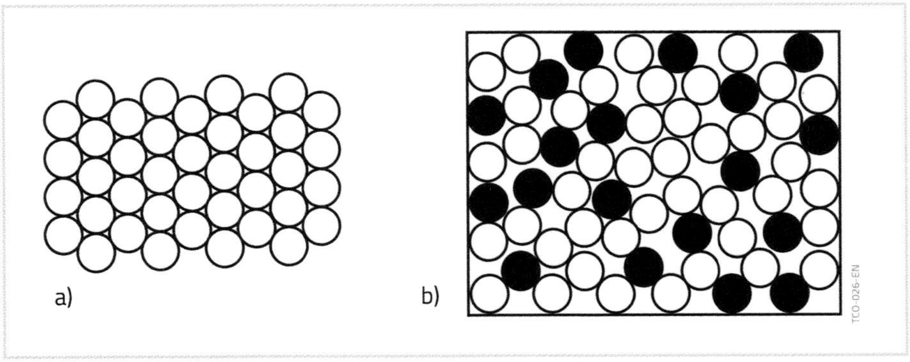

Fig. 1.26: Alloy addition and lattice distortion

[1] Since the IACS standard was established in the late 19th century. The processing of copper ores has been improved to increase the purity of copper resulting in an increase in conductivity from the original 100 % to 102 % IACS.

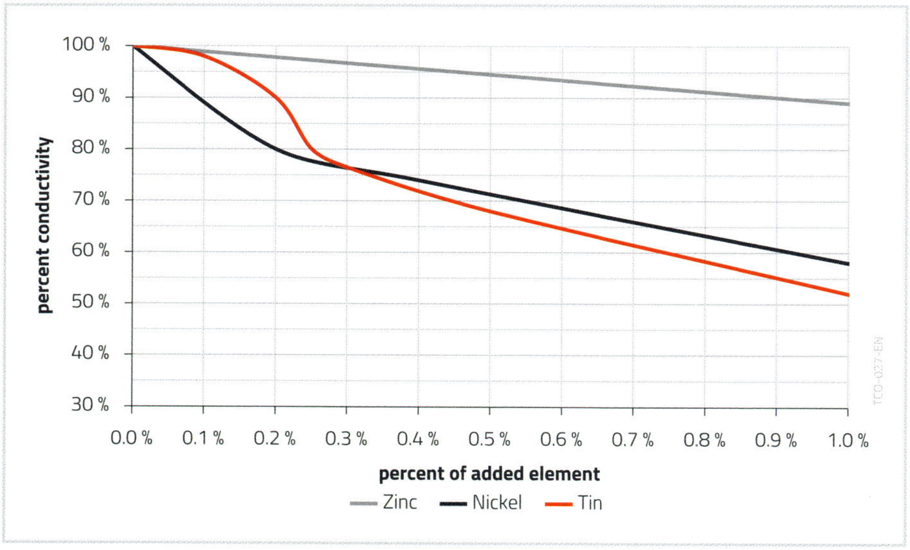

Fig. 1.27: Copper alloy addition and electrical resistance

Alloying and Mechanical Properties

Before discussing the effects of alloying on mechanical properties a discussion of pure metal mechanical properties is in order. Figure 1.28 contains a stress-strain curve for a metal under a tensile, uniaxial, load. The important parameters in the figure are:

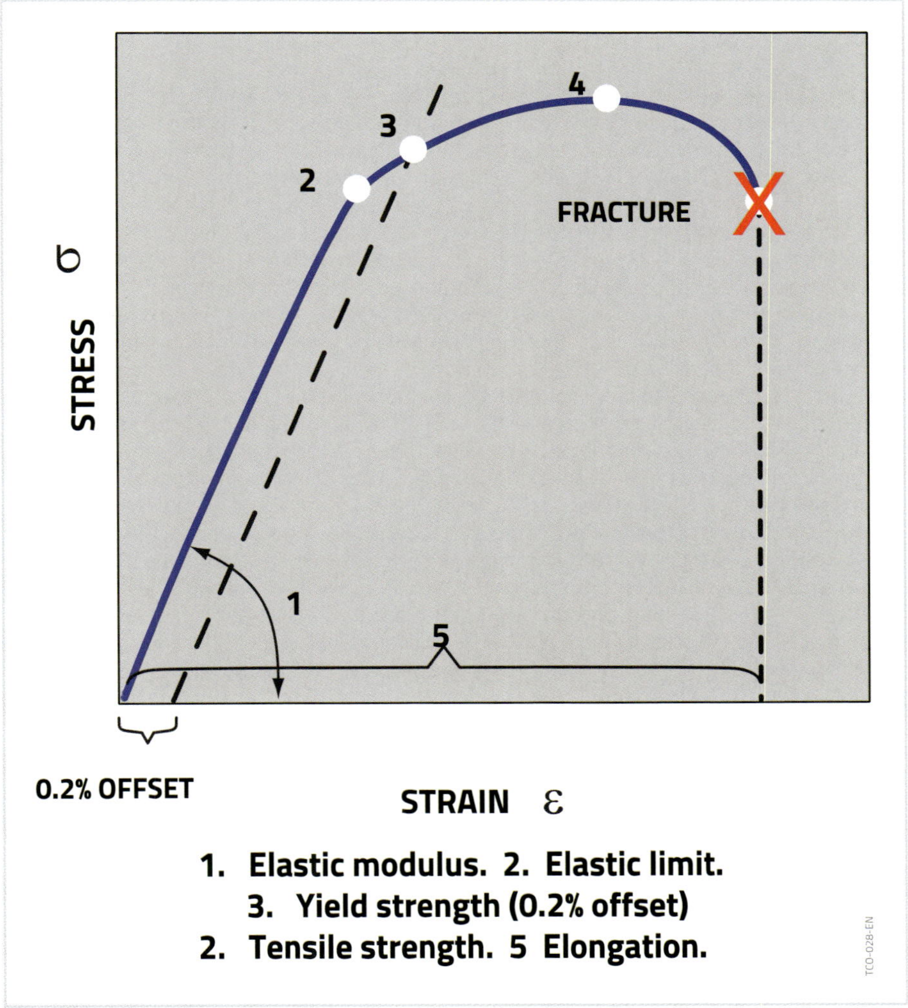

Fig. 1.28: Stress strain curve in tension

Elastic (Young's) Modulus, E [N/m²]

The Elastic Modulus is the slope of the stress versus strain curve in the initial stages of loading. Elastic means that if the load is applied and removed the sample will return to its original length. The elastic modulus depends on the material and, in copper alloys, is not significantly affected by alloying. The range in E across the commonly used copper alloys is between 16 to 20.106 psi. In section 1.10.1 Contact springs the elastic modulus will be related to the spring rate, the force versus deflection, of a cantilever beam, an important connector design parameter.

Yield strength, is the stress at which a given permanent set, a permanent increase in the length of the sample when the load is removed, is introduced into the sample. For copper alloys the yield strength is generally taken at 0.2 % offset, that is, the stress at which the length of the sample has increased by 0.2 %. The value is obtained from the stress-strain curve by drawing a line parallel to the Elastic modulus from the 0.2 % elongation point on the strain axis. Permanent set is due to plastic, non-reversible, deformation of the sample. Plastic deformation begins when the applied stress is sufficient to cause individual atoms to be displaced from their equilibrium positions in the metal lattice on an atomic scale distorting the local lattice structure. As the distortion increases under increasing stress, groups and planes of atoms, called dislocations, move relative to each other, generally along close packed planes resulting in the macro distortion of permanent set. The introduction of such defects into the lattice makes the lattice more resistant to additional deformation, strengthening the lattice. This strengthening is referred to as "work hardening" because the "work", the deformation, "hardens", strengthens, the lattice. All metals, pure and alloys work harden. There is an additional hardening mechanism relevant to both pure metals and alloys, grain size hardening. The copper and copper alloys used in connectors are polycrystalline. A cross-section through a contact would show numerous grains separated by grain boundaries. Figure 1.29 shows two grain size reference photographs. The number with the figure indicates the "average" grain size in the photo. The hardening mechanism of grain size actually comes about due to the grain boundaries. The orientations of the grains on either side of the boundary will be different, so the close packed planes will not be continuous and the dislocation motion will be stopped or inhibited, thus, the material will be strengthened. Clearly the relative amount of grain boundary compared to grain is higher as grain size decreases so the hardening effect is stronger. There are two other hardening mechanisms that come into play due to alloying. These mechanisms will be discussed when the discussion of the stress-strain curve is completed.

a) 0.045 mm   b) 0.120 mm

TCO-029-EN

Fig. 1.29: Grain size[2]

[2] 6 cf. ASTM E2-36

The final property of interest in the stress-strain curve is elongation, the increase in the length of the sample under load. The importance of elongation is its relationship to ductility, which, in turn, is a measure of the ability to stamp and form the material without unacceptable damage, i.e. cracking in the bend zones. Yield strength and elongation are inversely related, which is to be expected.

Alloy Hardening Mechanisms

Alloying directly affects yield strength in two ways. First, as noted in the discussion of electrical conductivity, alloying elements distort the lattice structure, creating localized strain fields which strengthen the lattice. This strengthening mechanism is called alloy hardening and depends on the alloying elements . The greater the disparity in atomic size and electron structure between the alloying element and copper, the greater the hardening effect will be. Second, the alloying elements affect the work hardening mechanisms and rates producing a different shape to the curvature of the stress-strain curve in the plastic region which is reflected in the determination of the 0.2 % offset yield strength. These two mechanisms are active in all alloys. Some alloying elements, however, introduce a new strengthening mechanism, second phase strengthening.

Second phase strengthening depends on the introduction of fine particles, the second phase, into the base metal matrix. There are two second phase strengthening mechanisms, precipitation hardening (PH), and Dispersed Second Phase (DSP). The hardening mechanisms are different in the two cases.

Precipitation hardening produces a distribution of small particles into the copper matrix that are very effective at inhibiting dislocation motion, that is, they strengthen the material. These second phase precipitates form because the solubility of some elements in copper varies with temperature. For example, beryllium has a solubility in excess of 1.4 % at temperatures above 600 °C and the solubility approaches zero at room temperature. If beryllium is taken into solution at high temperature and rapidly cooled, quenched, to room temperature the beryllium remains in metastable solution. Aging at an intermediate temperature causes the beryllium to precipitate out as copper beryllides, an intermetallic compound, CuBe. For this reason, precipitation hardening is often referred to as **"age hardening"**. Similar precipitation of intermetallic compounds, including some non-copper intermetallic, can be realized. This second phase strengthening mechanism can be both potent, effective at inhibiting dislocation motion, and stable, resistant to thermal degradation. It is the thermal stability of the dispersed intermetallic compounds that allows PH alloys to maintain their strength at elevated temperatures over time. This time – temperature stability, called stress relaxation resistance is an important parameter in high temperature connector performance and will be discussed again in section 1.9. In contrast, the residual stresses of work hardening, lattice distortion, that inhibit dislocation motion can be relaxed, diminished, by time at lower temperatures.

In dispersed second phase strengthening the hardening mechanism is an enhancement of the work hardening response of the alloy. The dispersed second phase causes more dislocation generation than would occur at the same deformation without the particles. The effect of the additional dislocations is accentuated as the dislocations interfere with one another, an additional strengthening mechanism. The second phase dispersion is dependent on thermo-mechanical processing with both temperature and working acting simultaneously to create the dispersion. The dispersed second phase can also result in a finer grain structure, again enhancing the strengthening.

1.5.2.2 Copper Strip Processing

The vast majority of connector contacts are stamped and formed from copper alloy strip. This discussion of copper strip processing begins with the cast slabs of the chemical composition of the desired copper alloy, say 70 % copper and 30 % zinc for cartridge brass The dimensions of the slabs vary with the alloy and the casting process and vary from a few cm to over 15 cm in thickness, 0.3 to 0.9 m in width and up to 8 m in length. The thickness of the strip used for stamping and forming of connector contacts is of the order of 0.04 to 1.0 mm so a significant decrease in the thickness of the cast slab is necessary to attain this thickness at a controlled value of yield strength and formability. This reduction is realized through two different rolling operations, hot and cold rolling. As with electroplating, the rolling processes are proprietary and only general comments will be provided.

Hot rolling is used to accomplish the major reductions in thickness, from the original thickness of, 10 cm down to 1 cm. The temperature of hot rolling is critical in that the metal must be, and remain, hot enough to avoid work hardening during the entire process which may include several passes through the hot rolling mill. By controlling the temperature, the final hot rolled strip will be in an annealed (soft) condition with a controlled grain size. The cold rolling process influences the grain size, which determines the final mechanical properties of the metal strip.

The **cold rolling** process takes the strip down from the hot rolling thickness of 10 mm to the final specified strip thickness, say 0.4 mm. During the cold rolling process the yield strength and elongation of the strip are also established. Recall that deformation of a metal, in this case reduction in thickness by rolling, work hardens the metal increasing its resistance to deformation or increasing its yield strength. Work hardening also decreases the ductility of the metal, the amount of deformation it can support without cracking. This means that only a limited amount of reduction in thickness can be realized in cold rolling without damage to the strip. In practice a specified reduction in thickness is taken and the strip is then annealed, held at an elevated temperature for a specified time, to return it to its soft condition when it can then be subjected to another reduction in thickness. This roll/ anneal process is repeated as often as necessary to get to the thickness required for the final annealing and rolling process which is when the mechanical properties of the strip are established. The final anneal establishes the final grain size of the strip, and the final reduction in thickness establishes the yield strength/elongation. The relationship between reduction in thickness and mechanical properties, in particular yield strength and elongation, is known for each copper alloy. A generalized example is provided in Figure 1.30. For a given connector design, the design yield strength is determined by the contact normal force requirement, as will be discussed in section 1.9 (Contact Springs). The elongation requirement depends on the forming necessary to realize the conformation of the contact spring. As an example, say a cantilever beam receptacle contact normal force requirement dictates a yield strength of 500 MPa (75 ksi) and an elongation of greater than 5 % is necessary. For the material indicated in Figure 1.30, a 30 % reduction in thickness on the final cold rolling pass will meet the yield strength requirement with an elongation of just under 10 %, which also meets the elongation/ductility requirements.

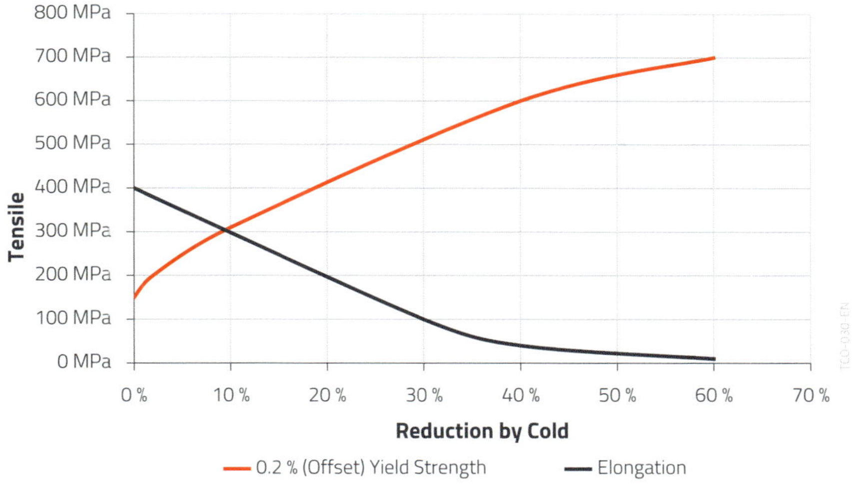

Fig. 1.30: Final pass: Reduction in thickness vs. properties

This description of the hot and cold rolling processes taking a copper alloy from a cast slab to a specified strip thickness and mechanical condition is, in principle, very straightforward. But the process controls, both thermal and mechanical, are very demanding as is the surface condition of the rolls and the supporting structures of the rolling mills that are used. Trade-offs among the various thermal and rolling process variables used by copper strip providers explains the variation in the comparative relationship between yield strength and formability among suppliers.

1.5.2.3 Connector Contact Stamping and Forming

The stamping process is used to convert blank metal strip into contacts or shields by cutting and forming it. The press brings the pressure to the die which performs cutting and forming steps.

Figure 1.31a illustrates the die into the press, when figure b shows inside the die (in between the two parts). On the c figure we can see the metal strip evolution along the process (from a blank metal strip down to a typical RJ45 metallic shielding)

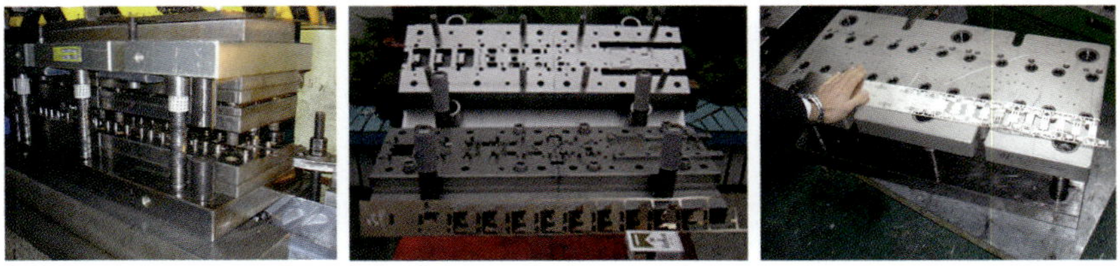

a) Die / Press

b) Inside the die

c) Metal strip evolution along the process

Fig. 1.31: Stamping progressive die

The vast majority of connector contacts are produced by stamping and forming processes using multi station dies that perform a variety of stamping operations including piercing, blanking, shearing and bending among others. The design of the dies and the sequence of the various stations depend on the complexity of the contact design with receptacle contacts being mostly more complex than those of plug contacts.

1.5.2.4 Copper Alloys

Copper alloys are commonly divided into several categories depending on the alloying element(s) and the amount of alloying. The dominant strengthening mechanism varies according to the same parameters as noted above. Some general comments on the alloy families, based on categories used by the Copper Development Association, follow. The explanation here is general due to the process dependence of mechanical properties.

Alloy data should be obtained from the manufacturers.

Coppers: Coppers are a minimum of 99.3 % copper. They are high conductivity alloys, greater than 80 % IACS, with varying strength levels and stress relaxation resistance depending on the alloying elements. Principal alloying elements are phosphorus, magnesium and zirconium. DSP hardening is realized by copper-zirconium and magnesium-phosphorus particles.

High Coppers: High coppers are a minimum of 96 % copper. Principal alloying elements include beryllium, chromium, cobalt, magnesium and iron. The beryllium alloys are PH strengthened. They have the highest strength, best stress relaxation resistance and lowest conductivities of the high coppers. The others are DSP strengthened by phosphide intermetallic compounds. They include a range of mechanical properties due to their different responses to second phase strengthening, moderate to good stress relaxation resistance and moderate conductivities.

Brass: Brass is a copper/zinc alloy. Cartridge brass, 30 % zinc, is the most highly alloyed connector material. Brass is unusual in that the addition of zinc to copper has a much lower effect on conductivity than any other alloying element, the conductivity of cartridge brass is 28 % IACS. Brasses have moderate strength, stress relaxation resistance and conductivity.

There are two categories of modified brass. Tin brass uses the alloy strengthening of tin to realize higher strength levels with lower work hardening which enhances the formability and stress relaxation resistance of the alloy.

The second category uses DSP hardening through additions of iron, cobalt, aluminium and silicon. Depending on the alloy, grain refinement, enhanced work hardening response, or combinations of both, provide additional strength for a given amount of cold working.

Bronze: Classic bronze is a copper-tin alloy with the tin providing strong alloy hardening and enhanced cold working response. Tin additions dramatically reduce conductivity with the commonly used 5 and 8 % tin alloys being less than 20 % IACS. These alloys are commonly called phosphor bronzes because of the residual phosphorus from the deoxidizing process remaining in the alloy. Bronzes have high strength capability and good stress relaxation resistance. Modified bronzes using copper-aluminium and copper-silicon base alloys are also used.

Copper-nickel-silicon: Copper-nickel-silicon alloys having a combination of high strength, good stress relaxation resistance and moderate conductivity have been developed recently and are seeing increasing use in automotive applications .

1.6 The Contact Interface and Contact Physics

1.6.1 Overview

The contact interface is arguably the heart of a connector system. The structure and properties of the interface which is created as the plug and receptacle contact springs come together determines both the electrical and mechanical performance characteristics of a connector.

The most important characteristic of the contact interface is that it is created when two rough surfaces come into contact. While a cursory look at the surfaces of contact springs suggests that they are smooth, on the microscale of the contact interface these surfaces are quite rough. The individual high points on the surface are referred to as asperities. Figure 1.32 schematically illustrates the contact interface that would be created if two spheres were brought in contact under an applied load. As the two surfaces approach each other the asperities that happen to be in proximity to each other come together first. As these asperities deform against each other under the applied load, the surfaces continue to approach each other bringing additional asperities into contact. Thus, a contact interface consists of a number of individual asperity contact points, called **a-spots**. The number is determined by the roughness of the surface and the applied load which determines the total area of a-spot contact. These a-spots are distributed over an apparent contact area that is determined by the geometries of the post and receptacle contact springs as they come together.

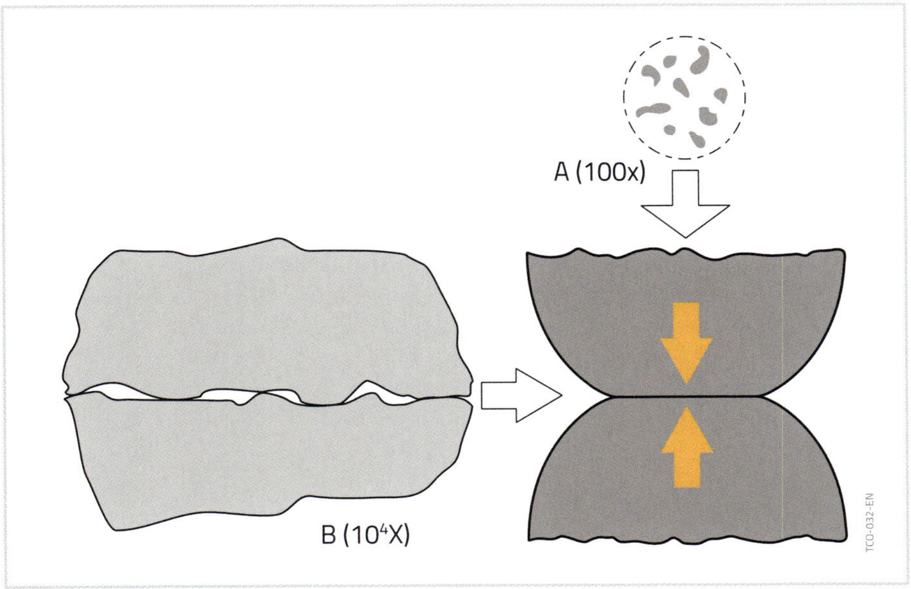

Fig. 1.32: Schematic contact interface

That this a-spot model is realistic is verified by consideration of Figure 1.33 which shows an actual contact interface between a stainless steel ball bearing and a glass plate. The ball bearing deforms slightly against the glass plate under the applied load and individual a-spots are created. Note the similarity to Figure 1.32. The photo on the right was taken after an increase in the applied load which brings the surfaces closer together. This, in turn, increases the contact area of existing a-spots as well as creating additional a-spots.

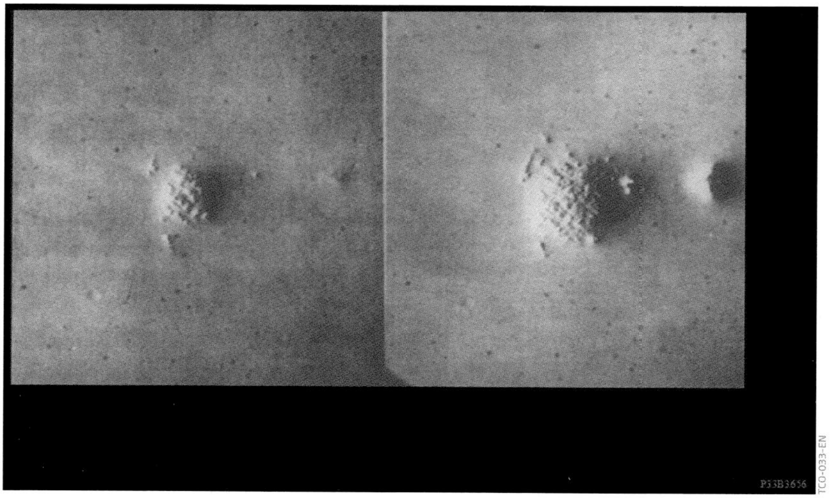

Fig. 1.33: Contact interface at increased load

Returning to Figure 1.32, a side view of the contact interface, also shows a distribution of a-spots across the contact interface. This view will lead to an understanding of why the contact interface structure affects both the mechanical and electrical performance.

Consider two mechanical effects. First, if a shear stress is applied to this a-spot contact interface the interface will remain stable until the stress is sufficient to break the a-spot interfaces. Second, when the interface does break a wear process comes into play. These two effects will be discussed in section 1.6.2

Electrical performance is affected by the fact that any current flow from the upper sphere to the lower, would have to flow through the individual a-spots. This condition leads to an electrical resistance called constriction resistance which will be discussed in some detail in section 1.6.3.

1.6.2 The Mechanical Interface: Friction and Wear/Durability

Friction and wear are two aspects of the same basic kinetic process. Friction forces must be overcome to cause motion, and motion must occur to cause wear. Consideration of Figure 1.34 will be used to explain this statement.

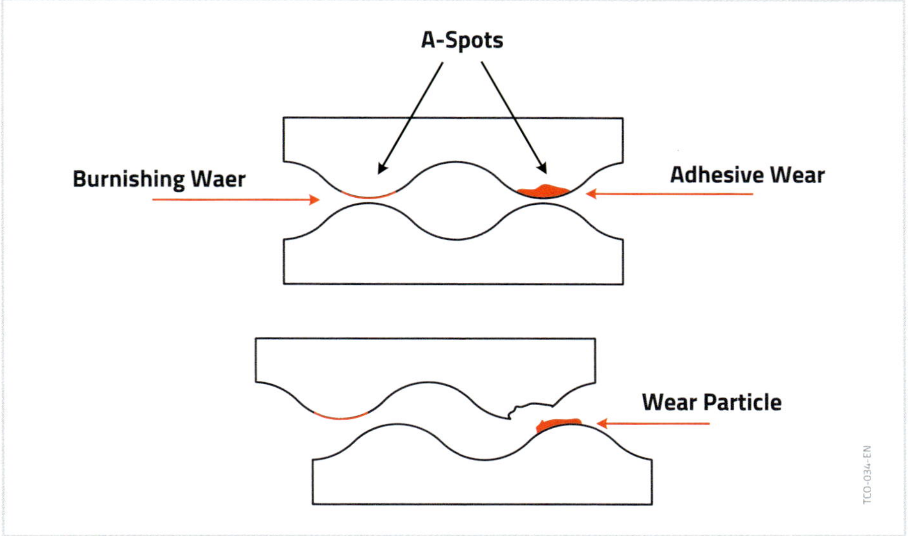

Fig. 1.34: A-spots and friction wear

Consider Figure 1.34 through a thought experiment where two a-spot contacts are created. The a-spot on the right is created first under a given applied load. The individual asperities deform against each other under the load. Four things happen during the deformation. First, the asperities deform plastically due to their small size causing material flow. Surface films and contaminants are disrupted or displaced on the microscale of the flow. Second, the clean metal surfaces exposed due to the microflow can form cold welded junctions3 when the microflow ceases. Third, the asperity surfaces work harden due to the deformation. Fourth, the a-spot contact area increases as the load increases.

While all this is happening at the first asperity contact the surfaces are continuing to come closer to one another and the second a-spot is created. The same process then begins to take place at this second a-spot. We now have two a-spots that have experienced different amounts of deformation. The difference in deformation means that the first a-spot is larger, due to more metal flow, and stronger – due to a greater work hardening response and more cold welding.

The increased area and strength of the first a-spot means that it will be mechanically more robust. Consider how the two a-spots will respond under an applied shear stress. This illustration is qualitative and only serves to explain the kinetics of friction and wear.

The applied shear stress will not lead to motion of the interface until the a-spots are broken. Given that the first a-spot is larger and stronger, it will determine the stress necessary to break the interface and cause motion. The necessary stress is related to the friction force of the contact interface. Thus the a-spot structure and distribution determines the friction force and, therefore, the mechanical stability of the contact interface.

The kinetics of separation at the a-spot interfaces will be different as well. Recall that the interface of the first a-spot is stronger than that of the second. In fact, due to work hardening and cold welding, the interface may be even stronger than the cohesive strength of the base metal. In such cases, the a-spot interface may break within the base metal rather than at the interface resulting in a wear particle as indicated in Figure 1.34. Such wear processes are commonly referred to as **adhesive wear**. In contrast, the weaker interface of the second a-spot may break at or near the original interface without a wear particle being generated. Wear of this type is referred to as **burnishing wear**. The wear track in a connector experiencing adhesive wear, sometimes called galling wear, will be rough, while that of burnishing wear will be relatively smooth.

The significance of wear processes in a connector is straightforward. **Wear results in the loss of surface material.** If a connector is designed to take advantage of the benefits of a contact finish, wear can lead to loss of the contact finish at the mating area and a consequent decrease in performance. This is particularly significant when the small thicknesses of contact finishes are recognized. **Susceptibility to wear is one reason connectors are rated for a specific number of mating cycles.**

Friction has two effects on connector performance, one positive and one negative. The positive effect is that friction provides mechanical stability of the contact interface against forces tending to drive motion of the connector. Disturbances of the contact interface can be a significant degradation mechanism for a connector for two reasons. First, the micromotions of a contact disturbance can induce wear at the contact interface. Second, micromotions can drive corrosion mechanisms, especially in tin finished contacts, and can lead to bring corrosion products and contaminants around the contact area into the contact interface. Thus high friction forces, generally due to high contact normal forces, can have a positive effect on connector performance.

There are two negative effects of friction forces. First, high friction forces correlate directly with wear mechanisms as has been discussed. Second, high coefficients of friction will increase the mating force of a connector. **Mating force varies directly with the coefficient of friction.**

These issues, friction, mating force, contact force and wear will be discussed in detail in section 1.10.1.

1.6.3 The Electrical Interface: Contact resistance

Attention now turns to the effects of a-spots on electrical performance, and, in particular on the electrical resistance the contact interface introduces into the connector. To address that issue requires a consideration of the sources of resistance in a connector.

Figure 1.35 shows a cross section of a connector with the various sources of resistance indicated. There are three types of resistances noted. There are two permanent connection resistances RP.C., the resistance of the crimped connection to the incoming conductor and of the pin connection to the plated through hole of the printed circuit board. Permanent connection resistances are dependent on the permanent connection technology used There are also two bulk resistance contributions RB, from the pin and receptacle contacts indivicually. Bulk resistances are determined by the resistivity of the material of the contact spring and its geometry. Finally, there is the resistance of the contact interface RC. Only one resistance is cited even though there are two contact interfaces electrically in parallel.

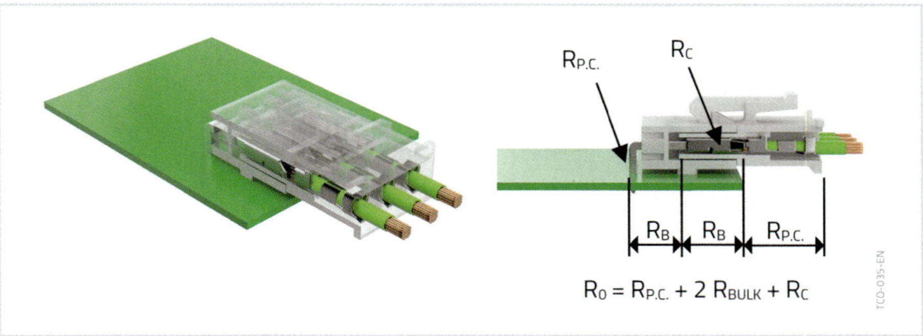

Fig. 1.35: Connector resistance

In order to get a perspective on the relative contributions of these three resistances, consider the following thought experiment. If a test probe is inserted through the insulation to the conductor of the wire coming into the crimped connection and a corresponding probe placed in contact with the pad on the plated through hole connection the overall resistance $R_{overall}$ of the connector can be measured. To provide a benchmark, it is assumed that the connector system shown uses a 0.64 mm (0.025 in) square post. For a connector of this size a reasonable value for the overall resistance is at a maximum of 20 mΩ. The permanent connections will be less than a milliohm each, the bulk resistances will total about 17 milliohms and the contact resistance will be of the order of a milliohm. Thus, for this connector system the contact interface resistance is of the order of 5 % of the overall resistance. **Why then is the contact interface resistance so important?** The answer is found when the connector system is subjected to a test program to assess its stability under a variety of conditions. If, after such testing, the overall resistance of the connector is measured and found to be, say, 200 milliohms virtually all that change will be in the contact interface resistance.

Bulk resistances are essentially constant. Permanent connection resistances tend to be stable because they have larger contact areas and contact forces than separable interfaces. **So, an understanding of the parameters that determine contact interface resistance is critical to knowing how to design and maintain its integrity over the broad range of connector applications and application environments.**

As noted previously, electron flow can only occur through an a-spot, the majority of the apparent contact area is non-conductive. This means that the current distribution through the contacts is constricted to pass through individual a-spots at the contact interface. This geometrical current constriction gives rise to constriction resistance as illustrated in Figure 1.36.

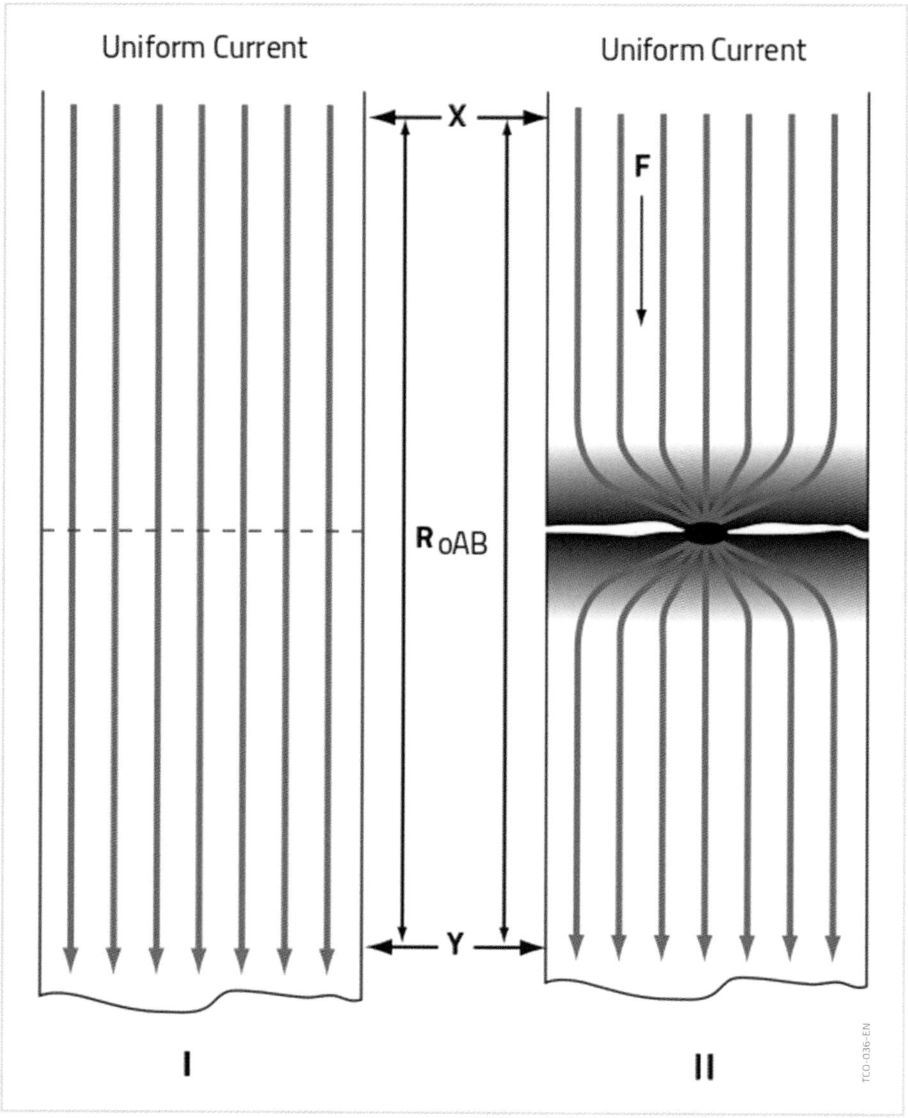

Fig. 1.36: Constriction resistance

Another thought experiment. The drawing on the left in Figure 1.36 depicts a length of round conductor. Voltage and current probes are connected to the conductor at points X and Y so that a direct current voltage can be applied between points X and Y and the resulting current measured. Ohm's law, U = I R, where U is the voltage drop in Volt (V), I is the current in Ampere (A) and R the resistance in Ohm [Ω], can be used to calculate the resistance.

An additional, independent, way to calculate the resistance R is using the below relationship: where ρ is the conductivity of the conductor material [1/(Ωm)], L the length between X and Y [m] and A is the cross sectional area of the conductor [m]. Naturally, these two methods result in the same value of resistance.

$$R=\rho\left(\frac{L}{A}\right) \qquad \text{(EQ 2)}$$

Now for the thought part. Assume that a zero thickness blade can be used to cut into the conductor midway between X and Y down to an infinitesimal distance from the center of the conductor. Given that change in conductor configuration the resistance is once again calculated using Ohm's law. The new calculated value will be much higher than the original value. To explain why we can use our general formula $R = \rho\,(L/A)$.

In the original conductor in Figure 1.36 the DC current is indicated by a uniform distribution of current flow lines across the conductor cross section. This, of course, is not what actually happens, but serves as a good visual representation for this discussion. The flow lines are all the same length and distributed uniformly over the conductor cross section.

In the reduced cross section conductor a very different flow pattern exists. If the cut to the center is small enough only one of the flow lines can pass along the conductor unimpeded. All the other flow lines must constrict to flow through the single continuous thread of conductor. This is the source of the term constriction resistance to describe this effect.[3]

Now consider the general formula. Recall that the length of all the flow lines but the central one must increase in length in the new configuration. R is directly proportional to L in the general formula which explains the increase in resistance. Alternatively, note that the current flow lines begin to constrict at some distance away from the reduced cross section. This effect can be modelled by taking a number of slices perpendicular to the axis of the conductor and noting that the cross sectional area of each current carrying slice is smaller than the original cross section. R is inversely proportional to

A so each of these slices has a higher resistance then the original slice, and all these slices are in series, thus, the increase in resistance.

Note that there is no interface in this model, the conductor is continuous. Constriction resistance is a geometric effect. All contact interfaces will include a number of a-spot interfaces, each of which will introduce a constriction resistance. Thus an interface must produce an increase in resistance to the connector. This resistance can be minimized by maximizing the a-spot contact area, but it cannot be eliminated.

[3] The current flow lines spread out in the other half of the conductor, but the effect on resistance is, of course, the same. In fact, in sem conductor technology this effect is called "spreading resistance".

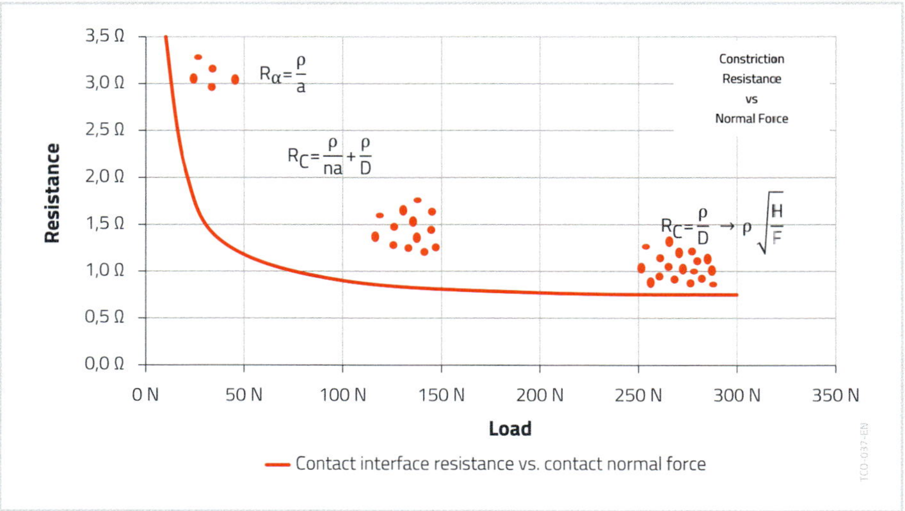

Fig. 1.37: Contact interface resistance vs. contact normal force

The resistance in Figure 1.37 is the constriction resistance , the resistance of the metal-to-metal contact interfaces. As the plug and receptacle surfaces are brought together under the contact normal force, a few asperity contact interfaces form, as illustrated previously in Figure 1.32, such contacts are referred to as a-spots. The resistance of each individual a-spot, R_a, is given by the formula shown in Figure 1.37:

$$R_a=\frac{\rho}{ad} \quad \text{(EQ 3)}$$

Where ρ is the resistivity of the material, and a_d is the **diameter** of the a-spot. As the force is increased and the surfaces deform against one another additional a-spots are created. The second equation now applies:

$$R_C=\frac{\rho}{na_d}+R_a=\frac{\rho}{D} \quad \text{(EQ 4)}$$

Where R_c is the contact interface resistance, n is the number of a-spots a nd D is the diameter of the apparent contact area, the area over which the a-spots are distributed at the contact interface. These two resistances are in series because the current constricts first to the dimensions of the apparent contact area, defined by D, and then constricts again to flow through the individual a-spots, defined by ad.

As the contact force continues to increase, the number of a-spots increases along with the apparent contact area. As n increases the a-spot resistance contribution decreases and the contact interface resistance becomes dominated by the resistance due to the apparent contact area. In other words, the interface acts like a single contact area with a diameter determined by the contact geometry.

At this point, the contact resistance can be calculated by the final formula:

$$R_C = \rho \left(\frac{H}{F}\right)^{1/2} \tag{EQ 5}$$

where H is the hardness of the contact material and F the contact force. The units of hardness are force/area so the square root of the inverse of hardness gives the diameter of the contact area. This equation shows why the contact force is a fundamental parameter of connector design.

1.6.4 Summary

A summary of the effects of the a-spot surface structure of contact interfaces follows:

- On the microscale of the contact interface all surfaces are rough.
- Surface roughness leads to the creation of small individual contact areas, called a-spots, at the contact interface.
- The number of a-spots created depends on three factors: the contact force, the surface roughness and the hardness of the material. The contact force determines the total a-spot contact area and the surface roughness determines how many a-spots will be included in that area.
- The geometry of the apparent contact area containing the a-spots is determined by the geometries of the plug and receptacle contact surfaces as they come together during mating.
- Current flow across the contact interface is constricted to flow through the a-spot distribution a geometric effect leading to an increase in resistance, constriction resistance, at the contact interface.

The distribution and number of a-spots provides a highly redundant network of resistances in parallel at the contact interface, a very important factor in the stability of contact interface resistance. It must be noted that permanent connection interfaces are also created in the same manner, and differ only in the amount of deformation, cold welding and the total contact area that is created.

1.7 Contact Finishes

1.7.1 Contact Finish Selection

The selection of a suitable surface treatment system for a particular connector application requires the consideration of a number of parameters. The parameters include the number of mating cycles during the service life, the required mating force and number of pins, and the current carrying capacity. Associated application requirements include thermal environment, hardness, corrosion, mechanical stress and intended field life. In the following section, these application-related parameters are considered.

1.7.2 Contact Finish Requirements

Contact springs are provided by connectors with a contact surface that fulfils two basic functions: Protecting the copper alloy spring elements from corrosion and optimising the mechanical and electrical performance of the contact interface.

Contact springs are generally made of copper alloys for connectors because they have high electrical conductivity and good formability with reasonable strength. Unfortunately, copper alloys are susceptible to corrosion in environments containing oxygen, sulphur and chlorine, which are typical operating environments for connectors. Therefore, the first task of a contact surface is to protect the copper alloy spring from corrosion.

Corrosion protection can be achieved by simply coating the contact spring with a material that is not susceptible to corrosion or, in the case of connectors, with a material that forms films that are self-limiting in thickness and can be easily mechanically displaced when a connector is mated. There are two general types of contact finishes, noble and non-noble Gold is the predominant noble metal finish material. The term "noble" refers to the fact that gold is not susceptible to corrosion. Tin is the most common non-noble finish material. Tin is susceptible to corrosion by the formation of a thin self-limiting thickness of tin oxide. This oxide protects the contact interface from further corrosion but is readily displaced on mating of the connector to provide the desired metal-to-metal contact interface.

1.7.3 Noble Finishes

Corrosion protection is necessary to ensure that metal-to-metal A-point contact surfaces can be created and maintained. A gold-to-gold contact interface is inherently metal-to-metal because gold does not corrode. A tin-to-tin contact interface becomes metal-to-metal when the surface tin oxide is displaced when the connector is inserted. "Optimising" the wear characteristics of contact surfaces is particularly important with gold contact surfaces because the wear that occurs with each mating operation causes the contact coating to wear away, exposing the underlying copper alloy, which is then susceptible to corrosion. The wear behaviour of gold coatings can be improved by increasing the hardness of the gold through alloying and the effective hardness of the gold by using a nickel backing.

A contact finish, C in Figure 1.58, consists of a surface coating, generally produced by electroplating, over the contact spring to provide two performance benefits: Corrosion protection for the contact springs and optimization of the electrical and mechanical characteristics of the contact interfaces. As mentioned, copper alloys are typically used as connector contact springs and are susceptible to corrosion in typical connector operating environments. The contact finish provides corrosion protection by covering the exposed copper to protect it from the operating environment. "Optimization" of contact interface performance is provided by influencing the corrosion and wear characteristics of the contact interface.

Connectors using noble metal finishes, usually gold, can be used in all applications and environments. Because they are more expensive than connectors of similar designs using non-noble finishes, due to the cost of gold plating, they are not generally used in commercial applications. They are, however, specified in most computer and telecom applications due to their performance advantages.

A noble metal contact finish is a system consisting of:

- a noble metal surface, usually gold
- a nickel underplate and
- the contact spring base metal (copper alloy)

All three elements influence both the electrical and mechanical characteristics of the contact interface.

Electrically the noble metal surface allows for the formation of the metal-to-metal a-spots that create the electrical contact interface; the nickel underplate provides benefits that protect the nobility of the contact interface due to external corrosion mechanisms; and the constriction resistance takes place in the contact spring material.

Mechanically it is the contact surface that will experience the wear process directly; the hardness of the nickel underplate improves the wear resistance of the system; and the hardness of the base metal contact spring influences the overall amount of deformation that occurs. This effect of the contact spring on deformation is due to the low thickness of the gold and nickel plating's, of the order of microns, so that the stresses applied to the contact interface penetrate into the contact spring.

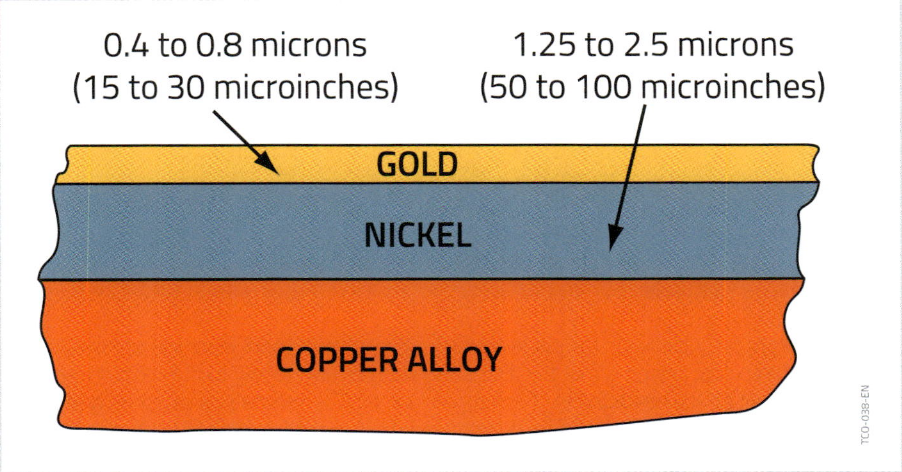

Fig. 1.38: Noble metal contact finish

Figure 1.38 schematically illustrates a cross section of a noble metal finish. The noble metal surface plating is usually gold. In computer/telecom applications the gold is typically electroplated to thicknesses of the order of 0.4 to 0.8 µm (15 to 30 micro-inches). The gold plating used in connectors is typically a "hard" gold. Hard golds are lightly alloyed, 0.1 %, with cobalt, nickel or iron, with cobalt being the most common. There is an alternative noble metal finish using palladium alloys. The most common palladium alloy is palladium (80)-nickel (20) with palladium (80)-cobalt (20) also used. In this case the bulk of the plating thickness is palladium alloy, but the surface plating may be a gold flash. A flash is a thin plating of the order of a tenth of a micron (4 micro-inches) in thickness.

The nickel underplate is typically electroplated to thicknesses of 1.25 to 2.5 µm (50 to 100 micro-inches) and provides several benefits with respect to corrosion and wear which will be discussed in chapter 1.8 "Contact Finish Degradation Mechanisms".

The nickel underplate:

- reduces the susceptibility to pore corrosion
- reduces corrosion migration
- provides a diffusion barrier to base metal migration
- improves the mating durability of the contact interface

Gold contact finishes can provide durability ratings from tens to thousands of mating cycles. This range is possible because gold contact finishes require normal forces in the range of 50 to 100 grams and hard alloy golds in the thickness range of 0.2 to 2.5 µm (15 to 100 micro-inches) are options. The combination of low contact force and a low coefficient of friction results in low mating forces and allows for high pin count capability. Gold contact finishes are suitable for currents ranging from milliamps to tens of amperes. Experience has shown those gold finished connectors are capable of maintaining acceptable resistance stability for both signal (milliamps) and power (tens of amperes) applications.

Hard gold contact finishes are suitable for application temperatures up to 125 degrees Celsius, beyond this temperature self-diffusion of the hardening agents to the contact surface my compromise the nobility of the gold. Soft gold and thicker nickel underplates should be considered for higher temperature applications. As noted, exposed copper alloy base metal, through porosity, wear or plating practice can lead to copper corrosion in environments containing sulphur and chlorine. Contact lubricants can be effective in enhancing performance in such environments.

Gold flashed palladium alloy contact finishes were developed as an alternative to gold finishes in the 80's when gold prices rose dramatically. Since that time the price of palladium has escalated to the point where it is now more expensive than gold.

Experience with gold flashed palladium alloy finishes has shown its performance is comparable to gold in most respects and the durability life is somewhat better due to the higher hardness of palladium compared to gold.

The contact spring is included in the noble metal finish system for both electrical and mechanical reasons. The electrical reason is that the distribution of the current flow to the contact interface is dependent on the conductivity of the contact spring. The mechanical reason is that the stresses applied to the contact interface during mating can extend down into the spring material due to the small thicknesses of the gold and nickel plating.

1.7.4 Non-Noble Finishes

Tin is the dominant non-noble contact finish due to its widespread use in connectors for commercial and industrial applications. Tin finishes are generally electroplated, though reflowed tin coatings are also used. The thickness of electroplated tin is generally in the range of 2.5 to 5.0 µm (100 to 200 micro-inches) with reflowed coatings being somewhat thicker.

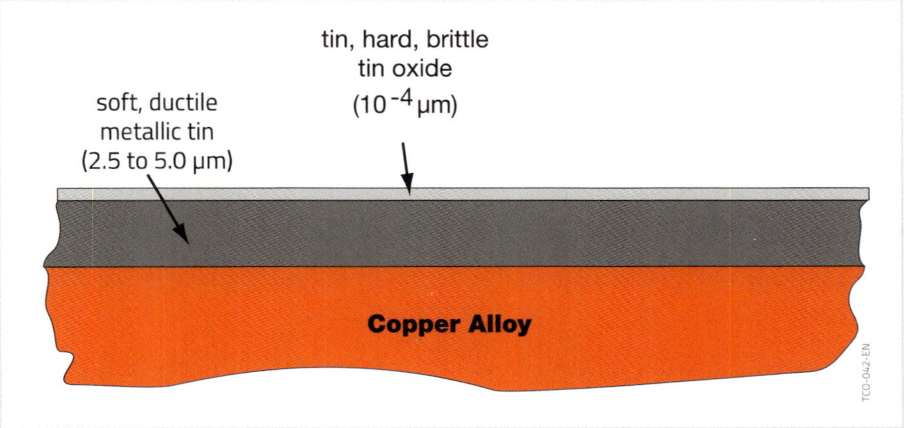

Fig. 1.39: Tin contact finish

Tin provides corrosion protection to the contact spring due to the self-limiting tin oxide film on the tin surface. As indicated in Figure 1.39 the oxide layer is very thin, hard and brittle and is readily displaced on mating of the connector resulting in tin-to-tin metallic contact areas. The major limitation to the use of tin as a general purpose connector contact finish is the fact that a tin-to-tin contact interface is susceptible to reoxidation if the interface experiences small scale movements due to mechanical or thermal stresses in the connector application environment. This degradation mechanism is called fretting corrosion.

Nickel and silver are also used in connector applications. Nickel has an even thinner oxide layer on its surface than tin. Nickel is most commonly used in battery contacts and also in high temperature applications. Silver surface films are primarily sulfides, tarnish films, which are also readily displaced on connector mating. Silver is used primarily in high current contacts due to the high electrical conductivity of silver and a better resistance to arcing damage compared to gold or tin contacts.

Tin and silver are the two most common used non-noble finishes. Tin finishes are used in a wide range of applications, from white goods to automotive. Silver is primarily used in power/high current applications.

1.7.4.1 Tin

Tin contact finishes are the most widely used non-noble contact finish. They are characterized by low durability (tin is a soft material) and high mating force (high friction coefficient and high contact force), and susceptibility to fretting corrosion.

These characteristics indicate that tin finishes are not suitable for connector applications requiring high mating cycles and high pin counts. These are intrinsic limitations though the range of mating cycles and pin counts can be extended by the use of contact lubricants.

Susceptibility to fretting corrosion suggests that tin finishes are questionable for use in demanding mechanical environments or those with large temperature variations. It must be noted, however, that tin finishes have been successfully used in under hood automotive applications; applications characterized by demanding mechanical and thermal requirements. This success has been realized using high contact forces to ensure the mechanical stability necessary to minimize fretting corrosion degradation. This is a good example of a guideline rather than a rule with respect to concern for fretting corrosion. But it is good practice to critically evaluate the mechanical and thermal characteristics of an application when considering tin as a contact finish. Apart from fretting corrosion, tin may be a good choice for harsh chemical environments due to the self-limiting passivating surface oxide. An additional temperature concern with tin finishes is the growth of copper-tin intermetallic compounds at the interface between the tin and the copper alloy contact spring. In worst case conditions the tin can be converted into intermetallic compounds with both mechanical and electrical effects. Mechanically the intermetallic compounds are brittle and electrically they are poor conductors. Tin finishes may not be appropriate in applications where the connector is exposed to temperatures more than 100 degrees Celsius.

The current carrying capacity and electrical resistance of tin finishes will be discussed simultaneously. High current applications, say tens of amperes, can be problematic due to the effects of Joule, or I^2R, heating. Joule heating and its consequences for connector degradation will be discussed in detail in chapter 1.11.1. At this point it is sufficient to note that Joule heating can create a positive feedback loop because contact resistance increases with temperature; Joule heating increases resistance and increased resistance increases temperature, and the loop continues. It is important, therefore, for connectors intended to carry high currents to have a low and stable contact resistance. Tin can readily realize low values of contact resistance. The stability of tin contact resistance, however, may be compromised by fretting corrosion driven increases in contact resistance as discussed earlier.

Once again, practice has shown that tin interfaces can provide reliable high current performance as long as the mechanical stability of the contact interface is maintained, generally through the use of high contact forces. The success of tin finished connectors in the white goods industry is an example of this capability.

To summarize:

Tin finished connectors are intrinsically limited to low durability, low pin count applications due to the fact tin is a soft material (high friction) and high contact forces are needed to ensure the mechanical stability of the contact interface to reduce susceptibility to fretting corrosion. Another intrinsic limitation is the application temperature due to intermetallic compound effects on electrical and mechanical performance.

When concerns for fretting corrosion can be minimized or controlled, however, tin finished contacts can be used in a broad range of connector applications. The difficulty in minimizing the potential for fretting corrosion, however, should not be under emphasized.

Whisker: Whiskers are single crystal fibres (when pure Tin plating is used) which can cause electrical failures. Lead was a good way to prevent them as there was no Tin whisker problem while Sn-Pb was the major plating used in the connector industry.

Whisker density, length and shape are quite different and unpredictable and can lead to transient or permanent shorts in electric and electronic circuits. Either directly on the component where they have grown or (if they break loose) somewhere else in the circuit.

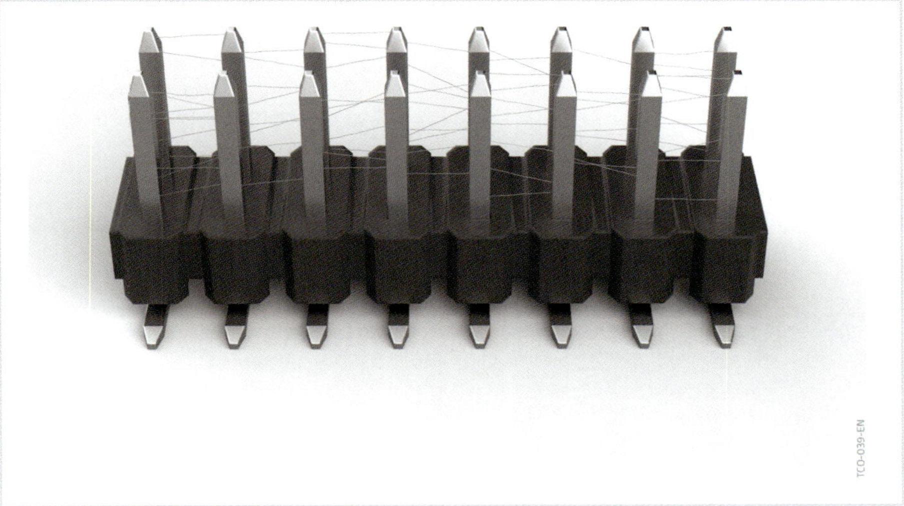

Fig. 1.40: Whisker effect on a pin header (drawing)

In the early 2000 years whiskers caused losses of over US$ 1 Billion, mostly in military, satellite and avionic applications, therefore Sn-Pb plating's remain active in these fields which are not concerned by the lead free considerations.

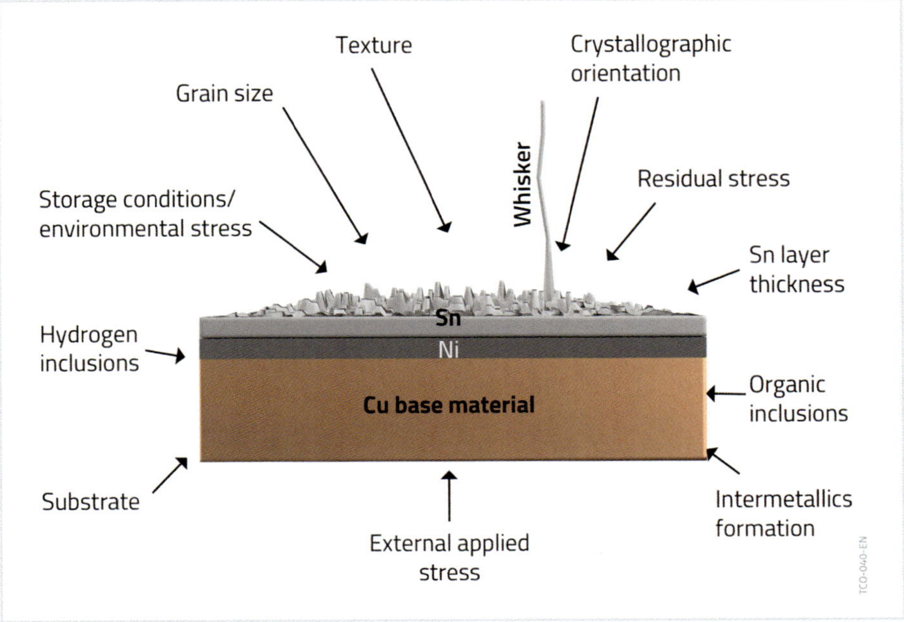

Fig. 1.41: Factors that influence the appearance of tin whiskers

The number of factors and their interactions are so complex that the whisker growth mechanism still hasn't been totally understood. As it's not the main purpose of this book to study in details the whisker formation principle, we will only state that it's mostly connector with small pitches, typically 0.5 mm or lower which are potentially subject to this problem.

In order to prevent Tin whisker creation, the most simple and efficient ways known are to:

- Use another plating than pure tin: like gold (full or selective gold ...)
- Use matt tin instead of bright Tin
- Use Ni under plating below the Tin

1.7.4.2 Silver

The major limitation of silver as a contact finish is its susceptibility to surface tarnish films. The tarnish films are more an appearance than a functional problem because they are readily disrupted on mating and the desired silver-to-silver contact interface is easily established. The use of silver finishes in typical connector applications is minimal due to resistance failures in low force non-wiping relay applications, in which the tarnish films were not effectively disrupted, highlighting the potential for tarnish related failures. The major application area for silver finishes is in high current connectors where the high electrical conductivity of silver takes on enhanced importance due to Joule heating concerns.

Silver is intermediate in hardness between tin and gold and can be used at contact forces lower than those for tin because the requirement for mechanical stability of the contact interface is not as demanding. Thus, silver finished contacts will have higher durability and lower mating forces than tin finished contacts. The pin count does not come into play because power connectors do not generally require high pin counts.

The current capacity, resistance and stability of resistance of silver finishes are good due to high conductivity and low susceptibility to fretting corrosion. High contact forces are typically used in power contacts, further adding to mechanical and electrical stability.

Environments containing sulphur promote tarnish formation. Sulphur plus chloride can result in more deleterious surface films. Silver plated contacts are often protected from tarnish by including special vapor phase inhibitor paper in the packaging.

1.7.5 Contact Finish Lubricants

Connector contact lubricants are used to provide two different performance benefits:

Reduction in friction and corrosion protection. Reducing friction will reduce connector mating forces and increase connector mating cycle life. Lubrication can also reduce the susceptibility to corrosion, both fretting and environmental, and enable the use of the connector in harsher environments.

For noble metal contact finishes, friction reduction is an obvious benefit with respect to mating durability allowing the connector to be specified for a higher number of mating cycles, or to meet a given mating cycle requirement with a reduced gold thickness. Reduction in mating forces is also important and can allow higher pin count connectors to be realized or allow the use of higher contact forces, enhancing mechanical stability, at a given mating force. Corrosion protection may also be important in reducing the potential for corrosion of exposed copper alloy in more demanding environments.

For tin contact finishes the benefits of lubrication in reducing the potential for fretting corrosion are dominant. The benefits of reductions in friction are not as significant for tin finishes because the intrinsic durability of tin is low, due to the low hardness of tin. The high contact force typical of tin contacts, to provide mechanical stability, results in a high mating force, which can be moderated by lubrication, but the pin count capability of tin connectors is still limited. Silver plated contacts benefit primarily from corrosion protection to reduce the potential for surface tarnish.

The components of a lubricant that reduces friction and a lubricant that provides environmental protection are different and, therefore, different formulations are necessary for noble and non-noble finish systems. Contact lubricant formulations are proprietary, but some general comments on lubricant components are possible.

The basic component is a solvent, or solvents, that can take all of the necessary active components into solution for ease and consistency of application. Application methods include sprays, dips and brushes/pads.

The friction reducing component of the formulation is an oil and/or wax. An oil may wet the surface and provide complete coverage. Waxes do not generally wet the surface yielding incomplete coverage. The waxes, however, may smear over the surface as the connector is mated enhancing the effective coverage.

Control of the lubricant viscosity is also important. The lubricant must balance an ability to cover the contact surface initially, be displaced out of the contact interface when the connector is fully mated and flow around the contact interface to provide an environmental "seal".

A contact lubricant may also contain one or more corrosion inhibitors, or surface wetting agents to inhibit access of the corrosive species to the contact surface.

Finally, ultra-violet tracers are often added to provide "after the fact" evidence that the contact lubricant has been applied.

While the potential performance benefits of contact lubricants are obvious, they also have limitations.

Application of the lubricant requires an additional process step. In some cases the lubricant can be applied at the end of the contact plating line by a dip or spray. In other cases secondary operations to apply the lubricant may be necessary. In either case there will be incremental costs of application.

Control of the quantity of lubricant applied may be necessary. While an appropriate amount of lubricant must be applied, excess lubricant may be a liability. This is particularly significant if the lubricant has a tendency to retain dust or contaminants in storage or in the field.

The lubricant may have a limited temperature dependent effective lifetime due to evaporation, flow from the contact interface or component degradation.

Despite these limitations, contact lubricants, when properly chosen and formulated, provide an opportunity to enhance the performance of connectors in more demanding environments, or to maintain the performance of connectors with reduced plating thicknesses.

1.8 Contact Finish Degradation Mechanisms

The two most significant contact finish degradation mechanisms are corrosion and wear. While wear is a degradation mechanism, its main effect on connector degradation is to make the contact interface more susceptible to corrosion due to wear through of the contact finish with resultant exposure of the copper alloy contact spring. Wear through of the finish can occur due to the multiple mating cycles the connector may experience during its application life time. That, however, is not the only active wear mechanism. Wear can also occur due to small motions of the contact interface due to mechanical or thermally generated stresses. **These small scale motions, fractions of a micron to a few microns, are called fretting.** While the wear that occurs during a mating cycle may be much more extensive, due to the larger sliding distance, than that of a fretting cycle, the number of fretting cycles a connector may experience in its application lifetime may be orders of magnitude larger than the number of mating cycles it will experience. Thus, fretting wear cannot be neglected in considering connector degradation mechanisms. In fact, as noted above, fretting wear leading to fretting corrosion is the primary degradation mechanism for tin contact finishes as will be discussed.

The discussion of corrosion as a degradation mechanism will be limited to noble metal and tin finishes. The corrosion related degradation of tin finishes is fretting corrosion as noted previously. In contrast, multiple corrosion mechanisms come into play for noble metal finishes. Recall that in the previous chapter it was noted that, "the term "noble" refers to the fact that gold is not susceptible to corrosion in most environments." This statement is true, the gold is not the source of corrosion products, exposed copper from the contact spring is the corrosion source. Due to this difference in corrosion mechanisms and susceptibility noble and non-noble degradation mechanisms will be discussed separately.

1.8.1 Noble Metal Finish Degradation Mechanisms

Noble metal finish degradation mechanisms include both corrosion and wear, with wear through of the noble metal leading to increased susceptibility of the contact interface to corrosion. Consider corrosion first.

As noted above, multiple corrosion mechanisms/kinetics are active in noble metal contact interfaces. These include pore corrosion, corrosion migration and surface corrosion. It is important to recognize the functional benefits of the nickel underplate in a noble metal contact finish system in mitigating the effects of all of these corrosion mechanisms. These benefits include:

- reducing the sensitivity of the interface to pore corrosion
- reducing the rates of corrosion migration at the contact interface
- reducing the rate of diffusion of base metal components to the contact surface.

A pore is a microscopic opening in an electroplated contact finish. As indicated in Figure 1.42, pores are microscopic in size and arise during the electroplating process. The number of pores in an electroplate depends on the plating thickness and the quality of the plating process.

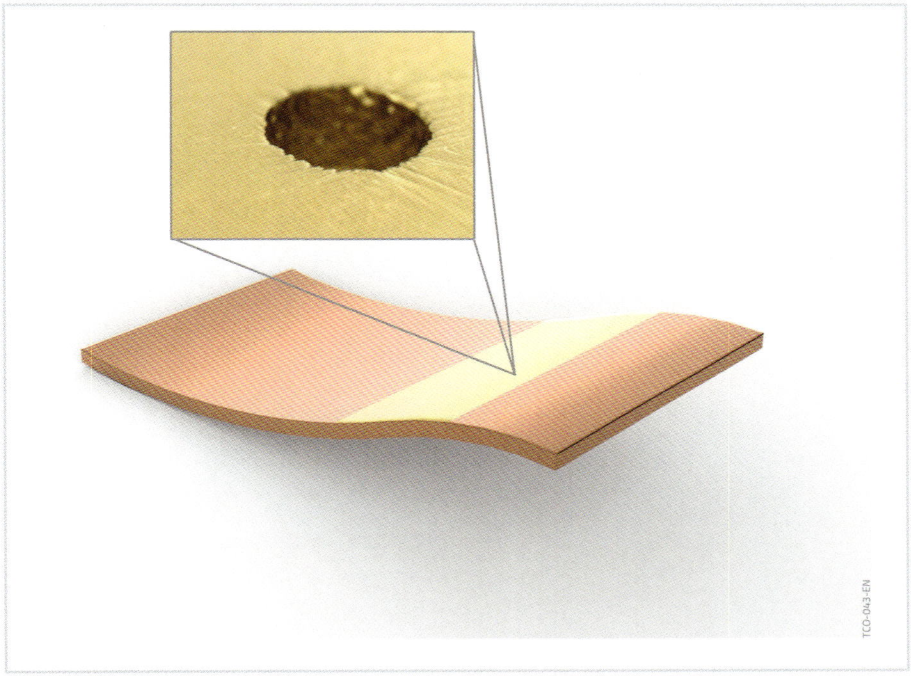

Fig. 1.42: Pore site

The corrosion potential of a pore site depends on the material at the base of the pore. Figure 1.43 schematically illustrates a passive and an active pore site.

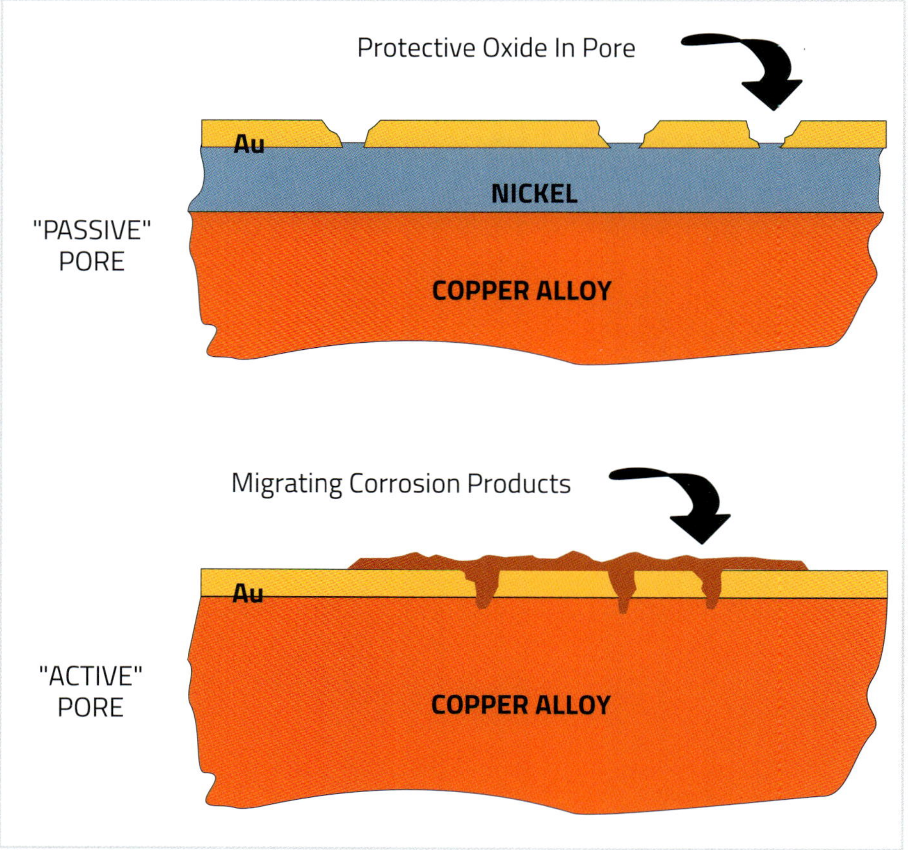

Fig. 1.43: Passive/active pore sites

The passive pore site ends at a surface that is not susceptible to corrosion, in the case of noble metal finishes, the nickel underplate. As noted earlier, nickel forms a very thin surface oxide which then "passivates" the nickel surface, that is, prevents further corrosion. Thus, a pore down to the nickel underplate produces no corrosion products that can make their way to the contact interface.

Without a nickel underplate the pore site would terminate on the copper alloy contact spring surface, a surface susceptible to corrosion. The copper corrosion products that form at the base of the pore site can migrate up the sides of the pore to, and over, the gold contact surface as indicated in Figure 1.43. These migrating corrosion products, of course, can interfere with the creation of the desired gold-to-gold contact interfaces.

Figure 2.7 shows a photomicrograph of pore corrosion migration over a gold contact finish. The rings around the pore site indicate different thicknesses of corrosion product. The sample in this case was a gold flash over palladium alloy over nickel over copper alloy.

The sample was exposed to a mixed flowing gas environment containing chlorine and sulphur at parts per billion levels. The carrier gas was air, thus the oxygen, also shown.

This test environment is known to simulate the corrosion observed in noble metal finishes exposed to industrial environments. Qualitatively it appears that the oxygen and chlorine remain in the vicinity of the pore site. The sulphur, however, clearly migrates, as indicated by the ring of enhanced sulphur intensity around the pore site. Thus, the rings around the pore sites shown in Figure 1.44 are primarily copper-sulphur corrosion products. This conclusion has been established in many studies of mixed flowing gas exposures.

Fig. 1.44: Pore corrosion

If pore corrosion occurs prior to the mating of the connector, during the storage life of the connector, and the mated contact interface happens to include a pore site of the kind shown in Figure 1.44, the contact interface resistance may be compromised. The probability of such an event depends on the pore density at the contact interface. This is why control of porosity is often a specification requirement for noble metal finished connectors. Both the pore density and a critical contact area may be specified.

Pore corrosion occurring after mating can still compromise the contact resistance. Corrosion product from a pore site near the contact interface can be taken into the contact interface as a result of fretting motions. In this case the fretting motions do not lead to the corrosion, as in fretting corrosion in tin. The mechanism in this case is that a fretting motion of the contact interface that moves the interface onto an area of corrosion product may cause an increase in resistance, either intermittent or "permanent". Further fretting motions may incorporate additional corrosion product into the contact interface with additional intermittent of "permanent" increases in contact resistance.

This description of contact resistance degradation highlights the importance of maintaining the mechanical stability of the contact interface. Corrosion product does not migrate into contact interfaces directly due to the small spacing between the contact surfaces. This restriction is sometimes referred to as a "gas tight interface". The required mechanical stability results from the friction forces at the contact interface due to the contact normal force as will be discussed in Chapter 1.9.

A brief discussion of the kinetics of corrosion migration is in order at this point. Consider the system schematically illustrated in Figure 1.45. The rectangular coupon in the center of the figure consists of a copper alloy base metal which has been plated with one of the finish systems indicated in the figure, gold, gold over nickel, palladium, palladium over nickel and nickel alone. The plating thicknesses were those of telecom applications, 0.8 µm (30 micro-inches) for the surface plating's and 1.25 to 2.0 µm (50 to 100 micro-inches) for the nickel. A disk was then stamped out of the coupon so that the top and bottom surface were plated with one of the systems and the sheared sides of the disk were primarily exposed copper alloy. The exposed copper alloy is the corrosion source for the following discussion.

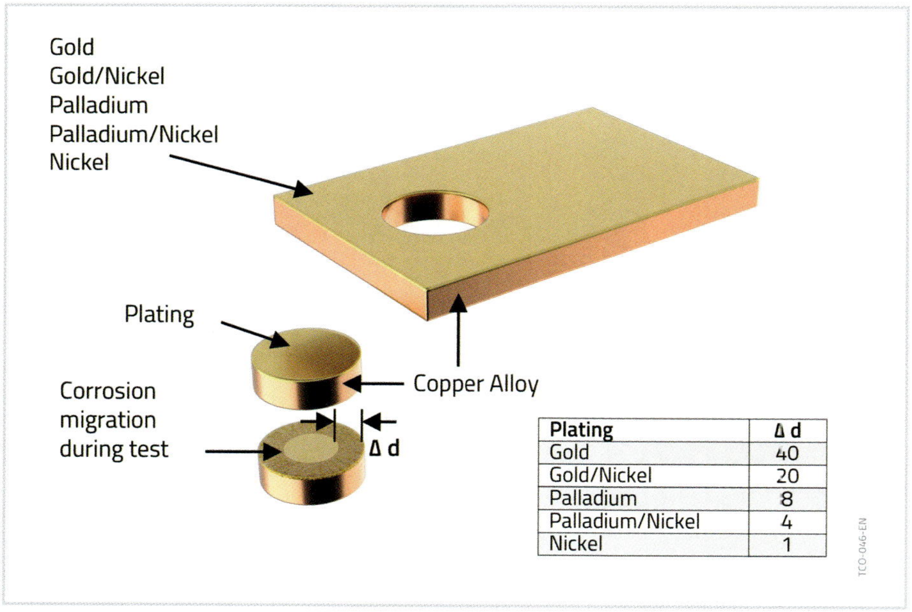

Plating	Δ d
Gold	40
Gold/Nickel	20
Palladium	8
Palladium/Nickel	4
Nickel	1

Fig. 1.45: Corrosion migration kinetics

The stamped disks were then exposed to a mixed flowing gas test environment intended to simulate an industrial environment. Details of Mixed Flowing Gas (MFG) environments will be provided in Chapter 1.18. The corrosion products forming on the exposed copper migrate up the sides of the coupon, across the plating system and then over the plated surface of the disk. The table in the figure gives the migration distance, Δd, for each of the plating systems. There are two patterns in the data that merit discussion. First, for the plating systems without an underplate, the corrosion products migrate further over gold than over palladium or nickel plating. Second, with the nickel underplate, the corrosion products migrate further over gold over nickel than over palladium over nickel.

These data can be explained, in a simplified but basically correct manner, by noting that corrosion products migrate faster over clean surfaces than over film covered or contaminated surfaces. Gold is the cleanest surface and shows the most migration, followed by palladium, which is not as corrosion resistant as gold, and finally, most slowly over nickel, which, as noted previously, always has a very thin oxide film on the surface.

The second set of data, which indicate that the presence of the nickel underplate reduces the migration distance by a factor of two, is similarly explained. In this particular case, the migration of corrosion products over the nickel underplate takes up about half the test time and the migration rate increases as the corrosion product reaches the surface plating. Clearly the corrosion migration could be reduced further by the use of thicker nickel underplates. This opportunity is mitigated by the fact that nickel plating's tend to become more brittle with increasing plating thickness and control of the surface roughness is also compromised. For these reasons nickel underplates seldom exceed four µm (160 micro-inches) in thickness.

The discussion of copper alloy corrosion product migration over gold and nickel surfaces demonstrate, as noted previously, the effectiveness of a nickel underplate in reducing the sensitivity of the interface to pore corrosion and reducing the rates of corrosion migration at the contact interface.

An additional corrosion related benefit of a nickel underplate, also noted previously, is to reduce the rate of copper migration from the contact spring base metal to the gold contact finish surface. Thus nickel underplates extend the storage and application lifetimes of gold plated contacts.

1.8.2 Nickel Underplates and Noble Metal Finish Wear

Nickel underplates in noble metal finishes provide another very important performance benefit that is not related to corrosion. The durability of noble metal finishes during connector mating and unmating is significantly improved by a nickel underplate. This improvement arises from the hardness of a nickel underplate as compared to that of a gold or palladium contact plating. Hard metals tend to be more wear resistant than softer materials and a nickel underplate increases the composite hardness of a noble metal finish system. This benefit is demonstrated by the data shown in Figure 1.46.

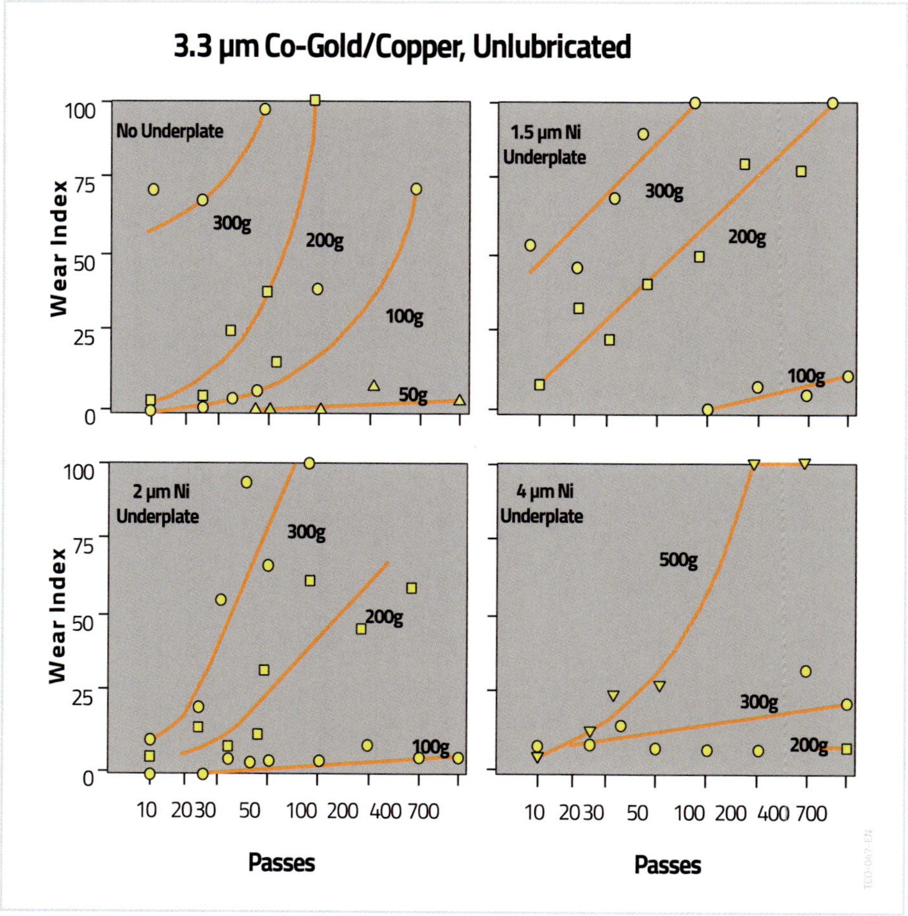

Fig. 1.46: Effect of nickel underplate on contact interface wear[4]

This figure is only one example of an extensive body of work on the durability of noble metal contact finishes carried out at Bell Telephone Laboratories in Murray Hill, New Jersey, USA during the 1970's. At that time telecom applications were using gold thicknesses in the range of 3 µm (120 micro-inches) as indicated by the 3.3 µm (130 micro-inches) thickness in the figure. Note that this example was not intentionally lubricated. Four nickel thicknesses were tested, 0 µm, 1.5 µm (60 micro-inches), 2 µm (80 micro-inches) and 4 µm (160 micro-inches). The data are plotted as Wear Index versus the Number of Passes along the linear mechanically driven wear track. A wear index of 25 means that 25 % of the length of the wear track showed exposed underplate or base metal, that is, all the gold was worn through in 25 % of the track.

Consider the data for 200 gram contact force as a reference for this discussion. With no nickel, the entire wear track, 100 % wear index, is worn through at about 50 passes. With 1.5 µm (60 micro-inches) of nickel, a 100 % wear index occurs at about 600 passes. Extrapolating the data for 2 µm (80 micro-inches) indicates a wear index of about 85 at about 700 passes. Finally, at 4 µm (160 micro-inches) of nickel the wear index at 700 passes is about 20. Clearly, as the

[4] Adapted from Antler, M: "Wear of Contact Finishes:", Proc. 11th Ann. Conn. and Interconn. Tech. Symposium, 1978

nickel thickness, and the composite hardness of the finish, increases the wear rate decreases significantly. As noted previously the 4 µm thickness approaches the upper limit of nickel thickness suitable for connectors.

Summary of Nickel Underplate Benefits

To summarize, a nickel underplate reduces:

- the sensitivity of the finish to porosity by passivating pore sites
- the rate of corrosion migration from exposed base metal edges and pore sites to the contact interface
- the rate of copper diffusion from the contact spring to the finish surface where the copper could react with the application environment
- the wear rate of the finish during mating or fretting cycles

Because of these benefits all noble metal contact finishes should include a nickel underplate with a thickness in the range of 1.25 to 2.5 µm (50 to 100 micro-inches).

1.8.3 Non-Noble Finish Degradation Mechanisms

Because tin is the most commonly used non-noble contact finish, it will be the focus of this chapter. Silver and nickel finishes are also used with silver being a good choice for power connectors and nickel commonly used as a battery contact

Fretting Corrosion of Tin Contact Interfaces

As noted previously the dominant degradation mechanism for tin contact finishes is fretting corrosion. Before discussing the mechanism of fretting corrosion a review of the kinetics of creating a tin-to-tin contact interface as schematically illustrated in Figure 1.47 is in order. The contact probe applies a load to the tin oxide covered tin surface. Because the tin oxide is very thin, hard and brittle, it cracks under the applied load. The load then transfers to the tin, which is soft and ductile, causing the tin to flow and the cracks in the oxide to spread apart. The tin then extrudes into the cracks in the oxide creating a tin-to-tin contact interface with the tin plating on the contact probe which has experienced a similar deformation and displacement of its tin finish. This mechanism is even more effective when the probe slides against the mating surface as it would in the mating of a connector. The sliding action effectively pushes aside the tin oxide layers creating the desired tin-to-tin interfaces. Creating the initial contact interface is easily accomplished.

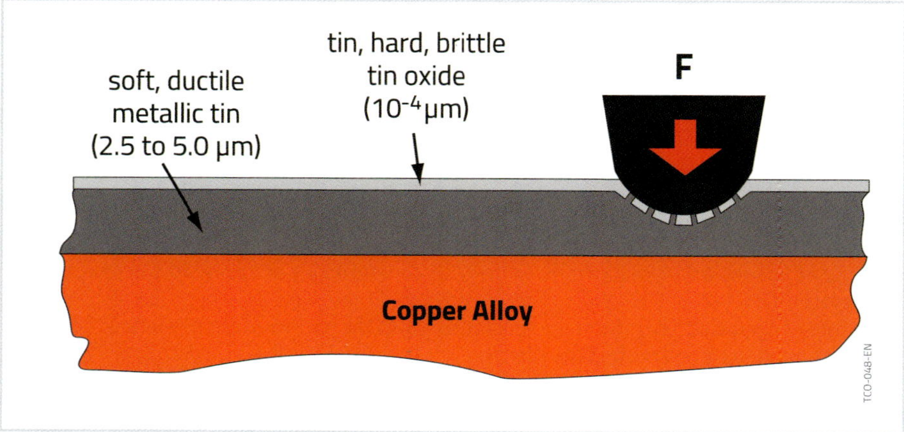

Fig. 1.47: Disrupted tin oxide surface on tin

The problem is that any exposed tin that does not make contact to a tin surface will reoxidize to create a new tin oxide layer over the tin. This is the mechanism of fretting corrosion as schematically illustrated in Figure 1.48. Figure 1.48a shows the original tin oxide disruption and displacement as just described. If, however, the contact interface moves under some applied shear stress in the application environment two things happen as indicated in Figure 1.48b. First, the sliding action during the motion disrupts and displaces the tin oxide at the new contact area creating a new tin-to-tin contact interface. At the same time the exposed tin at the initial contact area reoxidizes. After this single movement there will be little, if any, change in contact resistance. If, however, multiple movements occur, such movements are referred to as fretting; additional oxide debris will accumulate in the area of the contact as indicated in Figure 1.48c. This debris will not be as easily displaced as the original oxide layer because it is particulate rather than continuous and may be pressed down into the tin rather than being displaced. Eventually the entire sliding track consists of accumulated oxide debris, Figure 1.48d, and the contact may approach open circuit resistance values.

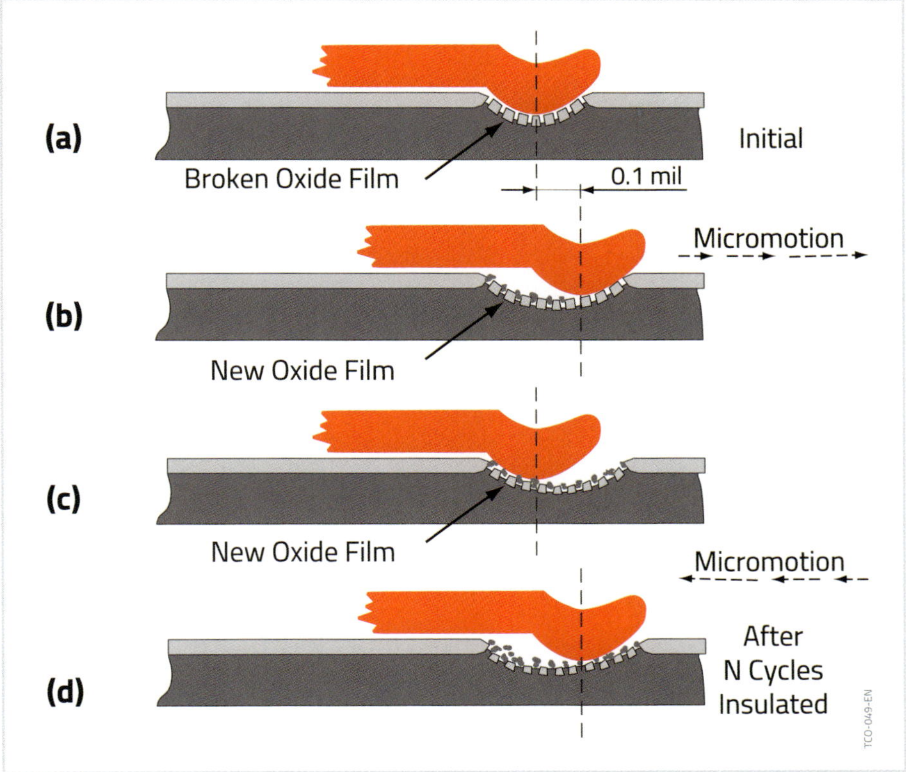

Fig. 1.48: Mechanism of fretting corrosion

Figure 1.48 shows a contact area after fretting corrosion. The oxide particulates are very small resulting in a black appearance of the fretted interface. Such black spots are typically the first diagnostic indication of fretting corrosion during failure analysis. A cross section of a fretted spot is shown in Figure 1.50. Note the mix of tin and oxide debris that makes displacement of the debris problematic after a number of fretting cycles.

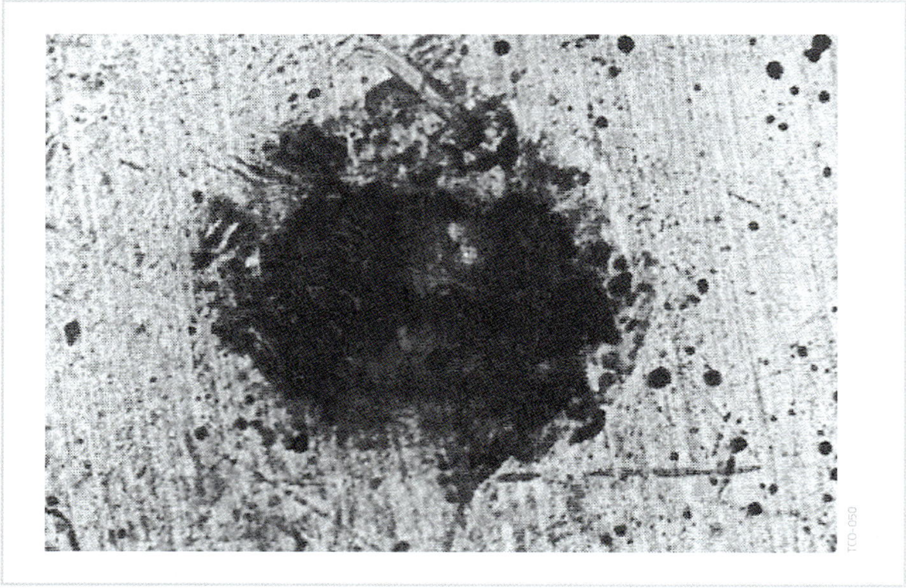

Fig. 1.49: Fretting corrosion spot[5]

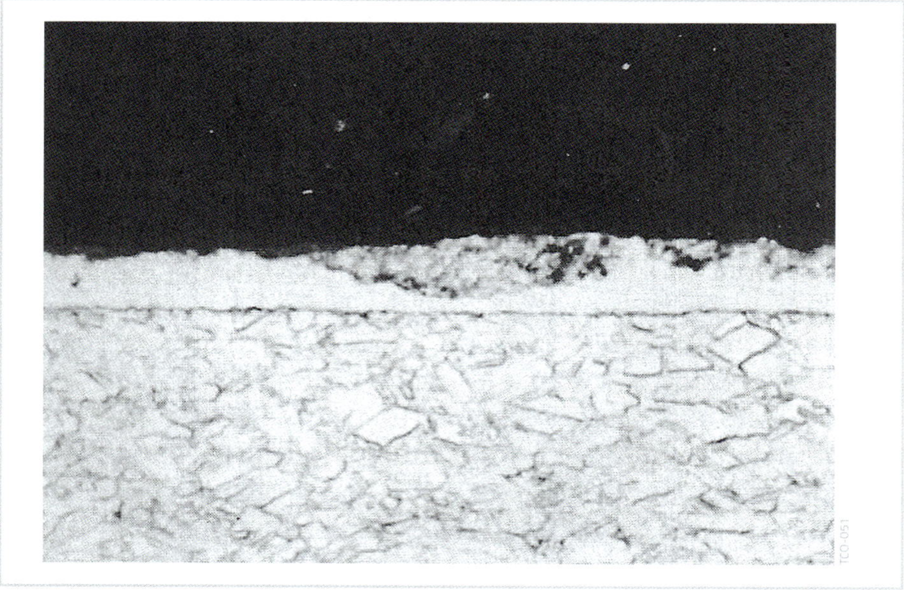

Fig. 1.50: Fretting corrosion spot cross section[6]

[5] Mroczkowski, R. S.: "Corrosion and Electrical Contact Interfaces", NACE, Corrosion 85, 1985
[6] Mroczkowski, R. S.: "Corrosion and Electrical Contact Interfaces", NACE, Corrosion 85, 1985

This accumulation of tin oxide debris, the corrosion part of fretting corrosion, will, over time, result in an increase in the contact resistance. The change in resistance may appear first as intermittent increases in resistance. Over time the average resistance will increase and the intermittent resistance swings will increase in magnitude and duration as schematically illustrated in Figure 1.51. The intermittent resistances may appear as electrical noise, but the increase in duration and magnitude may eventually disrupt system operation. Fretting resistance increases can increase to the tens or hundreds of ohms or even result in open circuits. Significant increases in resistance, tens of ohms, can occur after a few thousand fretting cycles.

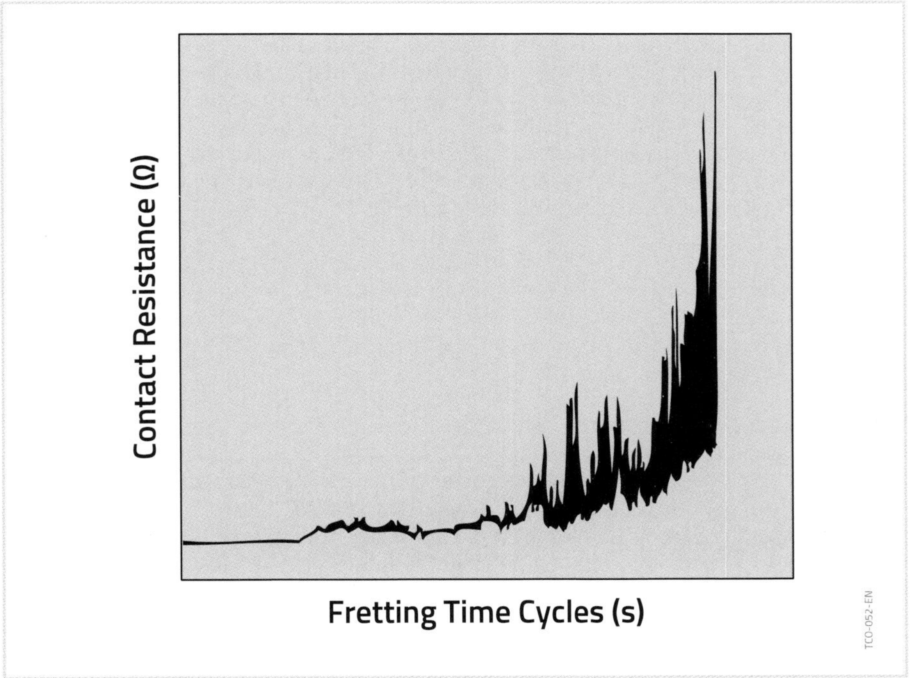

Fig. 1.51: Intermittent resistance during fretting corrosion

The rate of fretting corrosion degradation depends on several factors. The two most important are the contact force and the length of the fretting motion. The contact force dependence arises from the fact that increased contact force results in increased wear particle formation as previously discussed. The length dependence is due to the typical build-up of wear debris at the ends of the sliding track. A longer fretting track may take longer to accumulate the wear debris necessary to cause resistance increases all along the track. The length of the fretting track depends on the driving force for the fretting motion. It is important to note, however, that fretting motions of a few to several tens of µm can lead to significant fretting corrosion degradation.

The driving forces for fretting motions are both mechanical and thermal and are obviously application dependent. Mechanical drivers include both shock and vibration. Thermal driving forces arise from thermal expansion coefficient mismatches between connector housings and PCBs or other system components due to thermal cycling. Rapid thermal cycles are referred to as thermal shock. Slower cycles, for example temperature differences due to on-off cycling of equipment, are more frequent causes of fretting motions.

There are two approaches towards prevention of fretting corrosion degradation. One is to eliminate the fretting motions and the other to eliminate the corrosion, the reoxidation of the tin during fretting motions.

One means of eliminating fretting motions is the use of high contact forces to create high friction forces at the contact interfaces. This is why tin-plated contacts typically have contact forces of 100 grams and higher. There are, of course, limits to how much contact force can be used in a connector as will be discussed in section 1.9 "Contact Springs". At this point it is sufficient to note that tin plated contacts place additional limits on contact force because tin is a soft material with high wear rates (low mating durability) and high friction (high mating force) as a function of contact force. It is important to note, however, that if the increase in contact force is not sufficient to ensure mechanical stability the rate of fretting degradation increases with contact force, as noted previously, and the connector may be more susceptible to fretting corrosion.

The second approach, eliminating oxidation, can be realized by using a contact lubricant specifically formulated for this purpose. Contact lubricants will be discussed later but a few comments on anti-fretting lubricants are appropriate at this point.

Consider the data in Figure 1.52 which schematically illustrates contact resistance as a function of the number of fretting cycles for three different cases of surface treatment. The first case, dry, indicates a contact interface that has received no intentional lubricant. In the second case mineral oil has been applied to the contact interface. In the third case a specially formulated, anti-fretting, contact lubricant has been applied.

With no lubricant, dry, the contact resistance begins to increase at a few hundred fretting cycles with an increasing rate occurring at about 1000 cycles. Adding mineral oil to the contact interface results in a much longer stable resistance regime. But after a few thousand cycles fretting degradation begins to occur and quite rapidly. The reason for the increased degradation is loss of the mineral oil at the contact interface with time. Mineral oil has a low viscosity and flows away from the contact interface as well as evaporating. The anti-fretting lubricant (Figure 1.52) shows stable resistance out to tens of thousands of fretting cycles, demonstrating that appropriate contact lubricants can maintain contact resistance stability under fretting driving conditions.

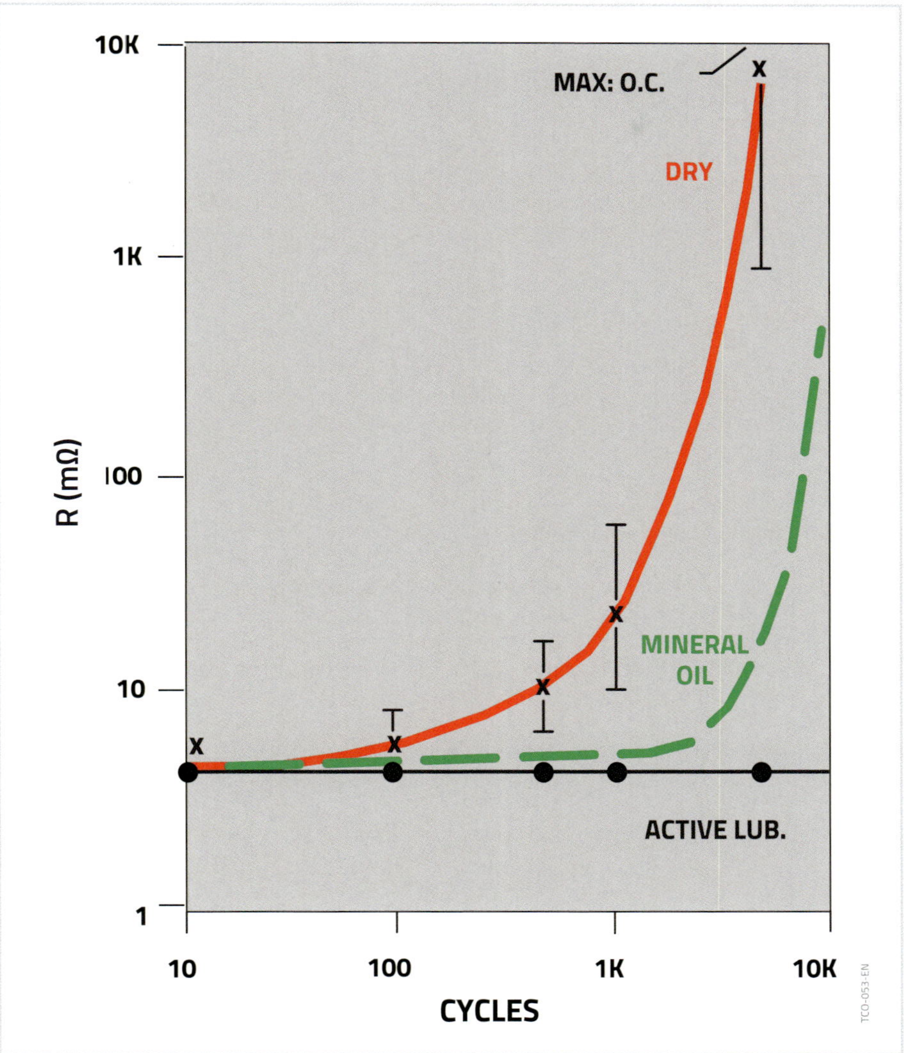

Fig. 1.52: Effect of contact lubricant on fretting corrosion

1.8.4 Fretting Degradation Mechanisms

For completeness it should be noted that there are several fretting degradation mechanisms:

Fretting Wear

All fretting motions will cause some degree of wear at the contact interface. The amount of wear will depend on the length of the fretting motion and the contact force. Fretting wear can become a concern when the amount of wear is sufficient to wear through the contact finish to the underplate or the base metal of the contact spring. Flash gold finishes, in addition to the earlier comment that they may not provide complete surface coverage, are particularly susceptible to fretting wear.

Fretting Corrosion

Fretting motions lead to fretting corrosion only in systems where the fretting surface is susceptible to corrosion. Examples of such systems include tin, nickel and palladium alloys. The palladium alloys are susceptible to fretting due to the 20% nickel or cobalt used in these alloys.

Fretting Polymer

Fretting or frictional polymer is a fretting degradation mechanism for palladium finishes. Palladium tends to catalyse organic contamination and during fretting motions the organics polymerize and create a polymer powder debris at the contact interface with the same effect as the oxide debris of fretting corrosion.

1.8.5 Nickel Degradation Mechanisms

As noted previously, nickel surfaces always consist of a self-limiting and very thin layer of nickel oxide, 0.01 µm (0.25 micro-inches). Nickel oxide is a P type semiconductor and will introduce a series resistance at the contact interface in excess of the metal-to-metal constriction resistance. Nickel contact interfaces, therefore, will have a higher constriction resistance than finishes in which a metal-to-metal interface can be realized.

Nickel is also susceptible to fretting corrosion similar to that in tin, a build-up of nickel oxide debris at the contact interface.

The most common application for nickel contact finishes is in battery contacts.

1.8.6 Silver Degradation Mechanisms

Silver surfaces react with sulphur and chlorine. Silver sulphide films, tarnish, are readily displaced on connector mating, but have been problematic in low force, non-wiping relay applications. Fretting displacement of silver tarnish does not result in fretting corrosion as described for tin fretting. Silver chloride is more tenacious, but less common as compared to silver tarnish.

Silver plating is generally applied in the range of 1.25 to 2.5 µm (50 to 100 micro-inches), and are soft compared to gold plating. The durability of silver finishes is lower than that of gold, but higher than that of tin.

Silver is recommended for power applications due to the high conductivity of silver and an increased resistance to arcing degradation compared to gold or tin finishes.

1.9 Contact Springs

The contact springs, B in Figure 1.3, are typically manufactured from copper alloys to provide three major functional capabilities:

Electrical conductivity

The contact springs provide the electrical connection between the two sub-systems being connected. Copper alloys provide the highest electrical conductivity of commonly used metals, considering the material price as well.

1.9.1 Nickel Degradation Mechanisms

As noted previously, nickel surfaces always consist of a self-limiting and very thin layer of nickel oxide, 0.01 µm (0.25 micro-inches). Nickel oxide is a P type semiconductor and will introduce a series resistance at the contact interface in excess of the metal-to-metal constriction resistance. Nickel contact interfaces, therefore, will have a higher constriction resistance than finishes in which a metal-to-metal interface can be realized.

Nickel is also susceptible to fretting corrosion similar to that in tin, a build-up of nickel oxide debris at the contact interface.

The most common application for nickel contact finishes is in battery contacts.

1.9.2 Silver Degradation Mechanisms

Silver surfaces react with sulphur and chlorine. Silver sulphide films, tarnish, are readily displaced on connector mating, but have been problematic in low force, non-wiping relay applications. Fretting displacement of silver tarnish does not result in fretting corrosion as described for tin fretting. Silver chloride is more tenacious, but less common as compared to silver tarnish.

Silver plating is generally applied in the range of 1.25 to 2.5 µm (50 to 100 micro-inches), and are soft compared to gold plating. The durability of silver finishes is lower than that of gold, but higher than that of tin.

Silver is recommended for power applications due to the high conductivity of silver and an increased resistance to arcing degradation compared to gold or tin finishes.

1.10 Contact Springs

The contact springs, B in Figure 1.3, are typically manufactured from copper alloys to provide three major functional capabilities:

Electrical conductivity

The contact springs provide the electrical connection between the two sub-systems being connected. Copper alloys provide the highest electrical conductivity of commonly used metals, considering the material price as well.

Separable Connections and Contact Force

As mentioned previously, the separable contact interface in a connector is created and maintained by the contact normal force generated by deflection of the receptacle contact springs by the plug contact on mating. **For this reason, the contact normal force is arguably the most important connector design parameter as will be discussed in separable connection chapter.**

Plug contacts, Figure 1.53, include pads on a printed circuit board and rectangular or round pins, respectively. Receptacle contacts come in a variety of spring geometries to serve different application needs and requirements.

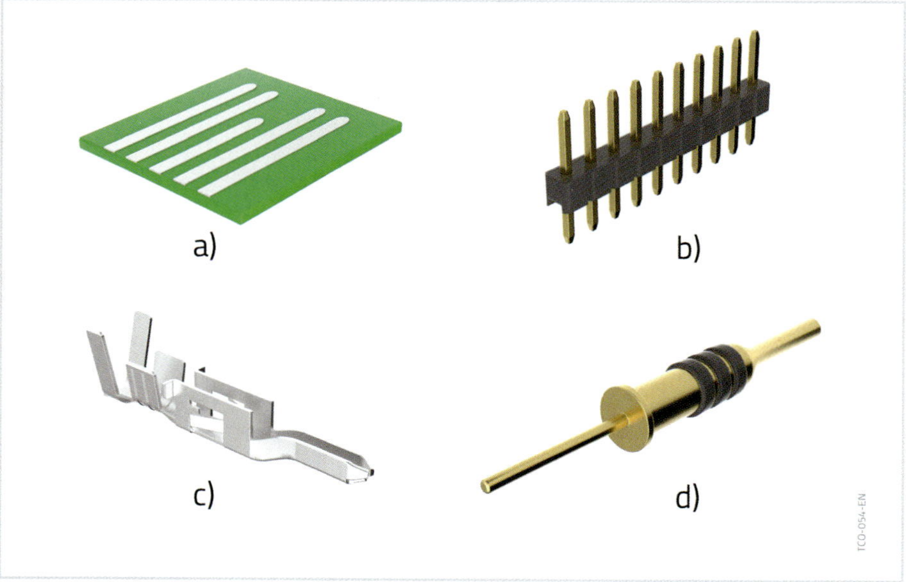

Fig. 1.53: Plug contacts

Figure 1.54 shows several receptacle spring geometries intended to provide functional capabilities, such as accommodating tolerances, and controlling the range of contact force in the connector. Copper alloys provide a good balance of spring characteristics and manufacturability. All of these contacts mate to a pin or post but require different amounts of forming to provide performance enhancements. Figure 1.54a is a simple flat stamping having twin cantilever beam contacts. The limitation of this contact is that the mating surfaces are the sheared edges of the stamping. This means a rough and uneven mating surface which is undesirable. If the contact was to be tin plated it could be used as it is. But if silver or gold plating is desired the sheared edge would have to be improved eliminating the cost advantage. Figure 1.54b and c are simple means for eliminating the sheared edge issue. In Figure 1.54b the contact beams are simply twisted by 90 degrees, so the mating surface becomes the rolled surface of the strip.

The spring system is no longer a simple cantilever beam due to the twist which stiffens the beam. In Figure 1.54c the contact beams are folded upward to make the contact surface a rolled surface. The spring system is no longer a cantilever beam. The beam deflection will also open the U shape at the base of the beams. Finally, Figure 1.54d includes multiple forming operations to provide a crimped permanent connection with an insulation grip, latching beams to retain the contact in the housing and a box entry, which provides an alignment feature as the pin or post is inserted into the contact.

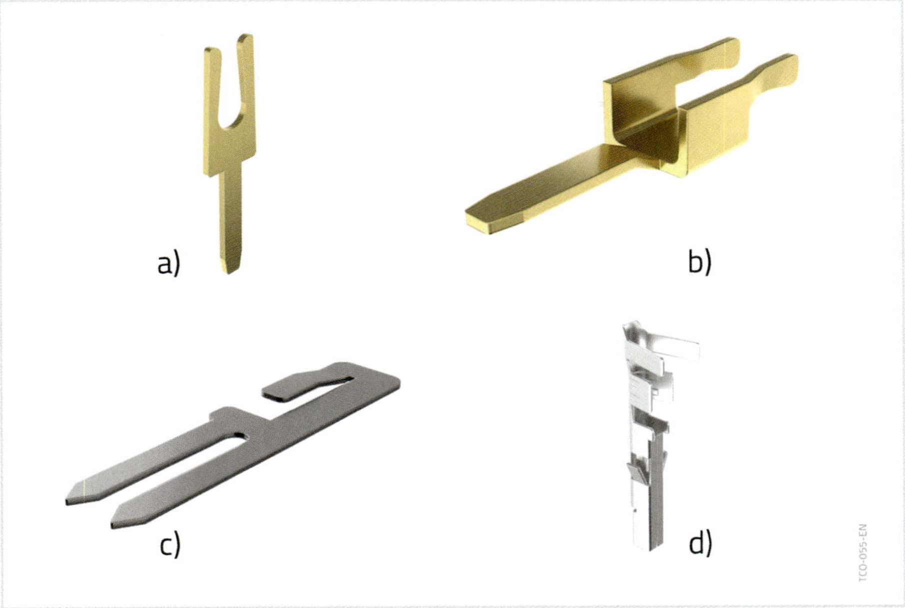

Fig. 1.54: Receptacle contacts

Permanent connections

A connector also includes permanent connections between the connector contacts and the sub-systems being connected. Permanent connection methods used in connectors include both metallurgical and mechanical technologies. The primary metallurgical permanent connection method is soldering. There are two basic soldering technologies used in connectors, through hole and surface mount, examples of which are shown in Figure 1.55 and Figure 1.56 respectively. The long solder tails in Figure 1.55 are inserted into plated through holes in a Printed Circuit Board (PCB) and wave soldered. The solder tails in Figure 1.56 are reflow soldered to surface pads on a PCB. Wave and reflow soldering will be reviewed in Chapter 1.14. The two major mechanical permanent connection technologies are crimping and insulation displacement, examples of which are shown in Figure 1.57 and Figure 1.58 respectively. Crimp and IDC technologies will be reviewed in Chapter 1.16 Permanent Connections. Copper alloys are advantageous for both metallurgical and mechanical permanent connections, because they are solderable/weldable and readily stamped and formed into the configurations needed to create mechanical permanent connections.

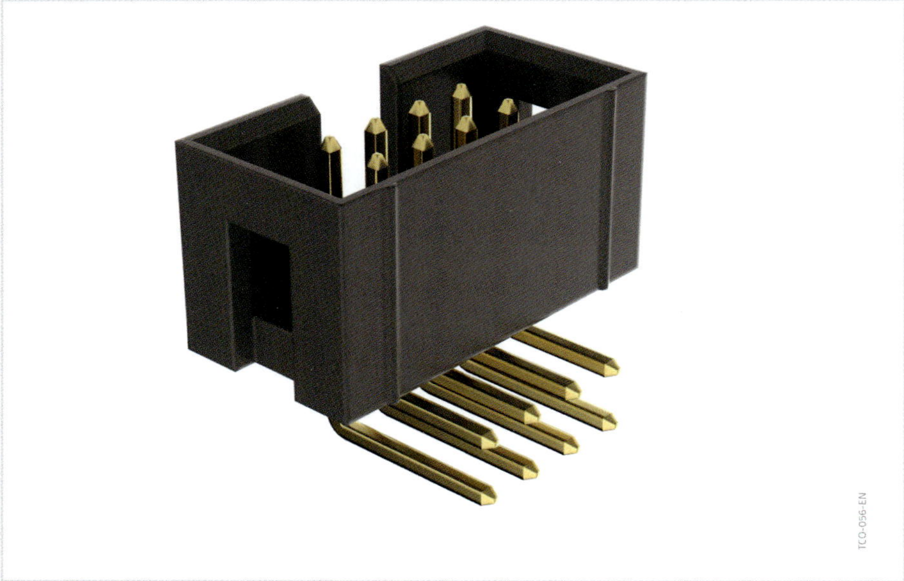

Fig. 1.55: Through hole box header

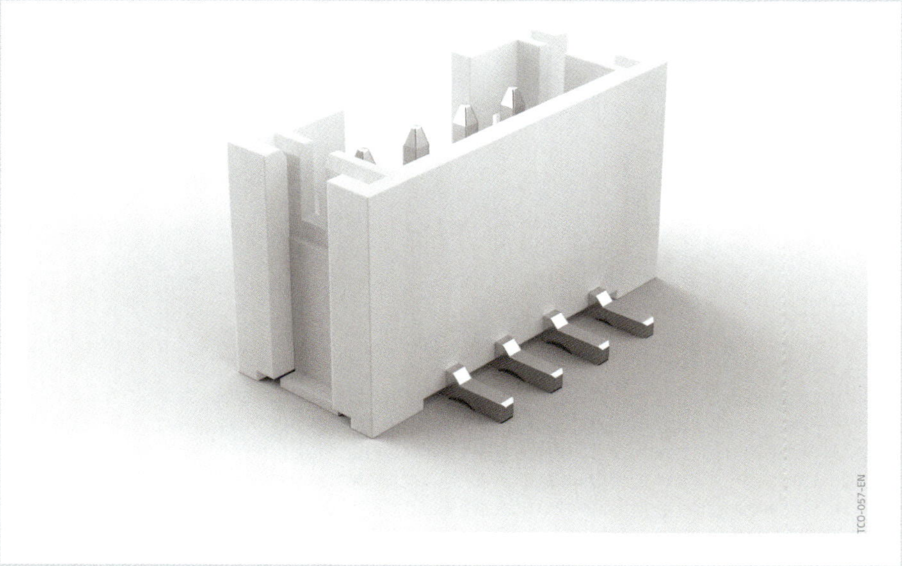

Fig. 1.56: SMT wire-to-board connector

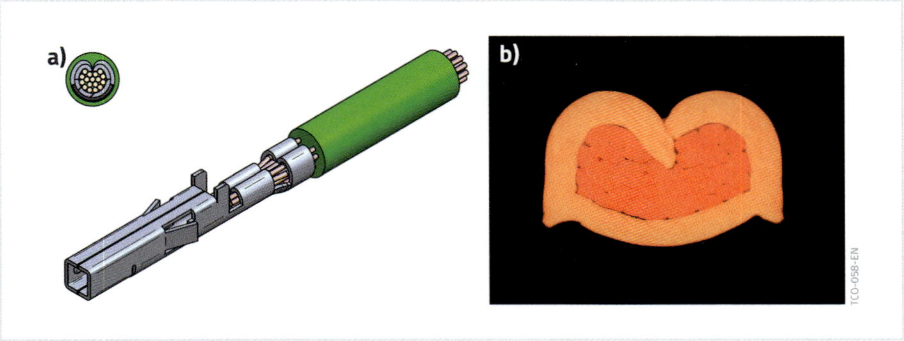

Fig. 1.57: Crimped permanent connection

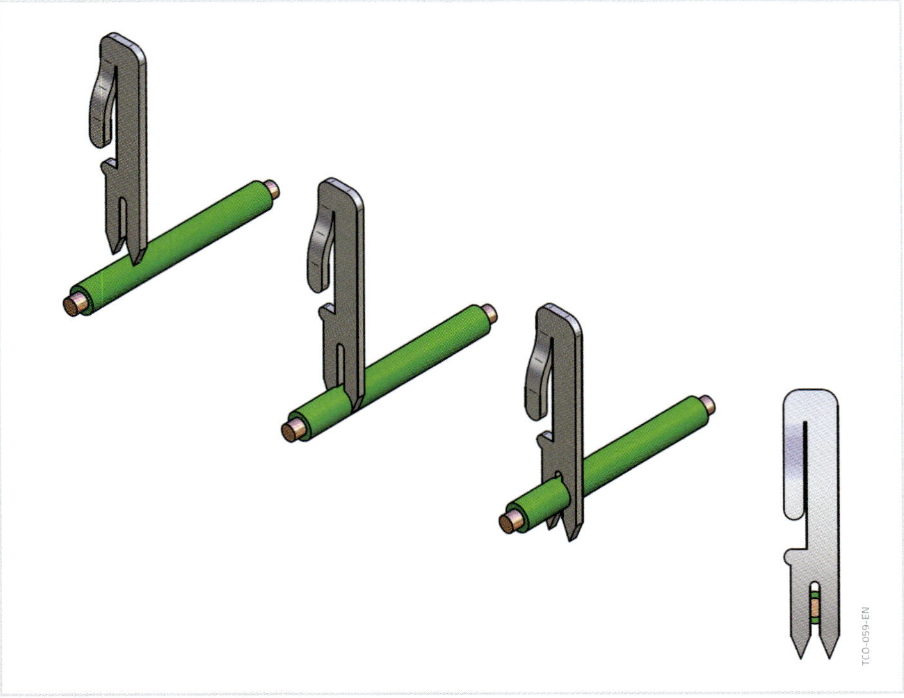

Fig. 1.58: Insulation displacement connection (IDC)

A connector is an **electromechanical** system in that it creates an electrical connection by mechanical means. There are two halves in a connector, a **plug** (Figure 1.59), and a **receptacle** (Figure 1.60). The deflection of the receptacle contact spring by the plug on mating of the connector develops a contact normal force that creates and maintains the integrity of the metal-to-metal contacts at the separable contact interface. This is arguably the most demanding requirement for the **contact springs** due to the many, and sometimes conflicting functions of contact force due to the separability requirement.

Fig. 1.59: Plug contacts

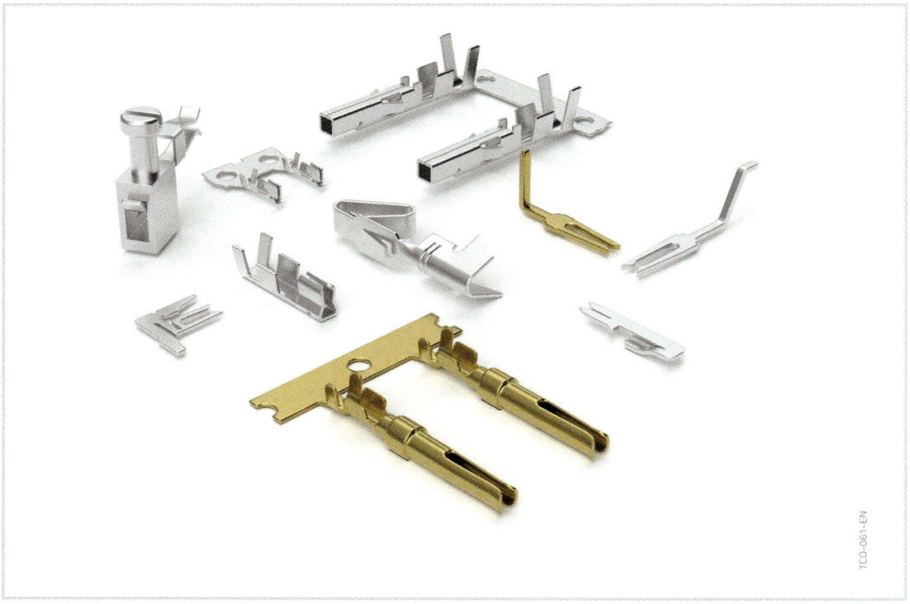

Fig. 1.60: Receptacle contacts

The contact springs must also have a balance of mechanical strength and formab lity to enable the creation of mechanical permanent connections between the connector and the subsystems being connected. These include crimped and insulation displacement connections to wire and cable and press-in connections to plated through holes in printed wiring boards. This chapter will discuss design and materials requirements for both separable and permanent connector contact interfaces.

1.10.1 Separable Interface Requirements

Separability requires that a connector has the capacity to be mated and unmated multiple times without any unacceptable effect on connector performance. In other words, the wear that occurs on each mating cycle, must be low enough to support the specified number of mating cycles. Contact interface wear rates depend on the contact force and the mating geometry. Wear rates increase as contact force increases and as the mating geometry becomes smaller. At the same time, the mating force of the connector must be low enough to allow for easy mating of multi pole connectors without mechanical assistance. Both requirements benefit from low contact forces.

This conflicts with the requirement for mechanical stability of the contact interface to maintain the integrity of the metal-to-metal contact interfaces, which benefits from high contact forces. The dependence of these and other connector performance characteristics on contact force will be discussed in this chapter.

1.10.1.1 Contact Spring Requirements

The three basic requirements for the contact spring are:

- Electrical conductivity
- Creating the contact normal force
- Formability sufficient for both separable and permanent connections

Consider each requirement separately.

Electrical Conductivity

The contact spring material must have a high electrical conductivity. In general, this means metallic conductivity. The conductivity of a contact spring depends on the intrinsic conductivity of the spring material and the geometry of the contact spring. The conductivity requirement is most demanding in power applications where Joule, I^2R, heating can limit the range of application of a connector.

Contact Force

The contact force is created by deflection of the contact springs, usually the receptacle contact spring, during mating of the connector. Material parameters that are important in generating the contact force include the elastic modulus, and the yield strength of the spring material. Examples of this dependence will be provided later in this chapter.

Formability

Formability is important for both separable and permanent connections. First, it is necessary to enable the various bending operations required to create the overall geometry of the separable connection portion of the contact spring. These geometries range from simple cantilever beams to complex curvatures to meet different application requirements. Second, it is necessary to support the large deformations and deflections needed for crimped connections as well as for insulation displacement connection geometries.

Fig. 1.61: Formability

These contact spring material requirements are best met by copper alloys. Copper itself excels in conductivity, but it does not have the balance of strength and formability to meet the required contact spring manufacturing and performance requirements. Thus, a range of copper alloys are used with the choice of alloy being dependent on the connector design and the performance requirements that must be met. For example, for springs requiring complex forming the balance between yield strength and formability may be dominant while in power applications electrical conductivity may drive the material selection process.

1.10.1.2 Separable Connections and Contact Force Optimization

Before the question of optimizing contact force is addressed it is necessary to ask a more fundamental question:

"What does contact force do in a connector?"

The two most important functions of contact force have been mentioned earlier. Contact force creates and maintains the integrity of the metal-to-metal areas of the contact interface. The next question is:

"How much force does it take to accomplish each of these functions?"

The question about the force necessary to create an "acceptable" contact interface has a more accessible answer.

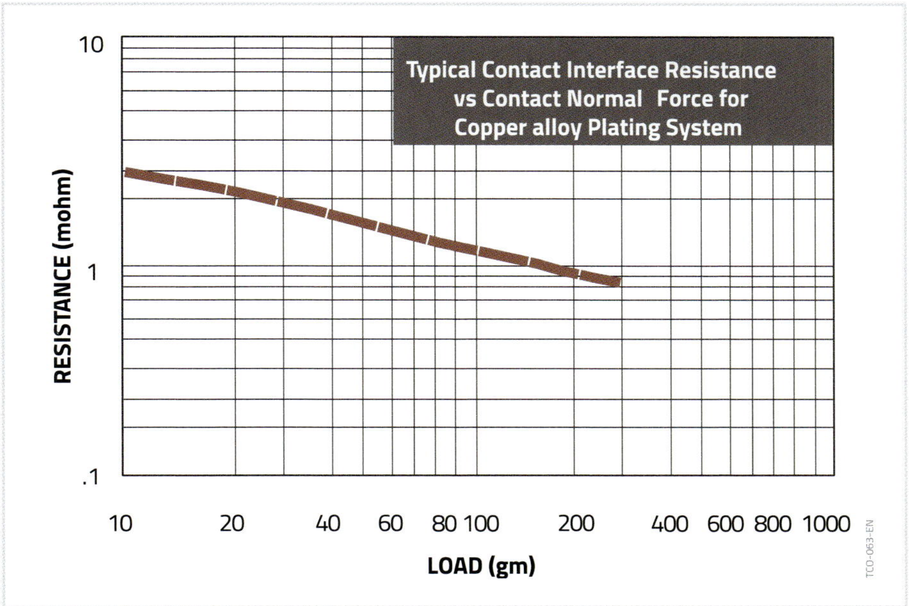

Fig. 1.62: Contact interface resistance vs. contact normal force

Figure 1.62 shows the relationship between contact force and the contact interface resistance, a reasonable measure of the acceptability of a contact interface. Note that contact resistance decreases continuously with increasing contact force, but at a decreasing rate. Between 10 and 100 grams the contact resistance changes from about 4 milliohms to about 1.2 milliohms, a decrease of about 70%, while between 100 and 200 grams the change is from 1.2 milliohms to about 0.7 milliohms, a decrease of about 40%. The most important data from this curve, however, is the fact that at a contact force of about 10 grams the contact resistance is about 4 milliohms, a value that, if it is stable, would be acceptable for most connector applications except for high current requirements.

The phrase "if it is stable" is the key point of that sentence. It is highly unlikely that the friction forces created at a 10-gram contact force would be sufficient to ensure mechanical, and, therefore, electrical stability under most application conditions. The conclusion to be drawn then is that the most demanding contact force requirement is to maintain the integrity of the contact interface. This force will be higher, possibly significantly higher than 10 grams. With this information in hand, the question of the optimization of contact force can be addressed.

As noted earlier in this chapter, contact force impacts several performance characteristics. Consider some of the major effects.

Increasing contact force will increase:

- the wear rate of the contact finish
- the mating force
- the stresses on the contact spring
- the stresses on the connector housing

These are all negative effects.

On the other hand, increasing contact force will:

- decrease the magnitude of the contact resistance
- increase the stability of the contact resistance

These are, of course, positive effects. The data in Figure 1.63 illustrate the first point. The second effect arises from the fact that increased contact force increases the friction forces at the contact interface which increases the mechanical stability of the contact interface, which, in turn increases the stability of the contact resistance.

Optimization of the contact force, then, reduces to determining the minimum contact force necessary to ensure mechanical stability of the contact interface in a given connector application. Unfortunately, the variability in mechanical and thermal driving forces for contact interface disturbance vary dramatically with the connector application so there is no single answer to that question, which is to say that there is no generally applicable minimum contact normal force requirement for a connector.

Contact Force and Mating Wear/Durability

The variation of mating wear/durability as a function of contact force is a good example of normal force optimization. The kinetics of the wear process were discussed in section 1.6.2 The Mechanical Interface: Friction and Wear/Durability. The following equation provides a parametric format for discussion of some of the major factors influencing the wear process.

$$V=K\frac{(F_nL)}{H} \quad \text{(EQ 6)}$$

where, V is the wear volume, F_n is the contact normal force, L is the mating length or fretting distance, H is the hardness, and K, the wear coefficient, is a parameter including effects of roughness, lubricity, contact geometry and normal force.

The wear volume, V, represents the volume of material removed, the wear particle in Figure 1.34 (Page 55). It can be related to the loss in thickness of the contact finish from the fact that V is the product of the contact area and the thickness removed.

Note that given this formulation, the thickness removed increases as the contact area decreases.

The dependence on L, the mating length or fretting length, is straight forward. Similarly for hardness, H, in that hard materials wear less than soft materials for a given contact configuration and contact force.

Note that the contact force, F_n, appears in two places in the equation. V increases linearly with contact force for a given value of K and F_n also appears, parametrically, in K itself. K is complex in its variation, with multiple parameters. For example the wear rate for a lubricated surface will be lower than that for a non-lubricated surface, but as contact force increases the effects of lubricity may be overcome by a transition from a burnishing to an adhesive wear regime. Thus, wear rates can vary significantly depending on the interactions among the multiple parameters included in. The key point about contact force and wear is that wear rates increase linearly with F_n and can increase discontinuously and dramatically if the wear kinetics transition from burnishing to adhesive wear. This is an important consideration at high contact forces.

1.10.1.3 Contact Force and Mating Mechanics

Consider now the effects of contact force on the mating mechanics of a connector. Figure 1.63 schematically illustrates the mating, or insertion, force of a pin as a function of the insertion depth of the pin into the receptacle contact. The receptacle is the spring contact in this system.

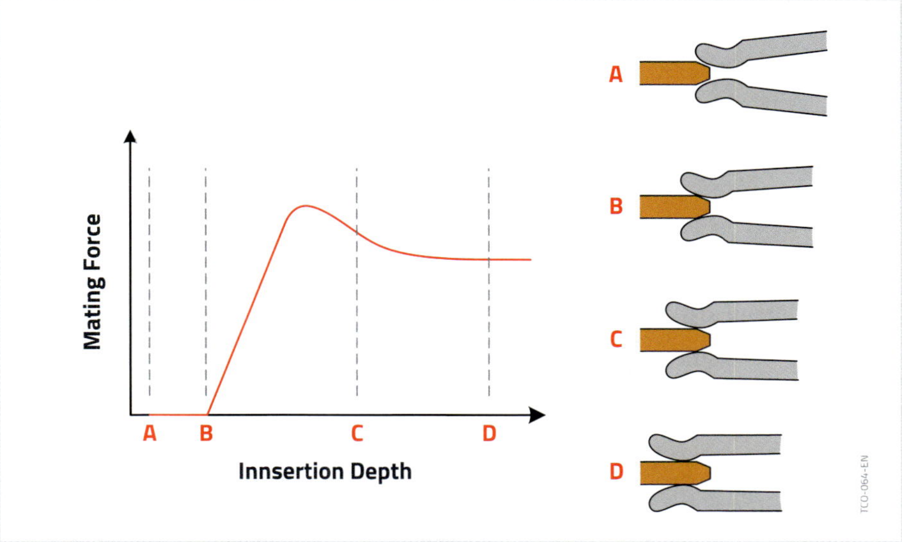

Fig. 1.63: Mating force as function of insertion depth

There are two phases to the insertion force. In phase I the receptacle contact beams are being deflected by the pin and in phase II the beams are fully deflected, and the receptacle contact is sliding along the surface of the pin contact. The insertion force in phase II is simply the product of the coefficient of friction, and the contact force because the friction force acts in opposition to the insertion force.

The insertion force in phase I arises from two contributions, the displacement of the receptacle contact beams by the pin and the friction forces created as the pin slides against the receptacle. The contact force and the insertion force increase as the beam is deflected dependent on the stiffness, or spring rate of the receptacle contact. The friction forces are always opposed to the direction of insertion and perpendicular to the tangent of the pin surface at the point of contact with the receptacle beam. Thus, the friction force contribution varies in direction as the receptacle contact slides up the contour of the pin contact. These variations are responsible for the magnitude and width of the peak in the insertion force curve.

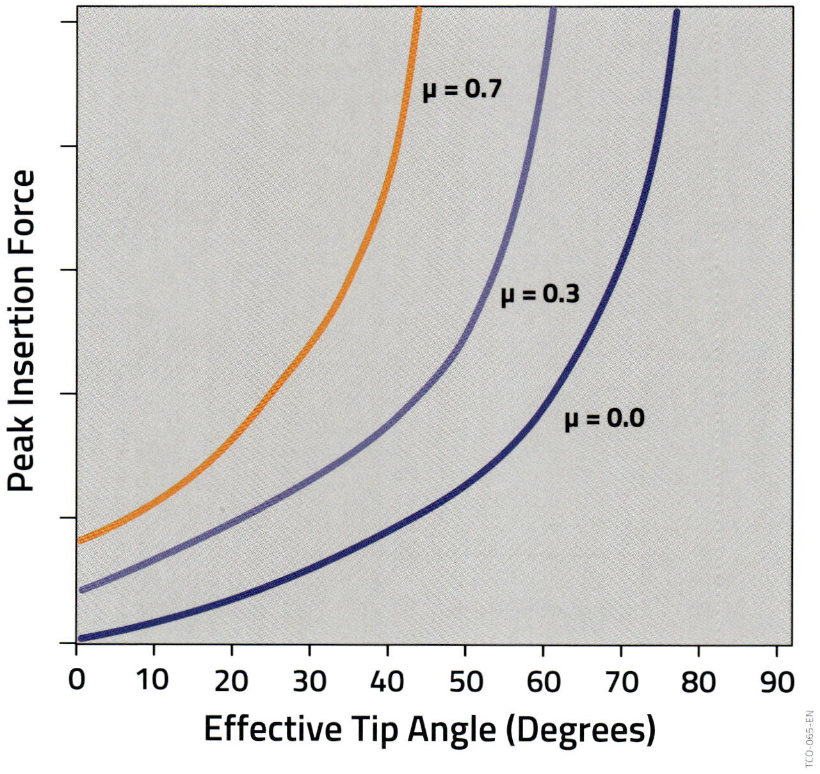

Fig. 1.64: Insertion force

Figure 1.64 illustrates the variation in **peak** insertion force as a function of the coefficient of friction, μ. The peak insertion force is the minimum mating force for the contact system. The contact force is a dominant contributor to the mating force and the same concern about a discontinuous change in friction coefficient, analogous to the wear coefficient discussed previously, applies. The values of μ shown are for an ideal frictionless case, μ = 0 and for nominal coefficients friction for gold, μ = 0.3 and tin, μ = 0.7. The abscissa shows the effective tip angle, the angle between the tangents to the pin and socket surfaces at the point of contact. Small effective tip angles will have a low and wide peak and the height will increase, and the width decrease as the effective tip angle increases.

1.10.1.4 Contact Normal Force Parameters

After this discussion about the effects of contact normal force, attention now turns to the design and material parameters that determine contact force. The following discussion will focus on a cantilever beam (supported on only one end) contact design for simplicity, but the same general principles, appropriately modified, will apply to other beam designs.

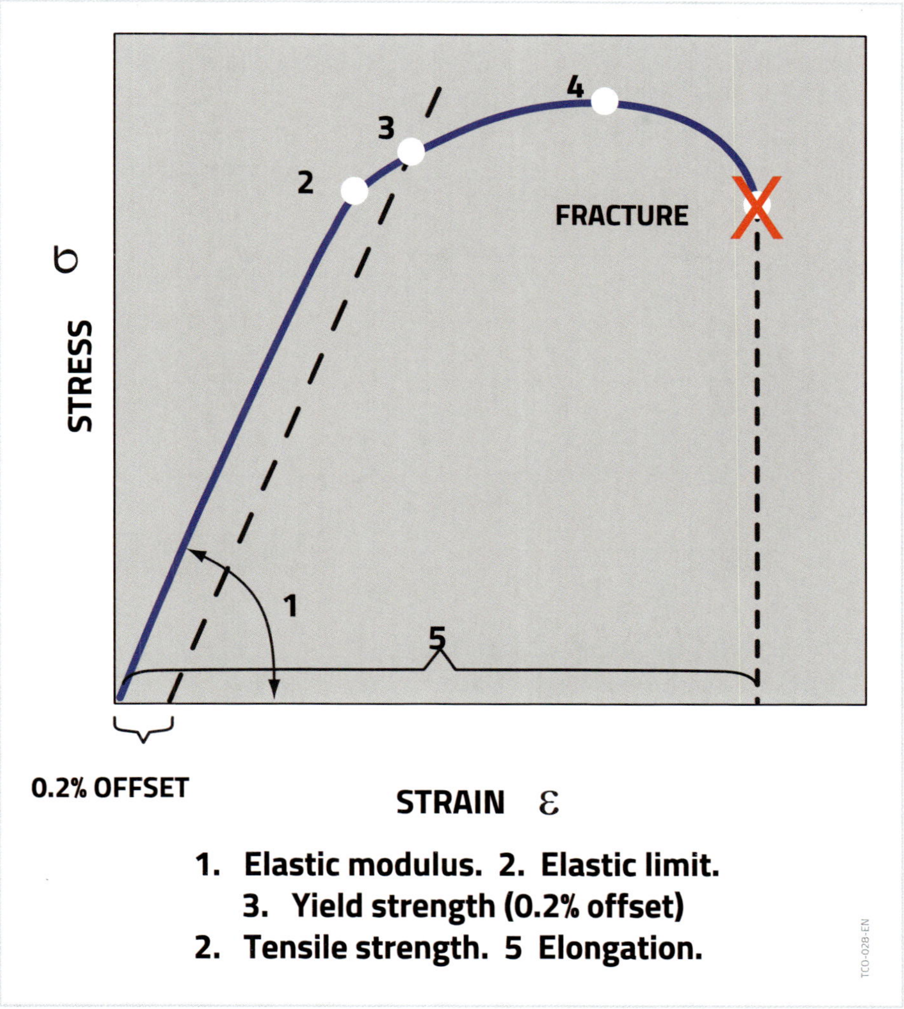

Fig. 1.65: Stress-strain curve in tension

Before considering the cantilever beam, however, a review of the stress-strain diagram is necessary to define some of the relevant parameters to be discussed. Figure 1.65 shows a representative stress-strain diagram obtained from a sample pulled in tension. The deflection of a contact beam occurs by bending, but the same parameters as noted in the tensile diagram apply to bending. The most important parameters in the diagram are:

1. The elastic modulus, E, the slope of the stress-strain relationship in the region where that relationship is linear.
2. The elastic limit is the maximum stress that can be applied without any permanent deformation. In the case of tensile loading, the elastic limit is the maximum stress the sample can support without increasing in length. In bending loading it is the maximum stress the sample can support without causing a permanent bend in the sample.
3. The yield strength is the stress that causes a specified elongation of the tensile sample. The 0.2 % offset is a commonly used criterion for specifying the yield strength of a metal. Using

this criterion, the stress that causes a 0.1 cm increase in the length of a 5 cm tensile sample is the 0.2 % offset yield strength.

4. The strain, or elongation, is given by ΔL/L where ΔL is the change in length of the sample and L the original length of the sample.

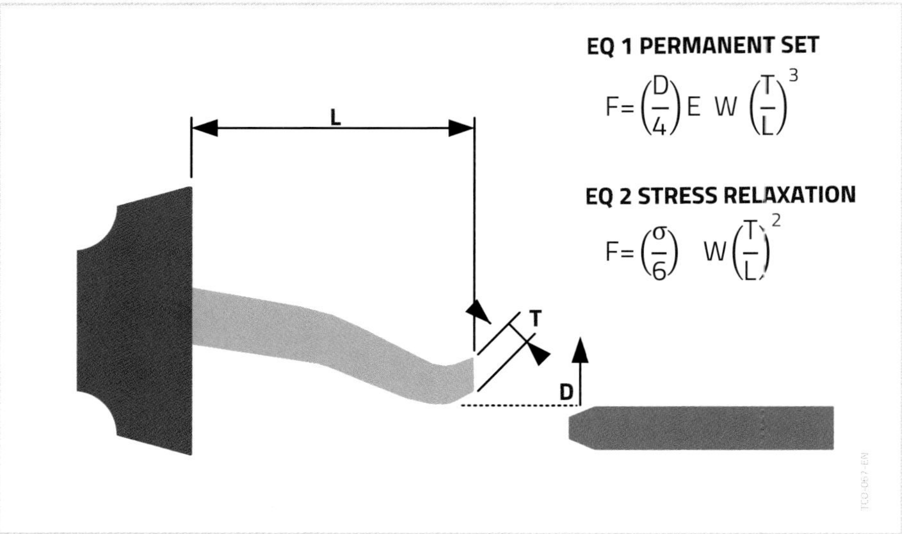

Fig. 1.66: Contact normal force and spring deflection

Figure 1.66 schematically illustrates a simple cantilever beam. The beam is fixed at one end and the maximum stress during deflection occurs at that end.

There are two equations for normal force in Figure 1.66. Equation 2.2 shows the relationship between contact force and receptacle contact beam geometry and deflection:

$$\left(\frac{1}{4}\right) E\, D\, w \left(\frac{t}{l}\right)^3 \qquad \text{(EQ 7)}$$

where F is the contact force, E is the elastic modulus of the base metal of the receptacle contact, D is the deflection of the receptacle contact beam by the plug contact, w, l and t are the width, length and thickness of the receptacle contact beam. Elastic modulus is a material parameter and varies with the contact material. The variation in elastic modulus over the typical range of copper alloys used in connectors is of the order of 110 to 140 kMPa (16–20 Mps). For example, consider three alloys used in connectors, brass (C26000), phosphor-bronze (C51000) and beryllium-copper (C17400) the elastic modulus of these materials is 110, 110 and 135 kMPa (16, 16 and 19 Mpsi) respectively. The beam deflection, D, is a design parameter. The contact beam geometry is also a design variable. For a given set of D, w, t, and l, and the limited range of elastic modulus for copper alloys, cantilever beam contacts will give a contact normal force within a range of about 20 % for the copper alloys typically used in connectors according to **Equation 7**. But that is not the whole story, as indicated in **Equation 8** Figure 1.66:

$$\left(\frac{\sigma}{6}\right) w \left(\frac{t}{l}\right)^{3} \qquad \text{(EQ 8)}$$

where is the stress introduced in the contact beam by the beam deflection, and w, t, and l, are as described above. The maximum stress a beam can support without plastic, or permanent, deformation is given by the elastic limit of the material. As mentioned previously, materials are generally specified in terms of the 0.2 % offset yield strength, a value roughly proportional to the elastic limit. The commonly used yield strengths of the example materials in connectors are of the order of 45, 55 and 65 kMPa (65, 80 and 100 ksi) for brass, phosphor-bronze and beryllium-copper respectively. These data show that, for constant geometry, and quasi-elastic deformation, beryllium copper can support a higher stress, that is a higher beam deflection and, therefore, generate a higher contact force than phosphor bronze, which, in turn, can generate a higher contact force than brass.

Equation 7 suggests that contact forces will be stable for beam deflections that leave the beam in an elastic state. Beams stressed beyond their elastic range will be plastically deformed and, for a cantilever beam contact, take a permanent set. In the case of dual beam contact systems, such permanent set will be evidenced by an increase in the gap between the two beams when the connector is unmated. This increase in the gap between the beams will result in a lower overall beam deflection and a lower contact force when the contacts are remated. Overstressing of contact beams may occur during mating of the connector due to improper alignment during mating. Many connectors have anti-overstress features such as alignment features or beam stops to reduce the potential for contact force reduction due to permanent set.

Similarly, **Equation 8** suggests that contact forces will be stable as long as the stress due to beam deflection remains stable. This may not be the case in practice due to stress relaxation. The metallurgical definition of stress relaxation is a time dependent decrease in stress at constant strain. For a connector this translates into a time dependent loss in contact force while mated. Stress relaxation, and the accompanying loss in contact force, is strongly temperature dependent. Figure 1.75 shows the loss in stress as a function of time for brass, phosphor bronze and beryllium copper alloys. Because Equation 8 shows that the contact force is proportional to the stress, the stress remaining axis could also represent the contact force remaining as a function of time at temperature. The designations on the right side of the figure identify the material and the temperature of exposure or heat aging. For example in the top three curves the brass and phosphor bronze were heat aged at 25 °C while the beryllium copper was heat aged at 105 °C. The stress relaxation resistance of be-copper is significantly better than that of phosphor bronze or brass (be-copper shows less stress relaxation at a much higher aging temperature). The other two curves indicate that the stress relaxation resistance of phosphor bronze is superior to that of brass. Stress relaxation limits the temperature capabilities of copper alloy. Be-copper can be used in the temperature range of +125 °C while recommended limit for phosphor bronze is +105 °C. While stress relaxation may not be significant in many connector applications, below, say +55 °C, it cannot be ignored at temperature above that value especially in power applications where Joule heating can increase the temperature of the contact spring up to 30 Kelvin above the application temperature.

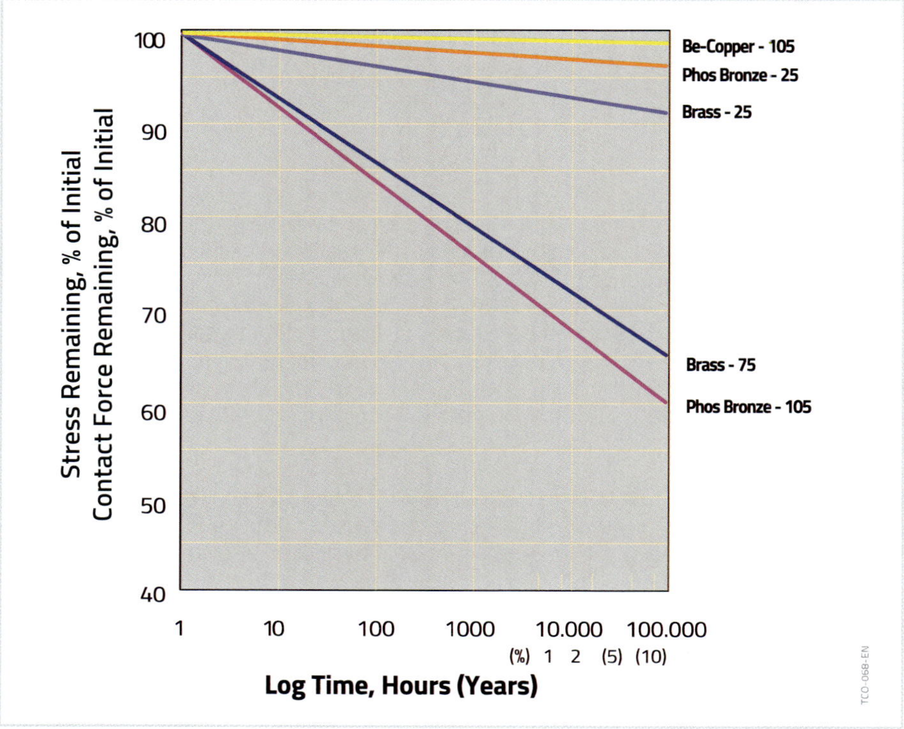

Fig. 1.67: Stress relaxation

To summarize, the contact normal force requirement for a connector is the force necessary to maintain mechanical stability of the contact interface under the stresses of the application environment. This value, typically in the range of several tens to a few hundred grams, is much larger than the force necessary to create a contact interface with an acceptable electrical resistance, a few milliohms.

The significance of the loss in contact force due to stress relaxation is due to the associated decrease in the friction force, which also means a decrease in the mechanical stability, of the contact interface. As noted previously, the mechanical stability of the contact interface is critical to preventing fretting motions and the effect of such motions on the corrosion susceptibility of the connector.

A second effect of stress relaxation is due to a change in the geometry of the contacts. Consider the cantilever beam contact of Figure 1.67.

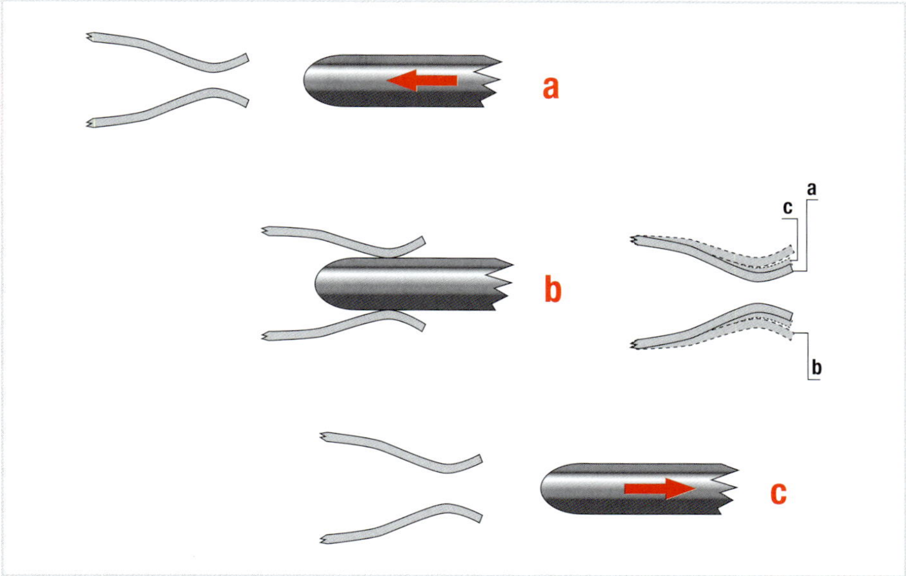

Fig. 1.68: Stress relaxation

The drawing shows the original beam position, a, the deflected beam position, b, and the beam position on unmating the connector, c, after stress relaxation has taken place. The reason for the change in beam position is due to the kinetics of stress relaxation. Atomic motion occurs on a microscale to reduce the localized stresses of beam deflection. Consider a cantilever beam contact where the deflection stress is concentrated near the fastening point of the beam. The atomic motions at this point effectively reduce the residual elastic stress so that when the beam is unloaded, the connector is unmated, the beam does not return fully to its original position. This means that subsequent beam deflections on mating will be reduced compared to the original dimension and the resulting contact force will be lower than the original value. If the stress relaxation is sufficiently large, the mechanical stability of the system may be compromised. This effect is analogous to the effect of overstressing and plastically deforming the contact beams due to pin misalignment or oversize pins.

1.11 Electricity – Basics

1.11.1 Derating

All resistors under current release heat. It is called Joule effect. The power P of the Joule effect is given by the formula:

$P = RI^2$

the parameters have the following meaning:

P: Electrical Power [W]
R: Electrical resisistance [Ω]
I: Current through the resistor [A]

Any connector could be considered as a simple resistor from a thermal point of view. The energy, caused by the finitely small contact transition resistance and the current flow heats the connector and this heat will be released in its direct environment. Even after thermal stabilization, the connector is hotter than its environment. That will give a temperature gap between the resistor and its environment.

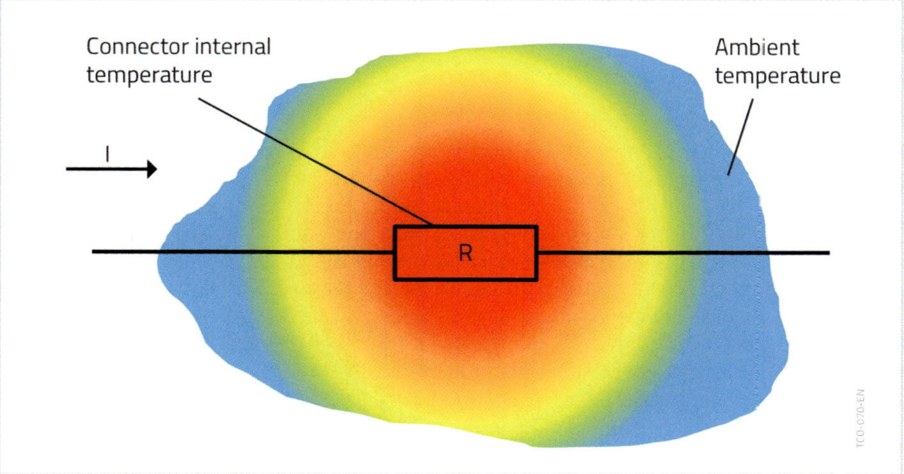

Fig. 1.9: Temperature rise of the connector (ΔT) is simply defined by

$$\Delta T\ (K) = \text{Connector } T\ (°C) - \text{Ambient } T\ (°C) \quad \text{(EQ 9)}$$

Ambient and internal temperature are given in unit °C but temperature rise is usually given in standards and literature in K (Kelvin). The goal is just to clearly make the difference between an "absolute" temperature and a temperature rise or gap.

ΔT depends on the energy lost in the connector and heat dissipation is obviously dependent of the environment. So ΔT is given by:

$$\Delta T = k \cdot R \cdot I^2 \quad \text{(EQ 10)}$$

ΔT temperature rise in Kelvin
k constant
R connector resistance in Ω
I current in A

The constant k depends of environmental factors like: type of plastic material, air flow and all factors that will improve or reduce thermal dissipation. This constant is product-specific and complex in its calculation. But it is possible to omit the factor k from the calculation when we compare values on the same system. If we measure ΔT_1 of a connector with a current I_1, ΔT_2 can be calculated at a different current I_2 without any measurement.

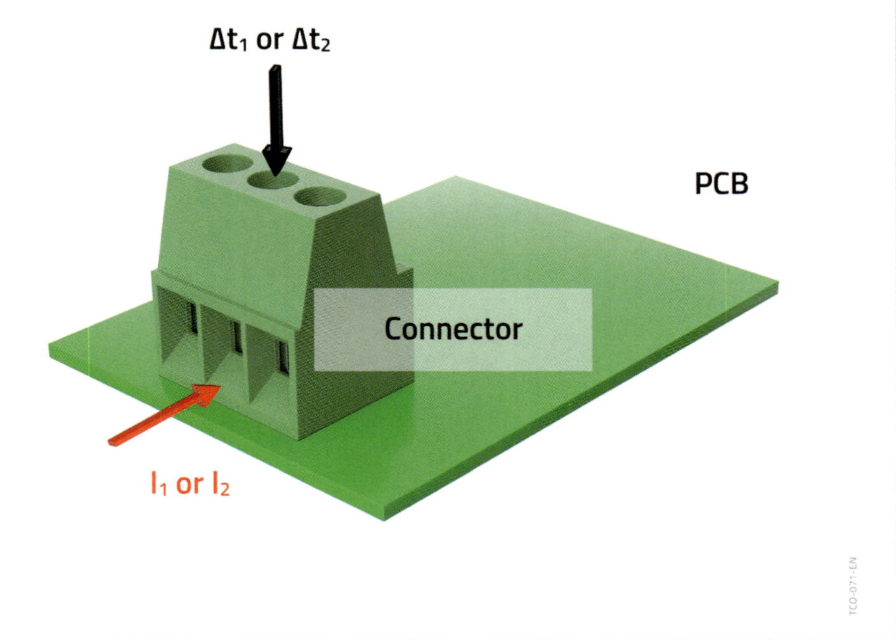

Fig. 1.70: Illustration on I influencing Δt

If we use Joule's formula, constant k and resistance R will be the same for these two lines.

$$\frac{\Delta t_1}{\Delta t_2} = \frac{I_1^2}{I_2^2} \quad \text{(EQ 11)}$$

Note: This formula is an estimation and therefore not precise because of the following reasons: injected current precision, measurement precision, environment, and temperature raw material resistance variation.

We performed the temperature rise test to some of our products to validate the theory. The set up is shown in Figure 1.71. The products were placed inside an enclosed space to avoid the influence of air flow. There was no temperature regulation.

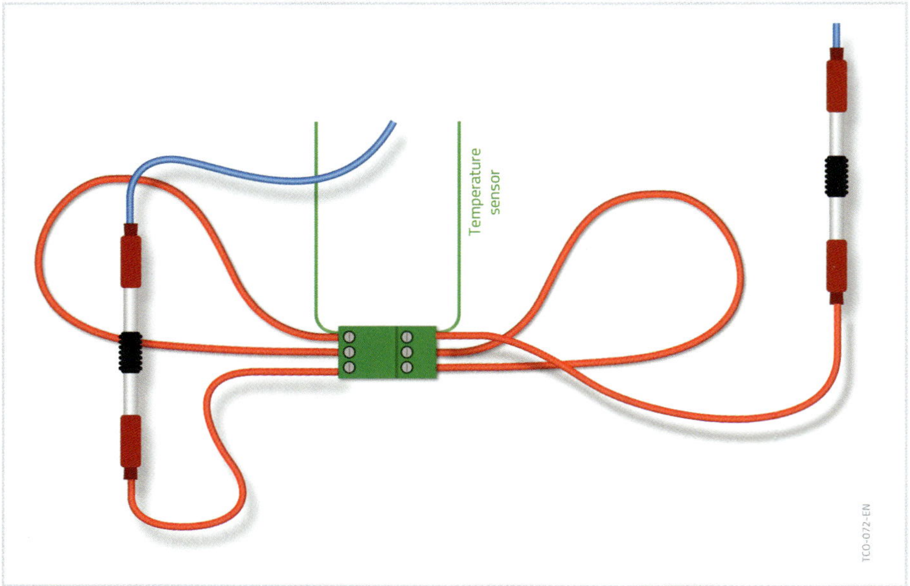

Fig. 1.71: Set up of the temperature rise test.

In the following example a PCB terminal block is connected with its three poles in series. Then a current of 20 A is applied using a 12 AWG wire. Thermocouple is placed on the central clamp of the connector, the hottest point, an additional thermocouple is used to measure the ambient temperature.

The result in Figure 1.72, at 20 A will be the reference and for the calculated curve (blue line), all other values will be calculated with it.

As an example, the estimation of ΔT at 10 A is calculated as follows:

$$\Delta t_{10A} = \frac{I_1^2}{I_1^2} \cdot \Delta t_{20A} = \frac{10^2\ A}{20^2\ A} \cdot 19.2\ K = 4.8\ K \qquad \text{(EQ 12)}$$

The following curve (Figure 1.72) presents a comparison between calculated curve and real measurements.

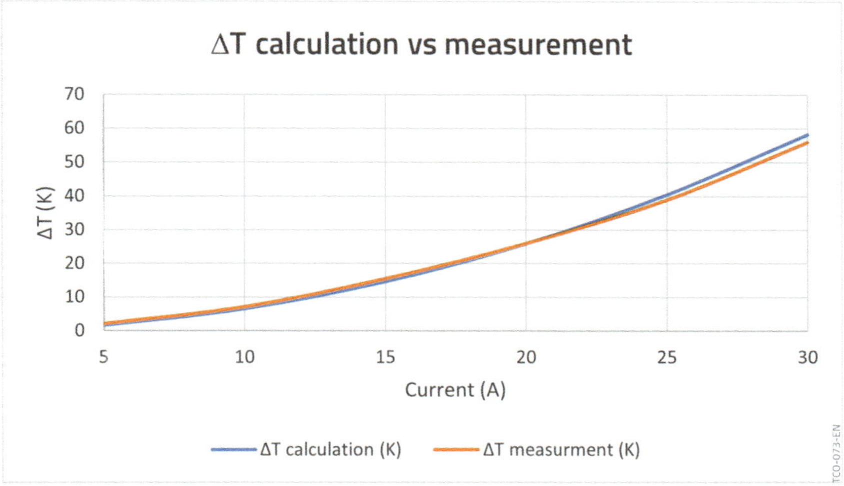

Fig. 1.72: Comparison between calculated curve and real measurements for a temperature rise test

The curves show that the calculation method is a valid approach to estimate the real ΔT value. It means that if you know ΔT of a connector at a current you can estimate ΔT at another current.

Be aware that this estimation is less accurate if there is a large difference between the two currents used (2 A and 50 A for example), and note that when you double the current, you multiply by 4 the temperature rise.

1.11.1.1 Electrical resistance and temperature

Würth Elektronik products are designed so that metal parts do not lose their efficiency across the working temperature range. However, we see a sensible Δt variation with temperature.

The main reason is that metal electrical resistances vary with temperature according to the following formula:

$$R_t = R_0 \cdot (1 + a \cdot (t - t_0)) \qquad \text{(EQ 13)}$$

With:

R_t resistance of a metal conductor at a temperature t in
R_0 resistance at a temperature t0 in
α temperature coefficient of resistance in K-1
t temperature in °C (or K)

α is a material dependant constant. Two examples of materials widely used as conductors:

α_{copper} ≈ 4 x 10-3 K-1
α_{brass} ≈ 1.5 x 10-3 K-1

Note: The symbol '≈' is used because the value varies slightly with material grade.

Obviously, the overall contact resistance is the addition of different parameters: different material conductors, contact between wire and clamp, soldering and contact between mated terminals.

To give an idea of the resistance variation on a connector with a copper alloy conductor, using as an example the temperature variation up to 100 °C of ambient temperature, Figure 1.73 shows the diagram.

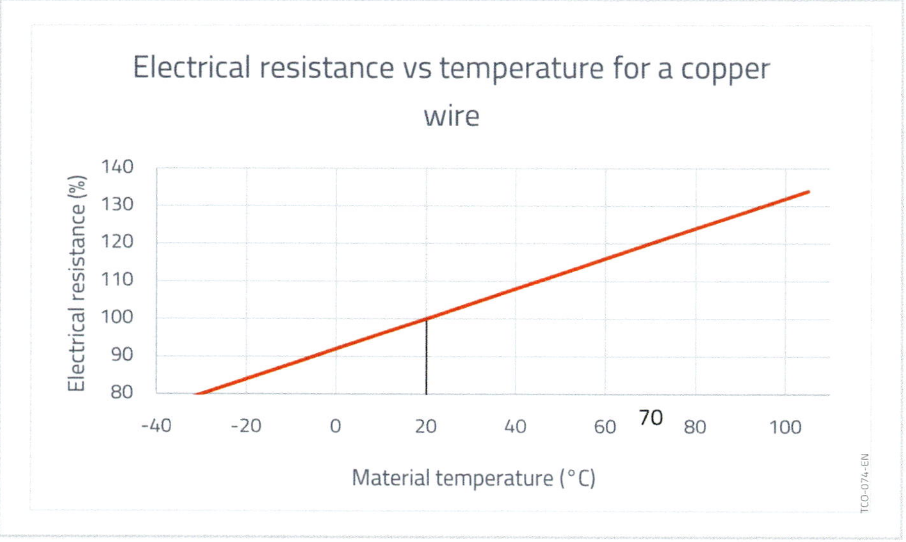

Fig. 1.73: Electrical resistance versus temperature of a copper wire.

For easy reference, remember that a 50 °C increase in ambient temperature results in a 20% increase in electrical resistance. Naturally, this will lead to greater power loss in the system, as electrical materials perform more efficiently at lower temperatures.

The operational current of a product is specified and tested with a maximum allowable temperature rise (ΔT). This value is determined according to industry standards, which typically specify the following guidelines:

- UL1059 standard: require ΔT <30K
- UL1977 standard: limit is the RTI of the plastic
- cULus (CSA C22.2 No. 182.3) standard: ΔT <30K
- VDE and IEC 60998-1 standards require ΔT<45K

As the size of the wires for different standards is given in AWG and mm^2 and ΔT are different, that explains why UL current and VDE current could be different.

1.11.1.2 Practical

Derating curves

We saw that in the previous chapter that working current allows a temperature rise of 30 K maximum according to the UL Standard used.

We know that all products have a working temperature range and especially a maximum operating temperature to be used under.

Now the question is: Can the product operate at the maximum working current under the maximum allowable temperature conditions?

The answer is that the current must be adjusted to prevent excessive temperatures, which can shorten the product's lifespan. It is essential to adhere to the derating curves provided for the product to ensure optimal performance and longevity.

They are designed as follows:

- Working current is of course allowed from the minimum operating temperature. The curves start from 0 °C to avoid a long flat curve area.
- For UL standard, from "t_{max} – 30 K" to maximum operating temperature, current will decrease according to the square of the current.
- For VDE standard, from "t_{max} – 45 K" to maximum operating temperature

When a connector is used close to the maximum temperature allowed, it is recommended to use derating curves.

Different curves are related to the maximum operating temperature of the product studied.

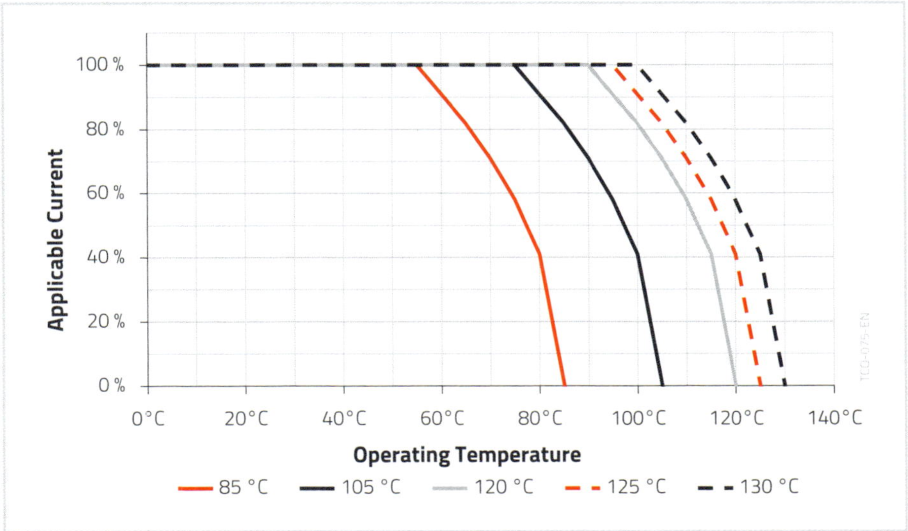

Fig. 1.74: UL derating curves for different operating temperatures

For instance, consider a product intended for use at an ambient temperature of 100 °C. This product is rated for a maximum operating temperature of 105 °C at a working current of 10 A. According to the black derating curve, the applicable current at 100 °C is 40 % of the maximum rating. Therefore, to ensure safe operation at 100 °C, the current should be limited to 4 A.

For a VDE current, it is recommended to use following VDE derating curve (t_{max} −45 K)

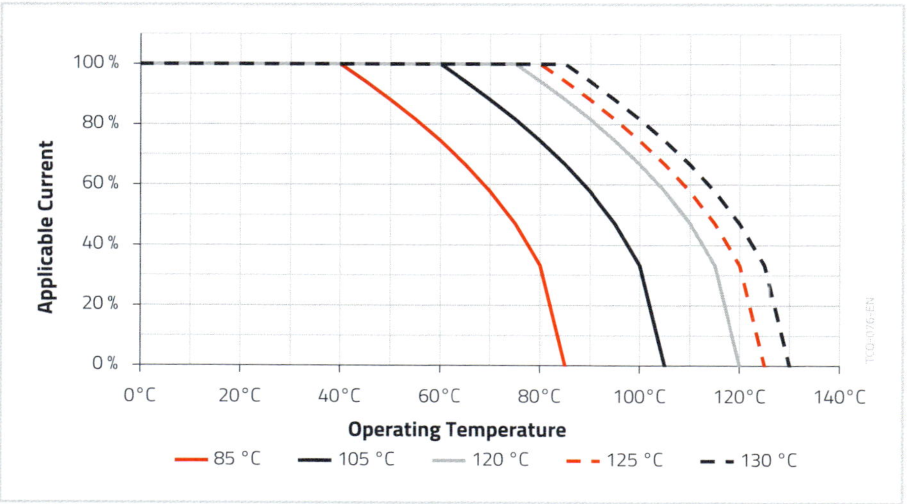

Fig. 1.75: VDE derating curves for different operating temperatures

1.11.2 Inrush-current

Previously, we observed that a connector carrying current generates heat. Product datasheets specify the maximum working current, considering the maximum operating temperature and the corresponding derating curves.

The rate of temperature increase depends on the connector's environment and the efficiency of thermal energy dissipation. The following curve illustrates the behavior of a connector under a stable working current in still air:

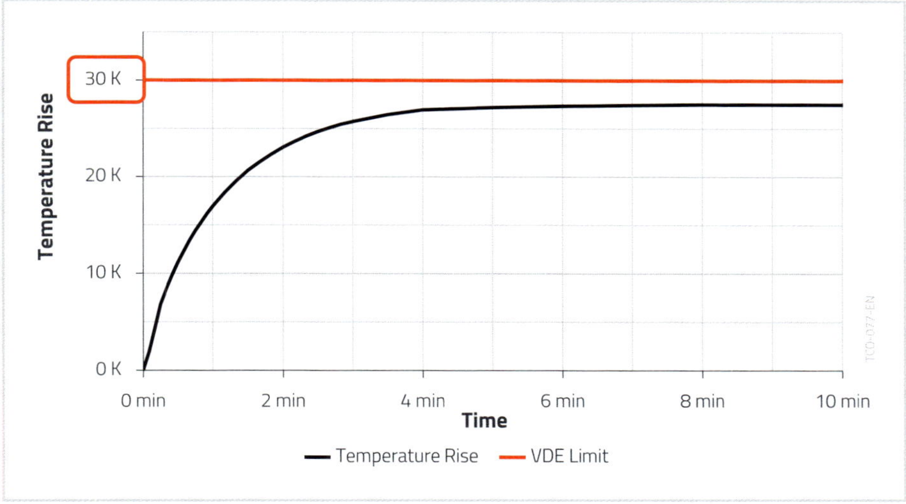

Fig. 1.76: Connector temperature rise in minutes

Temperature rise (ΔT) reach its maximum after several minutes and will keep stable in the time. Notice that ΔT <30K as given by UL rule.

Now let's focus on the first minute of the test (Figure 1.77). For example, at 10 s we can see that ΔT = 5K:

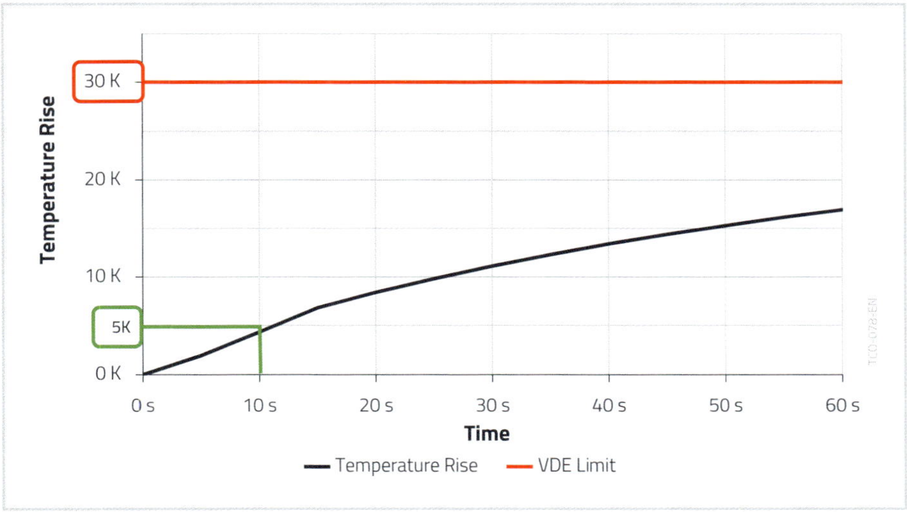

Fig. 1.77: Temperature rise in s

We can assume, that the current could be above working current for a short period of time without reaching ΔT = 30K.

Short time current temperature rise

Pre-test: inrush current duration that gives a ΔT of 30K

The question is: how can the current in a connector be estimated over a period of time without exceeding a temperature rise of 30K?

This can be determined through experimental methods. Connectors are tested beyond their nominal working current, with the temperature being closely monitored. Each test is stopped at 100 °C to ensure that the maximum operating temperature is not exceeded.

By systematically increasing the current and recording the corresponding temperature rise, a reliable relationship between current and temperature rise can be established. This experimental data allows for the estimation of the permissible current over a given period, ensuring that the temperature rise does not exceed 30K.

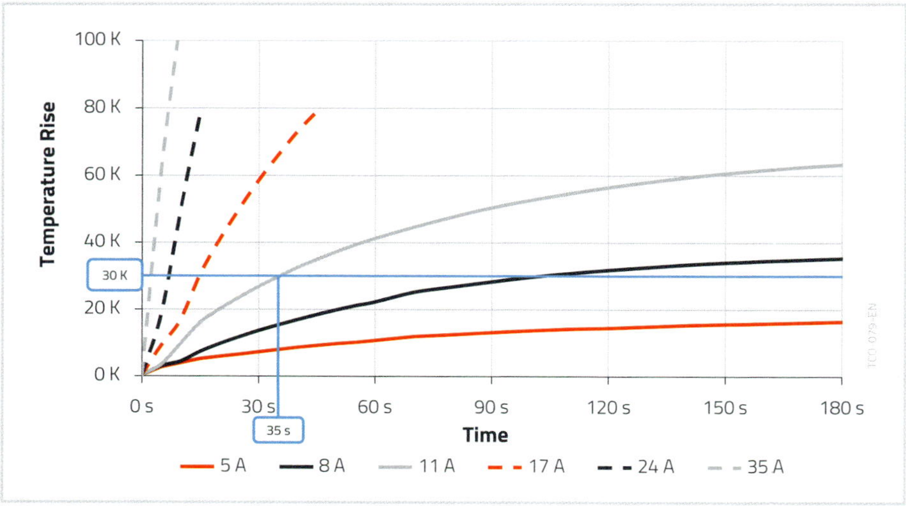

Fig. 1.78: Temperature rise above working current

Curves conclusions:

Working current gives a ΔT < 30K as requested by UL standard

Going higher than working current obviously exceeds 30K

If we limit the duration, it is possible to keep the connector ΔT < 30K. For example, with this connector, we are able to apply 11 A (2,2 times working current) during 35 s to reach 30K.

Inrush current curve proposal:

The objective is to determine the duration for which a current exceeding the nominal value can be applied to connectors, in order to test and validate the inrush current and time. A large safety factor of greater than 2 is selected, based on measured values, to ensure that the proposed limits for the user are both safe and reliable.

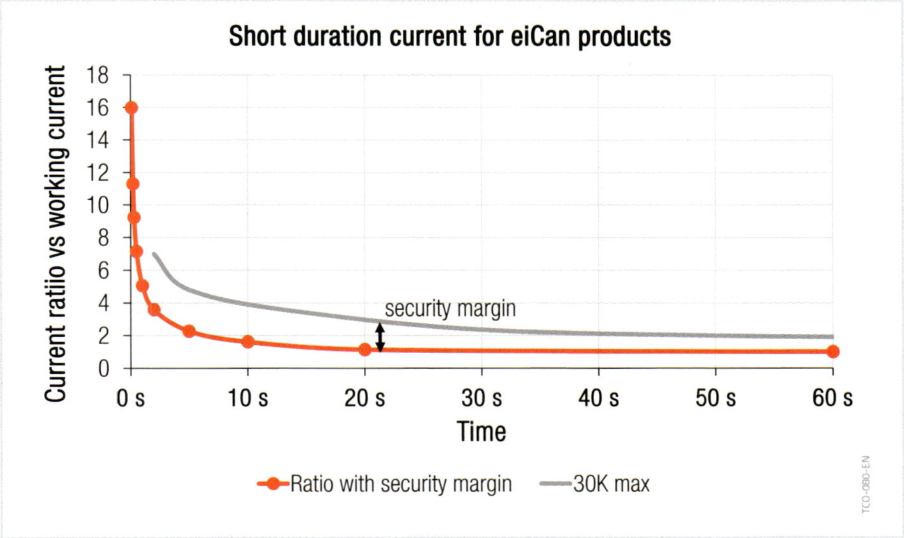

Fig. 1.79: Typical 30K temperature rise current and proposal with safety margin

Inrush current full tests

A series of connectors with a characteristic number of contacts was tested (red curve in figure 4) to check that the load from the increased current has no unacceptable effects on the connector. All values marked by the 10 red dots in figure 4 had to be tested on each connector (e.g. 1.1 times working current during 20 s). The aim was to verify that ΔT does not exceed 30K for each point.

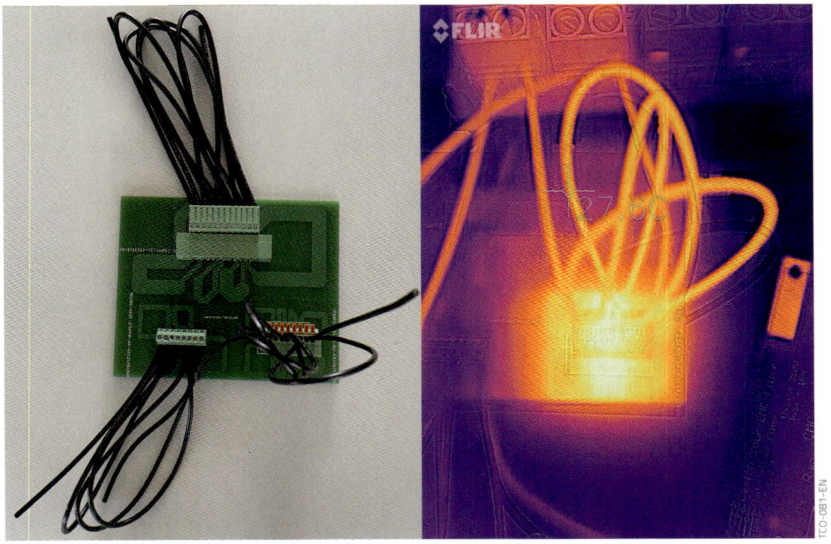

Fig. 1.80: Set-up of the inrush current test

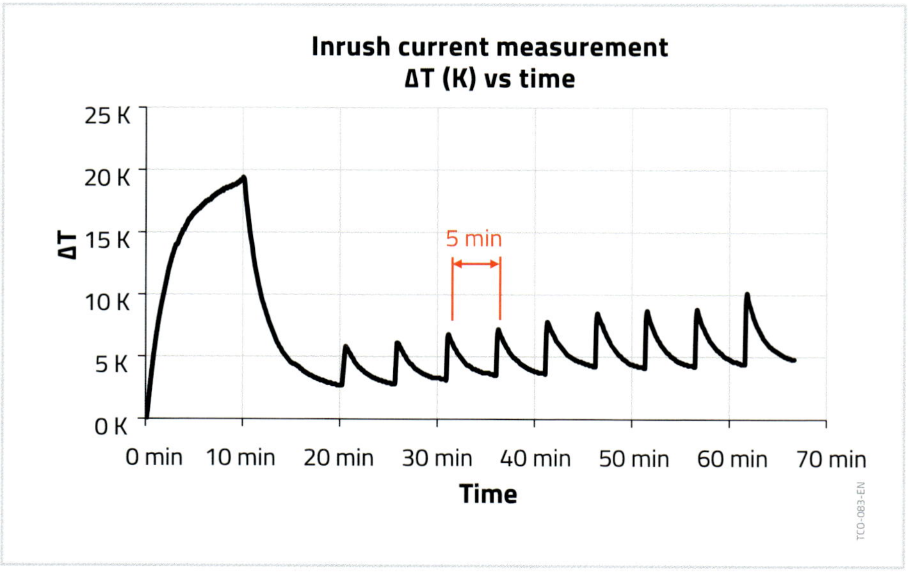

Fig. 1.81: Inrush current test results

10 tests were carried out with different currents, with an interval of 5 minutes to allow the temperature to slowly return to the initial value. The first ΔT-peak is the largest. It refers to the working current during a period of 10 minutes. All other "working points" have a very small effect on the ΔT, here about 5K maximum.

All tests carried out with different products showed roughly similar results. Time gap between peaks:

Thermal energy must be dissipated to the ambience of the connector. Otherwise, the energy is accumulated inside the connector. In that case, the temperature would continue to increase and put in danger the electrical parts of even worse the whole appliance.

Figure 1.82 shows a connector at 70 % of the nominal current with 10 peaks of current with a ratio of 7,1 and an interval of 5 minutes, that is necessary to let the connector coming back to its initial temperature.

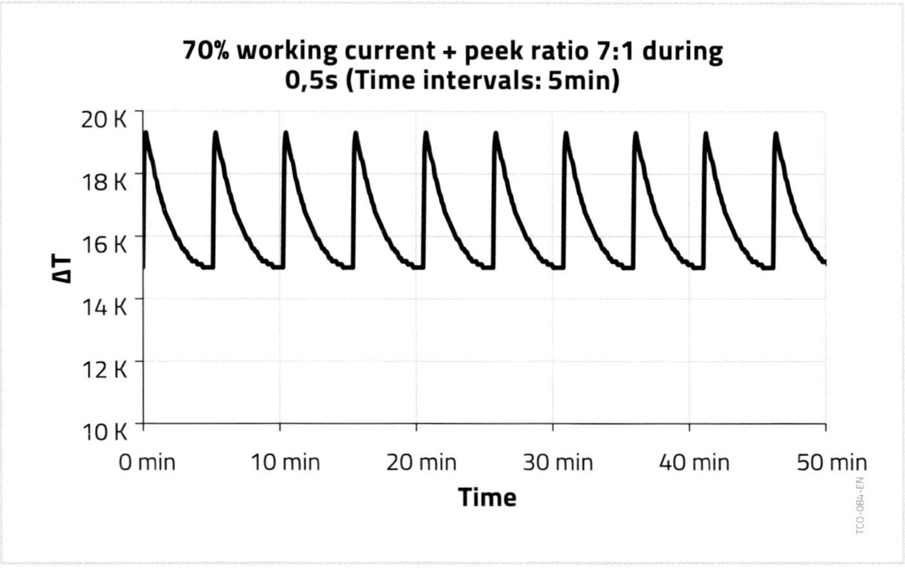

Fig. 1.82: ΔT curve with time intervals between peaks, enough for energy dissipation

Temperature rise always comes back to 15K, the system is stable.

But if the time gap is reduced, the thermal energy does not have enough time to be dissipated. Figure 1.83 shows, that the temperature continues to increase. The system is not stable. After a period, this will damage the connector so, this condition must be avoided.

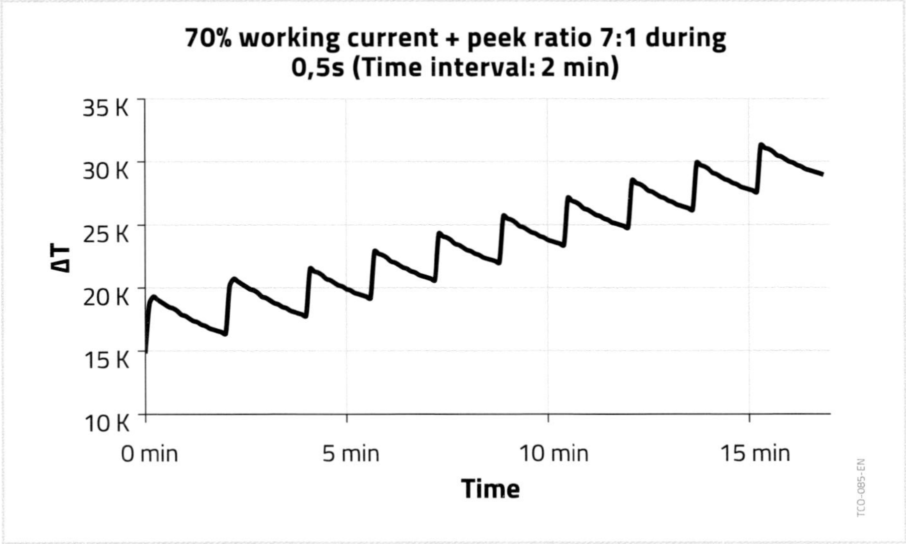

Fig. 1.83: ΔT curve with time gaps between peaks, too short for sufficient energy dissipation

Energy dissipation strongly depends on the connector's environment: size of PCB traces, size of wires, PCB put in a closed box or not etc. The amount of energy dissipation, at the 5 min intervals given here, varies with respect to each application. That is why it is necessary to run a temperature rise test under real conditions to confirm that the system is always stable.

Final applicable curves

Inrush current curve:

Finally, it is possible to have an inrush peak current every 5min. (minimum to be verified on real conditions) for each connector following the curves in Figure 1.84 and Figure 1.85.

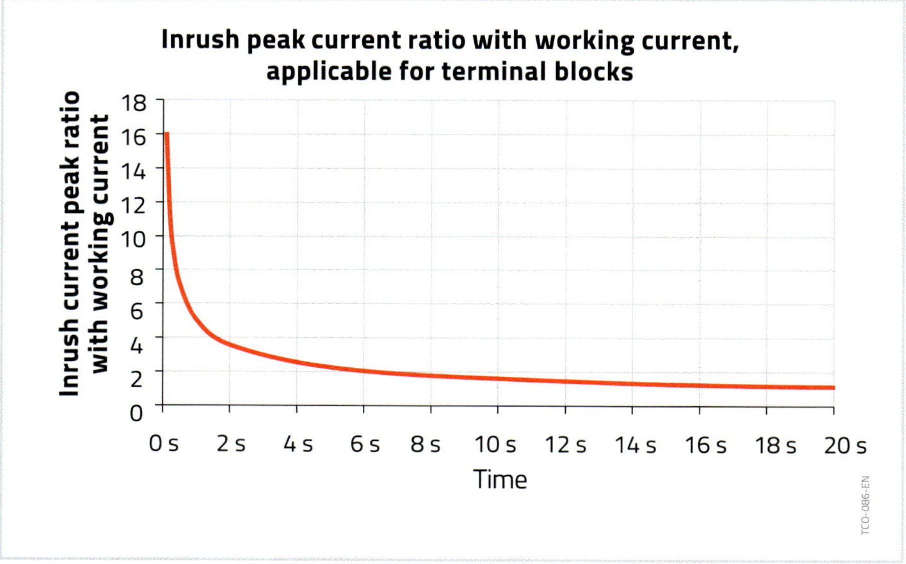

Fig. 1.84: Applicable inrush current for WE connectors. Different time scales

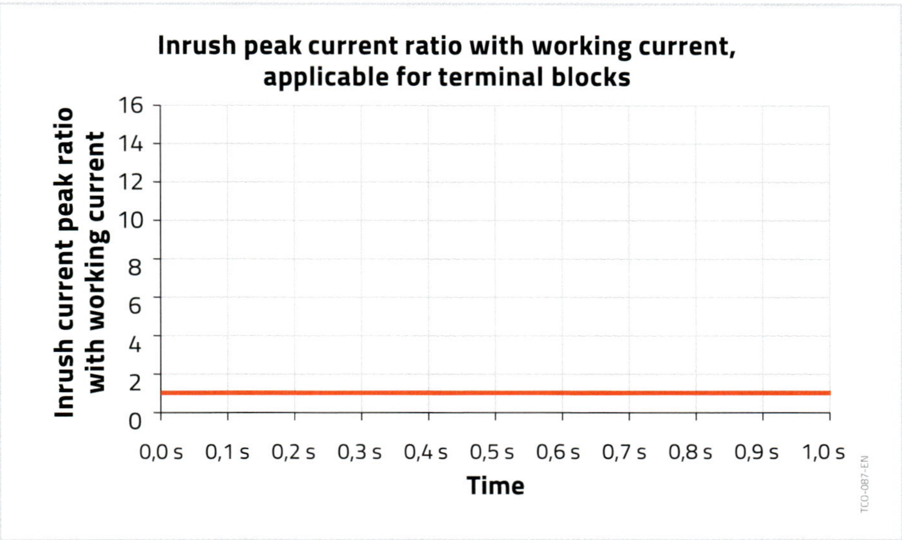

Fig. 1.85: Applicable inrush current for WE connectors. Different time scales

Example:

For MPC3 with a working current of 5 A, it is possible to use it at full current and have every 5 min. a peak current of, e.g., 2 times of the working current (10A) during 6s.

3-1-Derating curve – constant "inrush" current

If a WE connector is used, the following curve gives the current applicable related to the operating temperature (Figure 1.86).

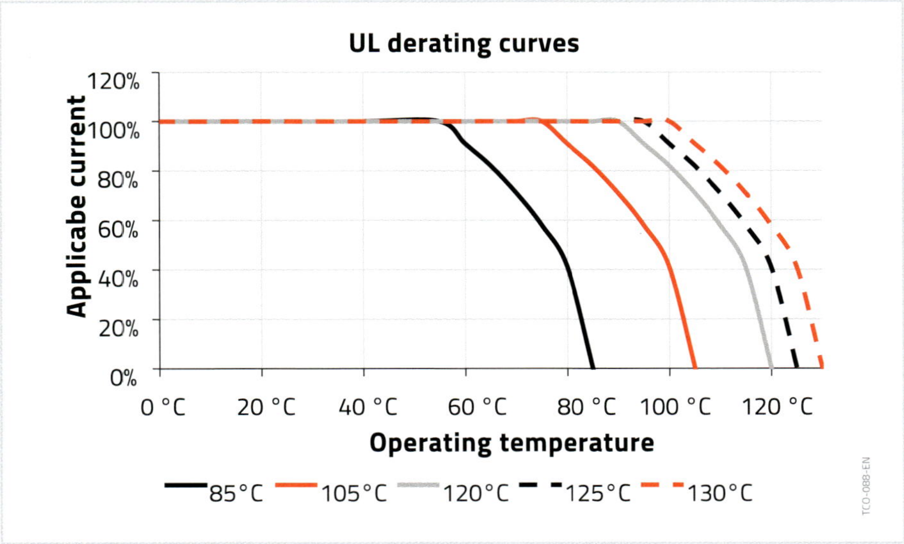

Fig. 1.86: UL derating curves for different operating temperatures

Würth Elektronik complies with UL regulations in every situation. A maximum temperature rise of $\Delta T \leq 30K$ is set, regardless of which inrush current is applied. It is therefore not possible to operate a connector with full current and additional inrush current at the same time.

For safety reasons, we assume that the maximum ΔT at inrush current is 15 K. This means that a maximum ΔT for the temperature rise at stable current of 30 - 15 = 15K is possible. We saw on previous chapter, for the same system, we know that ΔT is given by following formula:

$$\frac{\Delta T1}{\Delta T2} = \frac{I1^2}{I2^2} \qquad \text{(EQ 14)}$$

$\Delta T1 = 30K$ and $\Delta T1 = 15K$, then $I2 = I1/\sqrt{2}$

With inrush current, we must limit the working current to $(100\%)/\sqrt{2} \approx 70\%$.

So finally, if we want to use a connector with both, the continuous and the inrush current, we need to adjust the derating curves as follows:

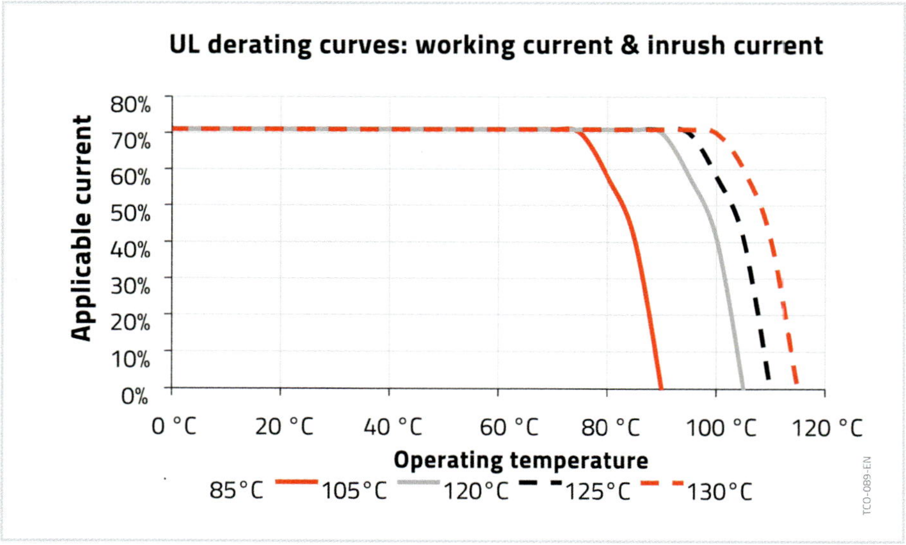

Fig. 1.87: Derating curves for continuous and inrush current at different operating temperatures. For WE terminal blocks, MPC3 and MPC4 only

Remember to always do a temperature rise test in real conditions to confirm that the time gap is long enough to keep system stable.

1.12 Connectors for shielded cables and their EMC consideration

1.12.1 How, why electromagnetic interference occurs?

There are two paths of electromagnetic interference that can enter a device from the outside and cause it to malfunction, and interference that leaves the device, emitted by the device: Conducted Interference and Radiated Interference. By law, a device must have a certain intrinsic immunity to withstand a minimum level of electromagnetic interference. Also, it may emit only limited levels of electromagnetic interference via cables, and radiated a limited amount of electromagnetic waves. These immunity requirements and emission limits are normatively defined and legally complied with as part of the CE requirements. But how can electromagnetic emission occur and why can an electronic circuit be affected by electromagnetic energy, i.e. "interference"?

Electronic circuits operate with alternating currents. It is due to the waveform, i.e. the change of the current over time. A digital signal has a certain time during which it is at 0 V, then a time during which it changes from 0 V to e.g. 5 V; then it remains at 5 V for a certain time until it changes back to 0 V within a certain time. So, in electrical terms, it is the period, the rise time and the fall time of the signal that play a role here. Depending on the impedance of the circuit, the change in voltage results in a change in current, which generates a changing magnetic field. Parts of the magnetic field will be radiated from the circuit, parts will also be coupled into adjacent circuits by inductive and capacitive coupling processes, and some of the energy will arrive at the receiver. Figure 1.88 shows the process graphically.

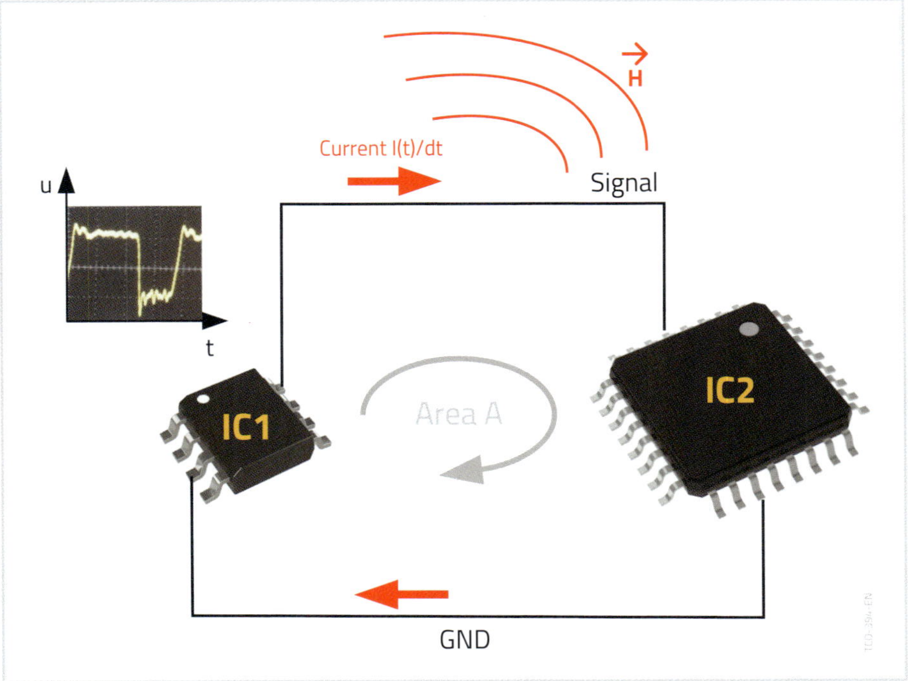

Fig. 1.88: Generation of an electromagnetic radiation by an alternating current.

Electromagnetic interference currents transmitted through power and signal lines can be attenuated on the electronic board by filtering or other circuitry measures such as impedance matching, reducing signal amplitudes, blocking voltage supplies, and others.

For radiated electromagnetic interference via the housing, shielding with metallic or electromagnetic material is essential. What remains is the interference that is radiated via the cables and also the coupling processes in which electromagnetic energy is conducted into the device via the cable, i.e. the case of immunity to electromagnetic fields, coupled transients and ESD (electrostatic discharge). Here, predominantly the cable, the connector, the interface filter and the associated ground system are of particular importance; these points are described in detail in the following article.

1.12.1.1 Fundamentals of the radiation of electromagnetic energy via cables

The purpose of cables is to conduct an electrical signal from one place, the signal source, to another place, the signal sink. This works to a certain extent. How well, however, depends on many influencing factors, these are:

- Signal parameter
  - Amplitude
  - Bandwidth (with consideration of the interface filters)
  - Balanced or unbalanced signal
- System impedances
  - Impedance of the source
  - Impedance of the cable
  - Impedance of the connectors
  - Impedance of the sink
- Cable properties
  - Wire symmetry (for balanced signal interfaces)
  - Cable attenuation (losses)
  - Shielding attenuation of the shield layers for shielded cables
  - Continuity of impedance over the cable length
  - Twisting characteristics of twisted signal cables
- HF suitability of the ground system

All these points influence the radiation characteristics of the cable, but also its immunity to external interferers, such as transient interferers coupled in by parallel cables, cell phone radiation and discharges of static electricity.

The electrical signal current, coming from its source, flows through the signal wires Depending on the signal topology, the return current flows back via a ground conductor or via a second signal conductor. The signal that, according to Figure 1.99 on top, flows in via a signal conductor and back via a ground conductor is called an unbalanced signal. The return conductor can be a cable shield or a housing ground. A signal, on the other hand, that uses another signal wire for its return path is called a balanced signal (Figure 1.89 below). This signal is shifted by 180° in polarity between one wire and the other (blue against red in Figure 2 below).

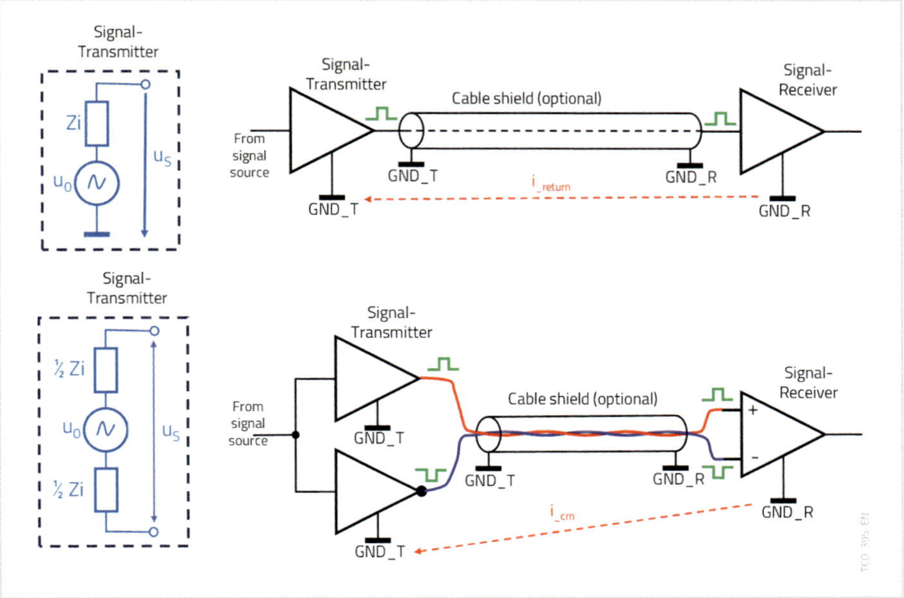

Fig. 1.89: Signal transmission of a balanced and an unbalanced signal by means of shielded cables.

Figure 1.89 also shows the principle circuit diagrams of the transmitters, the unbalanced one at the top and the balanced one at the bottom. The balanced transmitter does not require a ground reference. The return current does not flow through a ground path, as in the unbalanced signal transmission; therefore, the balanced signal transmission is not dependent on the system ground topology (e.g. metal housing, cable shield, etc.). However, symmetry properties of the transmission path have a significant impact on signal integrity, as any unbalance directly affects the overall balance of the signal, resulting in common mode noise, the current of which then flows uncontrollably back from the receiver to the transmitter, as shown in Figure 1.89 below.

Well, the same that applies to the useful signal also applies to the interference signal. Differential interference signals (DM, Differential Mode) are generated by current circuits which have a defined forward and return conductor and thus span a current loop area. This can be, for example, the current loop of a circuit between the signal conductor of an unbalanced interface and its ground return conductor, or the circuit between two integrated circuits, as shown in Figure 1.88. However, if the interference current on both conductors of a balanced interface flows back in the same amplitude and phase via a ground, e.g. the protective earth ground connected between two devices, we speak of common mode interference signals (CM, Common Mode, Figure 1.99 below). The electrical current of the common mode interference is in most cases weaker than the differential interference current formed by signals, but common mode currents often form large loops and can therefore still cause interference problems even from a greater distance.

Cable to cable or interference PCB element to cable coupling are two common and well-known phenomena. There are other types of coupling that occur primarily on a printed circuit board and are caused by mutual inductance or mutual capacitance. This coupling is referred to as crosstalk. Crosstalk is caused by near-field coupling between signals that are in close proximity to each other. Capacitive crosstalk is most common and occurs due to very short slew rates of signal voltages (dv/dt). The rapidly changing voltage waveform of the first signal induces a voltage on the signal running parallel to it (differential, push-pull coupling). Over 90 % of the RF energy radiated by cables is caused by push-pull interference from the electronics, which is then radiated through the cable in the form of common-mode interference. Figure 1.90 illustrates this "coupling mechanism".

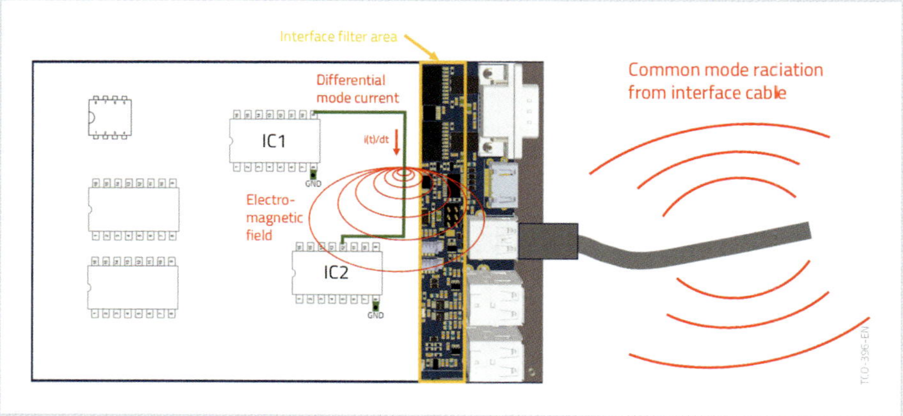

Fig. 1.90: Principle of symmetrical and asymmetrical spurious radiation and differential common mode conversion.

A signal current (useful signal) runs between IC1 and IC2 in Figure 1.90 via a conductor track, and the current loop is closed via the ground plane. The current loop (see Figure 1.88) generates a magnetic field which couples into the neighbouring areas of the interface. That is, portions of the useful signal are coupled into conductors and filters in the interface area. The coupled-in currents will flow into the cores of the interface cables and thus manifest themselves as an interference signal, measurable on the cable. How much of the interference currents are radiated through the cables depends on numerous conditions, which are enumerated at the beginning of this section. On the way from the signal source to the interface of the connector, the following measures can be taken to reduce the coupling into the cable cores:

- Reducing the current loop area between IC1 and IC2
- Relocating the signal conductor path to a different position than the interface filters
- Shielding the interface area

Another possibility to reduce the coupling is the analysis of the useful or interference signal, because the magnitude of the coupling depends essentially on the frequency and the signal current. Here we are not only concerned with the fundamental frequency of the signal (e.g. 10 MHz clock), but also with the harmonics of the signal. The relation will be explained in the next section.

1.12.1.2 Influence of the signal parameters (symmetrical/asymmetrical, amplitude, bandwidth) on the radiation

The radiation efficiency of a loop according to Figure 1 is proportional to the square of the frequency (+40 dB/decade). This characteristic holds for conductor loops that are no larger in circumference than ¼ of the wavelength, e.g., about 12.5 cm in an FR4 PCB at 300 MHz. If this characteristic is superimposed on the harmonics of a digital signal (e.g., clock signal), it can be seen, that the differential mode emission shifts toward high frequencies, and thus the period and rise time of the signal are significant parameters. Figure 1.91 shows the relationships.

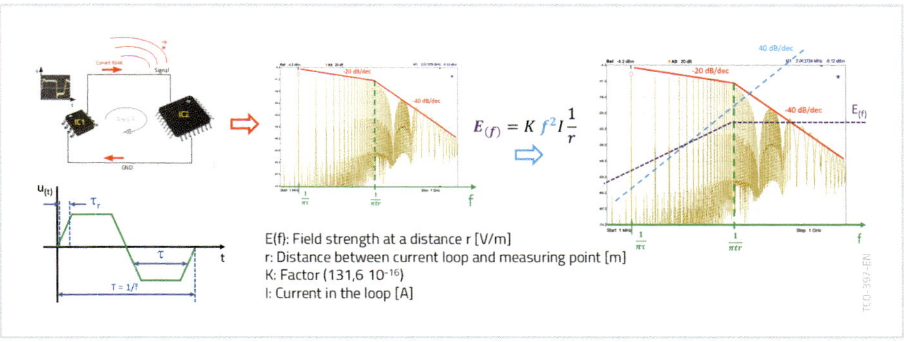

Fig. 1.91: Influence of the signal parameters of a signal circuit on the radiation characteristics.

The digital signal, with rise time τ_r and pulse duration τ is composed of many individual harmonics, whose amplitude spectrum can be seen in the middle diagram. Up to the frequency $f = 1/\pi t_r$ the harmonic spectral components fall in their amplitude with 20 dB/dec, above that with 40 dB/dec. Due to the coupling function of the current loop, the harmonics are shifted in their amplitude to higher frequencies, since the frequency in the formula occurs in the square, the coupling function is drawn in the figure on the right with 40dB/dec. This results in a shift of the amplitudes as shown in purple in the diagram on the right. What does this mean for the EMC at the interface?

- Not the fundamental frequency of the signal, but its harmonics are to be considered. An important parameter is the signal rise time and thus the frequency point $f = 1/\pi t_r$.
- High frequencies couple more easily than low frequencies.
- If the high frequency components in the circuit can be reduced by circuitry measures (e.g. filter, matching), the signal harmonics at the interface can also be reduced in their interference potential.

1.12.1.3 From the trace to the interface filter and to the connector on the housing

In most cases filters are provided between the signal source, e.g. a USB controller, and the connector. Filters are supposed to attenuate certain parts of a signal spectrum over the frequency. And this is where the problem starts, because a filter is thus a frequency-dependent voltage divider that requires a ground reference. If transient overvoltage is to be filtered along with the filtering of high-frequency signals, the ground system around the interface area is decisive for the desired function. Figure 1.92 shows the principle function of a filter.

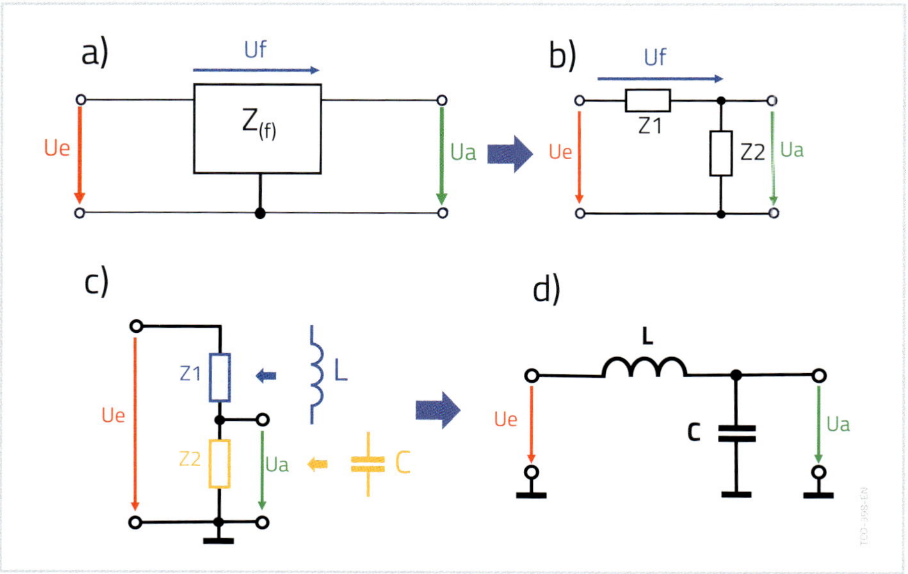

Fig. 1.92: Principle function of an asymmetric filter.

From the quadrupole diagram (Figure 1.92a), which can often be found in the literature, it can be seen that the impedance is a function of the frequency. In the more detailed schematic (Figure 1.92b) it can be seen that the low-pass topology chosen here consists of two impedances, at least one of which must change its impedance as a function of frequency. Figure 1.92c clearly shows the voltage divider principle, here it can also be clearly seen that for a function of the voltage divider, the ground, as a reference, must not have an interference potential. Every grounded interference potential will reach the interface output as a part of Ua. Figure 1.92d now shows the second order low pass with an attenuation of 40dB per decade.

The same principle applies to common-mode filters, which achieve an attenuation of the common-mode interference only by their arrangement of the windings, but additionally have an asymmetrical attenuation component due to the leakage inductance, which in turn forms the frequency-dependent voltage divider via capacitors to ground.

If the filter is to have additional protection to limit transient interference on the signal lines, a voltage-limiting component must be connected to capacitor C (varistor, TVS diode, ...). If the parasitic capacitance of the component is sufficient to achieve the required attenuation, capacitor C can be omitted. A circuit diagram of an interface filter (low-pass filter) is shown in Figure 1.93.

However, Figure 1.93 also shows the problem of the ground issue: the aim is to obtain an "EMC-proof" interface so that cables connected to the interface neither couple interference into the electronics nor radiate interference generated by the electronics. Let us first consider only the interface area itself. For the filter in Figure 1.93, top left, to work, both capacitor C30 and SMT varistor D1 must have a ground reference, ground connection (SGND in the schematic). Both, the high-frequency interference currents from the board electronics and transient overvoltage coming from the interface side must be dissipated via these ground connections. The reference point is the metal housing and, if present, the cable shield of the interface cable. Every mΩ of contact resistance and every nH of additional parasitic inductance will impair the filtering effect. This ground connection can be made through the board attachment, as shown in the upper right of Figure 1.93, which usually results in additional parasitic impedance (mΩ, nH), or the connection can be made through the connector jack, as shown in the bottom two images in Figure 1.93. This ground connection allows a low impedance connection between the housing, cable shield and filter.

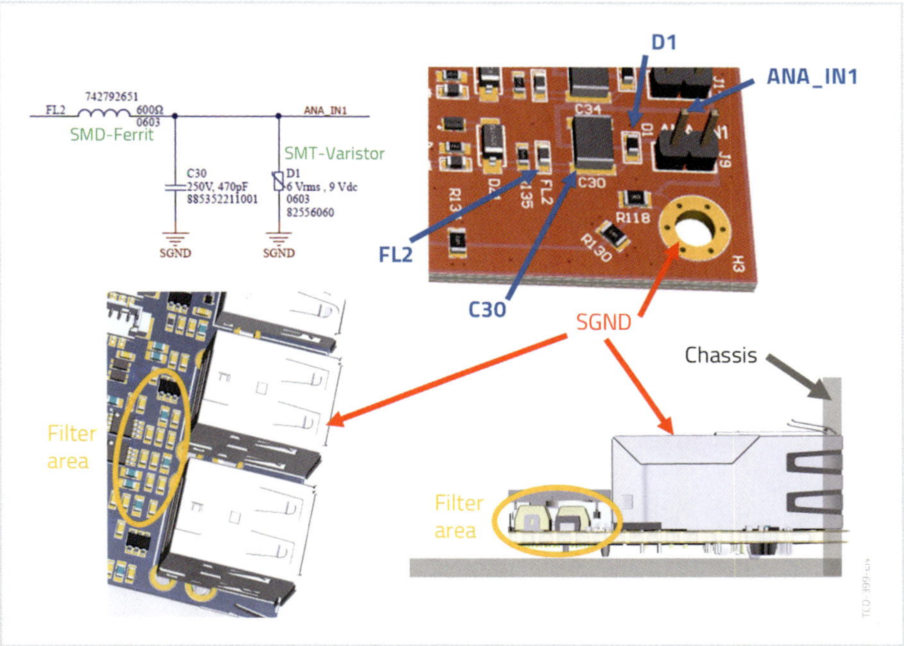

Fig. 1.93: An essential prerequisite for the proper functioning of the filter is the ground reference

If no metal housing is available, a metal connection panel or extensive ground shielding on the PCB will still be necessary for "functional" EMC and "clean" signal integrity in the interface area, which cannot replace a metal housing, but can offer a compromise depending on the electronics. Why the low impedance between filter and ground is so important is shown in the estimation in Figure 1.94.

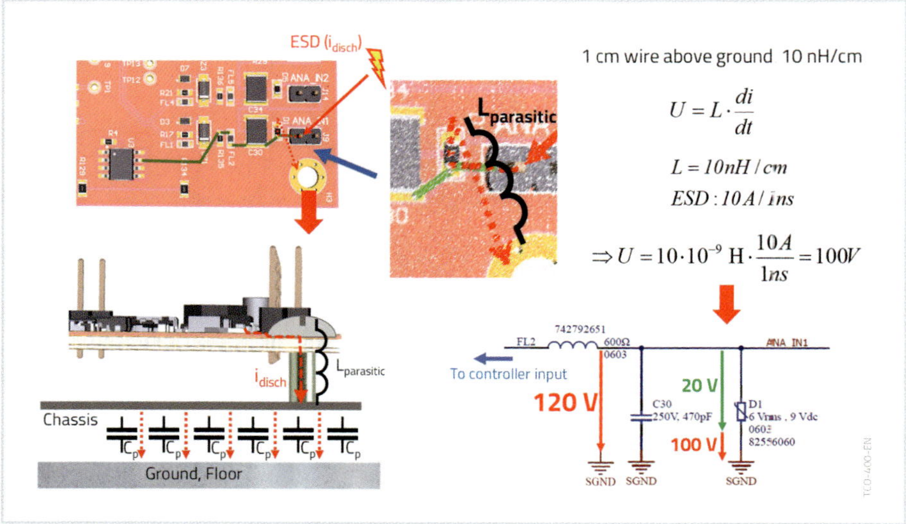

Fig. 1.94: Voltage drop at parasitic impedances between transient arrester and ground

An electrostatic discharge at the ANA_IN1 interface connection generates a discharge current of 10 A (ESD test in the laboratory up to 30 A!). The SMT varistor limits a voltage drop to approx. 20 V (clamping voltage of the varistor). Due to the parasitic inductance between the varistor and the case bottom of here only 10 nH, a voltage drop of 100 V occurs. Together with the varistor voltage, there is thus a voltage of 120 V at the input of the controller. In most cases the controller will be destroyed by the too high voltage! Not included here are contact resistances between board and screw connection, which cause additional voltage drop.

One more explanation regarding ESD: The reference, i.e. the reference ground for the electrostatic discharge is the floor. The energy discharged at the electronic device flows capacitively to the floor via the housing floor of the device. For a device without a metal floor, the device ground is therefore the GND layer of the printed circuit board!

If now instead of the ground connection selected in Figure 1.94 a concept via interface connector is chosen, as shown in Figure 1.93 below, a much better constellation results. The connector can create a low-impedance connection to the overvoltage-limiting component due to its multiple ground connection between the housing wall and the printed circuit board. However, the prerequisite is that the overvoltage-limiting component is placed directly on the connector and that there is a low-impedance conductive connection between the connector housing and the housing wall. Figure 1.95 shows the setup.

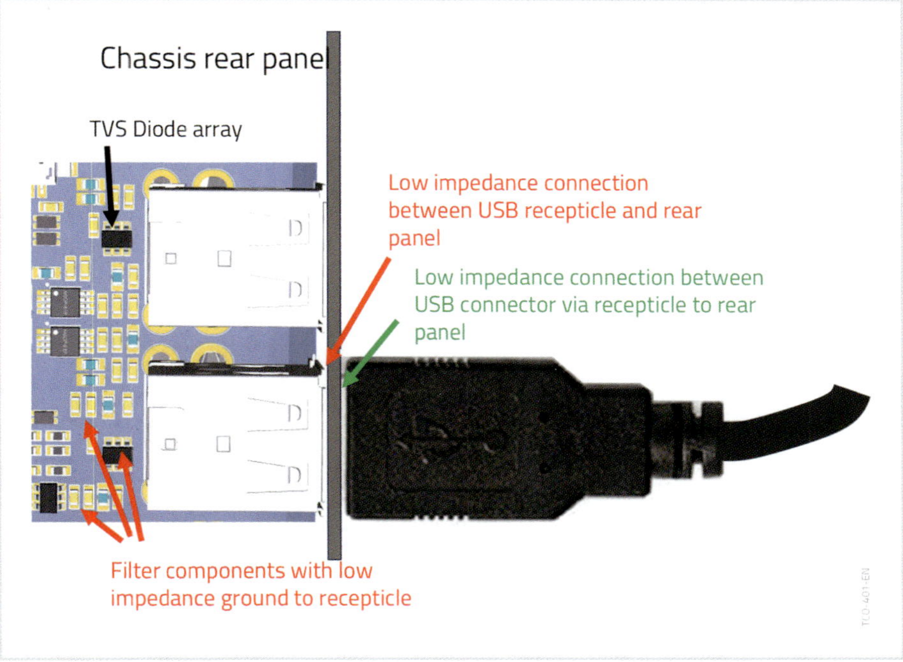

Fig. 1.95: Ground connection via interface socket to the transient protection and to the HF-dissipating capacitor

A prerequisite for a good ground connection is a high conductivity of the backplane of the housing. If the material of the rear panel causes a high contact resistance due to a surface treatment (varnish, passivation, anodizing), the filter capacitors and the voltage-limiting components will lack the ground reference. The corresponding filter effect against RF interference is missing and transient interference, coupled into the interface cable, is conducted directly into the electronics! Mechanical deficiencies, which do not allow a permanent contact to the interface socket, lead to similar effects. A ground improvement by additional conductive sealing to the base plate, as shown in Figure 1.96, can improve the situation here.

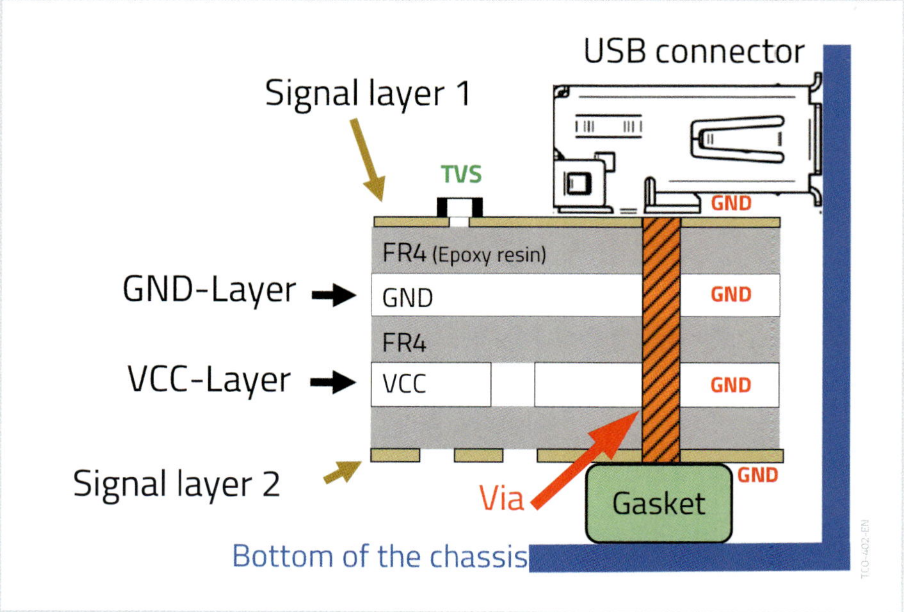

Fig. 1.96: Ground connection of the interface area by an additional conductive gasket or spring to the metal base plate.

A high surface conductivity of the housing base must also be ensured when the filter mass is connected to the base plate.

1.12.1.4 The interface with coaxial connector, RF interfaces with matching.

In most cases, high-frequency signals are carried over coaxial cables. These signals have one or more of the following characteristics:

- a high frequency, typ. > 100 MHz
- a high bandwidth, typ. > 500 MHz
- a high power, typ. > 1W
- a very small amplitude, typ. < 1 mV

All properties have in common that they are "EMC sensitive", be it because of the radiation of high-frequency power or because of coupling of interference signals due to very small amplitude. A prerequisite for the functionality of these interfaces is not only a high shielding effectiveness of the connectors and cables, but also the matching of the complete signal path to the system impedance. There is a lot of literature on this subject, so only the most important points will be summarized here. Figure 1.97 shows a typical transmission path for an unbalanced RF signal graphically.

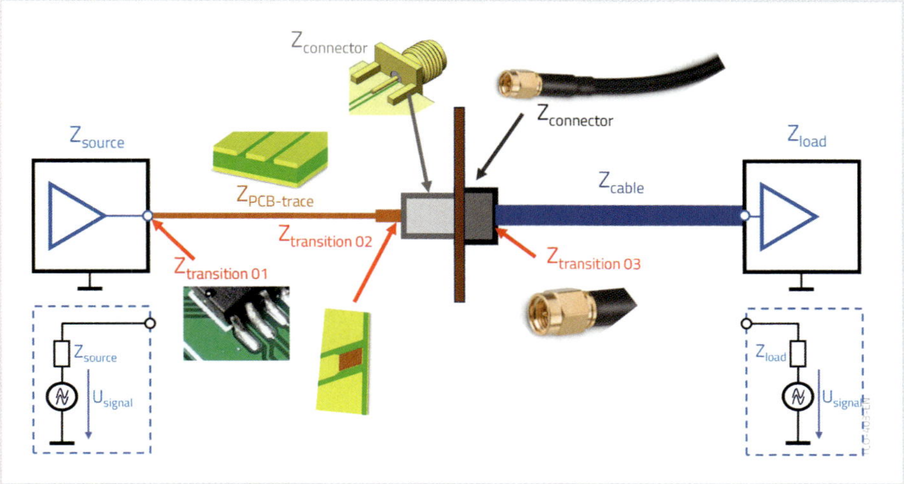

Fig. 1.97: Signal path with impedances for an unbalanced RF signal

The signal, coming from the source, is related to the source impedance Z_{source}. This source impedance determines all other impedances of the system, which, in short, must all be equal. This means,

$Z_{source} = Z_{PCB\text{-}trace} = Z_{connector} = Z_{cable} = Z_{load}$. Also, the transitions, i.e. the connections of the cable ends to the connectors and the connections of the traces to the connector sockets must have this impedance and here the problems start. In practice, one cannot achieve one hundred % equality of impedances, i.e. one hundred % matching. Mismatches remain, depending on the "quality" of the design. Even if the connector, cable, and PCB trace have the required system impedance, the transitions from one system section (e.g., trace) to another often cause mechanical problems, resulting in mismatches. Another point that requires special attention is the ground reference of the various system sections: The reference ground for the signal source (Z_{source}) is the printed circuit board, and likewise for the trace to the connector at the interface. Assuming that the housing of the system in Figure 1.97 is made of metal, the connector must be connected with its ground all around the sheet metal housing and so must the ground of the PCB with the defined trace impedance! If the connector is not "cleanly" connected to the housing ground, the situation shown in Figure 1.98 results.

1.12 Connectors for shielded cables and their EMC consideration

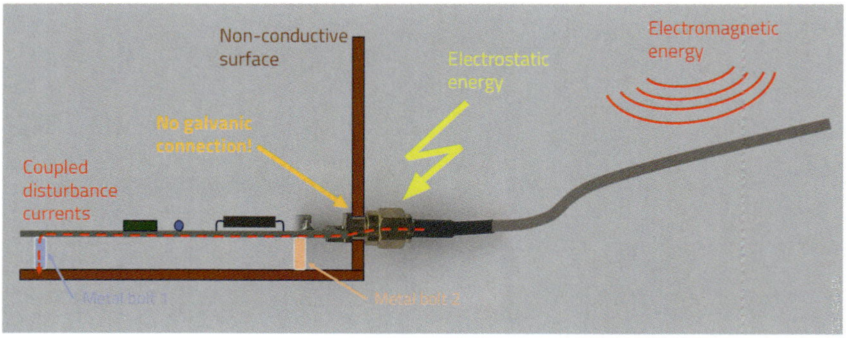

Fig. 1.98: Interference coupling in electronic module in case of insufficient ground connection of the plug connector

If the cable shielding ground is not galvanically connected to the housing with all-round contact, interference energy of any form (ESD, electromagnetic radiation, ...), which is coupled in by the cable, will enter the housing and thus the electronics as interference current. The effects of the parasitic current depend on the more or less random ground conditions between the PCB and the bottom of the enclosure. If, as shown in Figure 1.98, a metal bolt 2 is present, it can dissipate part of the interference current against the bottom of the housing, provided the impedance is low enough. If, however, the interference current should flow via the metal bolt 1, the interference current, depending on its frequency and its current amplitude, will cause a voltage drop which will lead to the malfunction, or even the destruction, of the electronics. A simple rough calculation based on Figure 1.99 will illustrate this:

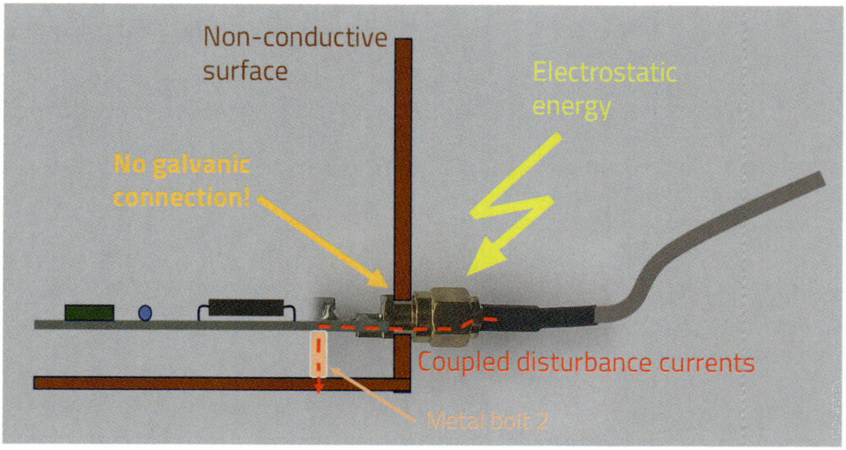

Fig. 1.99: Interference current path via metal bolt 2 during ESD discharge at the SMA connector.

The contact resistances are roughly estimated:

- SMA connector to SMA connector on PCB: 30 mΩ
- SMA connector to PCB via vias to metal stud: 50 mΩ
- Metal stud to sheet metal base: 20 mΩ

If an ESD discharge current of approx. 20 A is assumed, which is 10 A below the normative maximum discharge current, a voltage drop UESD of only 0.1 Ω x 20 A = 2 V results.

If the parasitic inductance is taken into account as a parameter instead of the contact resistances, a completely different picture emerges:

The parasitic inductance of the distance between connector and sheet metal base is approx. 10 nH/cm. An estimated distance of 3 cm results in an inductance of 30 nH. We assume a current rise time of 20 A/ns as the ESD discharge current (standard: 30 A/0.6 ... 1 ns). The voltage drop is now:

$U = L\, di/dt = 30\ nH \times 20\ A/ns = 600\ V$

The potential difference between the ground point and the base plate, the reference ground for the electronics, is 600 V! This should be too high for most electronic circuits at the interface input and leads to immediate destruction.

Now back to the cause, this is the missing galvanic connection between the cable shield or the connector ground and the housing. If the connection is made, the protection against the overvoltage is also guaranteed. It should also be mentioned here that the missing ground connection not only massively impairs the noise immunity, but also significantly increases the noise emission through common mode coupling to the cable shield.

1.12.1.5 The system properties and their influence on the cable radiation

Why can a cable connected to a device couple or radiate energy, which factors play a role here? If you look at the overall system, i.e. the device, the factors are:

- Signal parameters and impedance of the interference source
- "Antenna" or radiation characteristics of the cable
- Electromagnetic properties of the environment (around the device)

We will focus here to the cable's radiation characteristics, because this is where the connector has the greatest influence.

In the frequency range from approx. 30 MHz to 300 MHz, radiation coupling is dominated by cable emissions rather than direct radiation from the PCB. This is for the simple reason that cables resonate in this frequency range and their radiation efficiency at these frequencies is higher than that of PCB structures. The emission current is generated in common mode. With regard to the common mode radiation of shielded cables, there are various causes, these are illustrated in Figure 1.100.

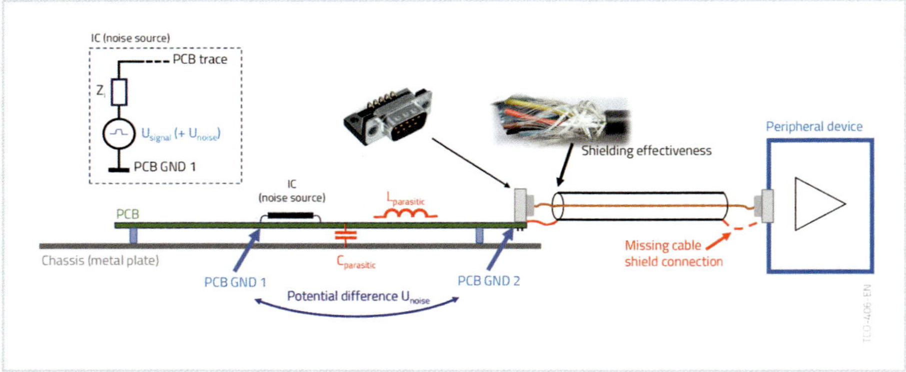

Fig. 1.100: Causes of common mode radiation from shielded cables

a) A Part of the useful signal or an interference signal is as RF potential on the device ground.
A Part of the useful signal (U_{signal}), or an interference signal (U_{noise}) can be measured as interference potential on PCB GND 2. This signal will appear as interference current on the cable shield and will be radiated as electromagnetic interference. Causes are the parasitic inductance between the interference source (IC) and the point PCB GND 2, as well as insufficient galvanic connection of the points PCB GND 1 and PCB GND 2 to the housing. The interference signal appears as a potential difference U_{noise} between the two points of the electronic assembly. The interference potential U_{noise} increases in amplitude with increasing frequency, typically with approx. 20 dB per decade.

b) Shielding effectiveness of the cable shield too low
Insufficient shielding effectiveness of the cable shield is in most cases due to the following causes:
- Insufficient area coverage of the shielding braid (not enough strands)
- Insufficient conductivity of the shielding strand material
- Only one-sided or no conductivity of the shielding foil
- Broken shield foil
- No HF-compliant connection of the shield to the connector

All causes result in two effects:
- Some of the signal energy passed through the cable will be radiated over/through the shield.
- The sensitivity of the interface to coupling of electromagnetic interference via the cable increases considerably.

c) Cable shield not connected to the ground of the peripheral device at the cable end.
The shielding effect of a cable shield is due to the effect that the inner conductor(s) builds up a mutual inductance with the shield and thus the signal and the interference return currents flow back almost completely via the cable shield. This is the only way the shield attenuates both electric and magnetic fields. If one side of the cable shield is not connected (Figure 1.100 right at the "Peripheral device"), the cable shield does not attenuate magnetic fields, e.g. induced by other cables running in parallel.

Another property that contributes significantly to a cables "radiating ability" is the ratio between the wavelength of the interfering signal and the electrical length of the cable. The wavelength of the interfering signal can be calculated with the following equation:

$\lambda = c/f$

Where λ is the wavelength [m], c is the speed of light [m/s], and f is the frequency [MHz].

From the above equation, as the frequency increases, the wavelength decreases. As the wavelength gets smaller, the cables appear electrically long, which means that several wavelengths can fit into the cable length. Thus, the cable must be considered as a "monopole" in its electrical property, a "rod antenna" that has its best radiating property when multiples of $\lambda/4$ fit into its electrical length. The electrical length of the antenna (of the peripheral cable) is essentially dependent on the ratio between the length and diameter of the cable and on the conductivity of the shielding surface; thus, in "EMC practice" it can hardly be influenced. As a rule of thumb, it can be assumed that the electrical length corresponds to approx. 90 % of the physical length of the cable. Figure 1.101 illustrates the representation of an electronic product, with connected peripheral cable and its "electrical equivalent circuit" as an interference source with antenna (monopole). Figure 1.101 on the right shows the radiation system as an interference source with source impedance $Z_{\_electronic}$ and antenna radiation impedance $Z_{\_cable}$.

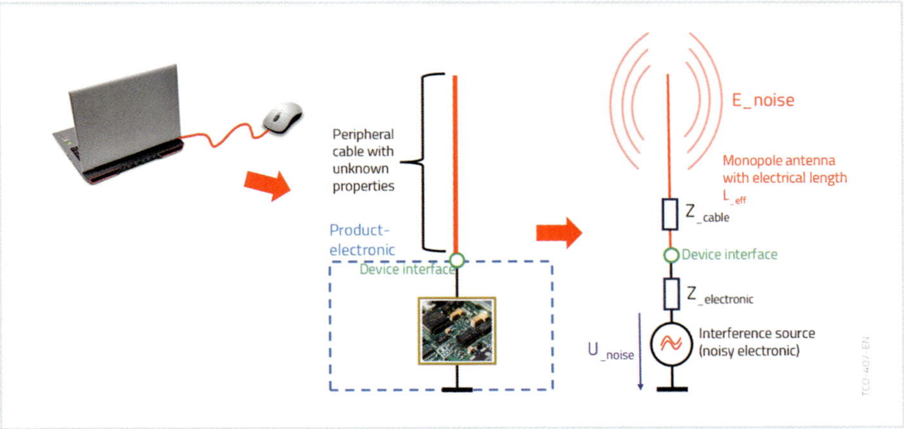

Fig. 1.101: Illustration of an electronic system (notebook with cable mouse) and the associated -equivalent circuit as a radiation source.

Depending on the two impedances $Z_{\_electronic}$ and $Z_{\_cable}$ and the amplitude of the interference source $U_{\_noise}$, the interference field strength $E_{\_noise}$ results, which is radiated via the antenna. $U_{\_noise}$ and $Z_{\_electronic}$ are essentially a matter of good electronic design, which will not be discussed here. What the length of the cable has to do with its radiation characteristics is made clear in Figure 1.102.

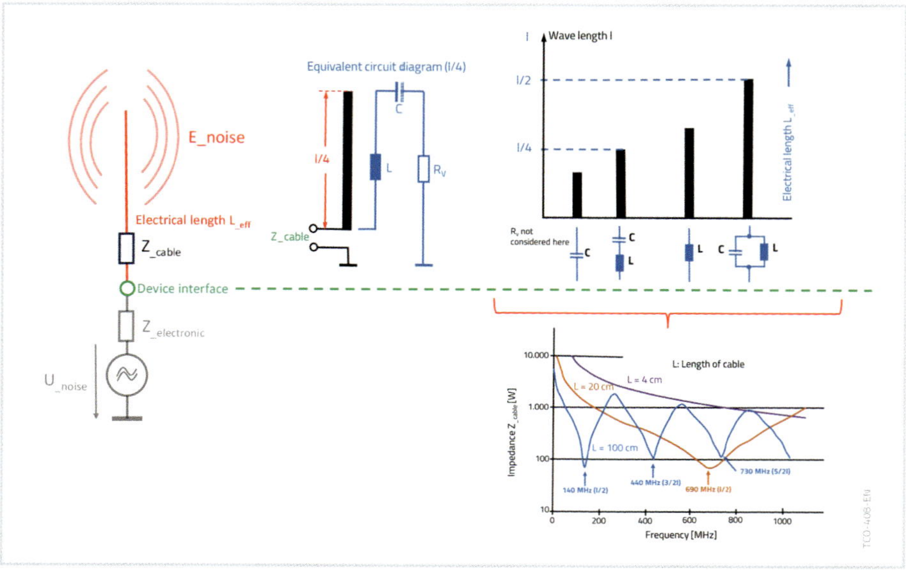

Fig. 1.102: Graphical representation of the radiation characteristics of a monopole (peripheral cable).

The figure on the top left again shows the equivalent circuit of the system consisting of interference source (e.g. notebook), interface and connected cable. As already described, this cable behaves like a monopole antenna. The electrical equivalent circuit of a monopole antenna is shown in Figure 1.102 center/right. At resonance, electrical antenna length = multiples of $\lambda/4$, the structure is in series resonance and has its lowest impedance here (37.5 Ω). Other lengths result in other, higher impedances. Figure 15, bottom right, shows the impedances of various cable lengths versus frequency. As can be seen in the lower right of Figure 15, the impedance of the cable can take on several kΩ. Short cables generally have high impedances at low frequencies, i.e., long wavelengths. Long cables show multiple resonances, which leads to an "up and down" of impedance jumps. In practice, after all, the cable length is not tuned to a specific frequency. The peripheral cable is fed with a broadband mixture of different useful and interfering frequencies and the length (electrical length) of the cable is generally predetermined. Thus, the radiation impedance $Z_{\_cable}$ varies with frequency and, together with the impedance $Z_{\_noise}$ of the interference source, forms a voltage divider. Thus, in summary, the following results with respect to the "radiating ability" of the peripheral cable:

- The amplitude of the radiated interference field strength depends on the impedance ratio $Z_{\_electronic}$ to $Z_{\_cable}$. If $Z_{\_electronic}$ is small (typ. < 200 Ω), the cable will radiate at multiples of $\lambda/4$ the wavelength of the interference spectrum.
- If $Z_{\_electronic}$ is large (typ. > 2 kΩ), most of the interference energy will fall off at the source at antenna resonance. However, in the frequency range where the antenna has a high impedance, spurious energy may still be radiated.
- **Note:** This is the reason why clap ferrites over cables sometimes help and sometimes not!

Since the properties of the cable obviously play a major role, the next chapter describes the shielded cable with its most important properties.

1.12.2 The shielded cable and its most important properties for EMC

There is no cable that meets the "EMC limits", but there are cables with a frequency-dependent shielding effectiveness and a defined impedance. The shielding effectiveness can be higher or lower and the impedance can be more accurate or not. There are other important properties that determine the quality of a cable, but they should not be considered here, where EMC is concerned. Cable shields must cover their entire length and around their entire perimeter, and this cover must have high conductivity and may have slits/openings. The problem with shielded cables is that

- external disturbances must be kept on the outside of the braided shield (skin effect) and compensated by eddy currents and a closed current loop, i.e. shield contacting to ground on both sides. Thin shielding materials (vaporized foils, etc.) cannot fulfil this task, nor can braided shields with a surface coverage of < 85 %.
- Coaxial cables require a "shielding sleeve" that is continuous over the entire length and as low in loss as possible in order to build up a negative feedback inductance between the center conductor and the shielding sleeve, which establishes the frequency-independent characteristic impedance and the shielding property as such. Only in this way can the signal carried in the coaxial cable be transmitted with the lowest possible losses and reflections. That is why professional coaxial cables are rigid and thick, and even the best glossy brochure cannot explain away these general physical preconditions.

For a better understanding of the "RF suitability" of a cable, the most important parameters are explained in more detail below.

The first coaxial cable had a solid copper shield, which had very good shielding properties, but was not flexible. They still exist, the semi-rigid coaxial cables, for applications in the RF range where very high shielding attenuation is required. For flexibility reasons, braided shielding is used in most cases. However, the shielding properties of the metal braid are not as good as those of solid tubes. Over time, various methods have been developed for assessing the shielding quality of coaxial cables.

1. Transfer impedance of Schellnukoff.

He defined the concept of transfer impedance (Zt [Ω/m]) inherent in the cable itself. This parameter is calculated using the following equation:

$$Z_t = \frac{\Delta V}{L\, I_s} \quad \left[\frac{\Omega}{m}\right] \qquad \text{(EQ 15)}$$

The method measures with ΔV [V] the potential difference between the shield and the cable core, which occurs at one end of the cable when a current I_s [A] flows on the shield of the cable under test. L [m] is the length of the cable sample. The transfer impedance parameter illustrates the coupling between the external magnetic wave with the transverse electromagnetic wave propagating inside the cable. The purpose of shielding is to achieve the lowest value of transfer impedance to minimize this coupling.

2. Leakage of electromagnetic waves through the shielding braid according to Vance.

Detect the coupling through the holes in the braid, as shown in Figure 1.103. Vance said that the magnetic coupling through the holes can be represented as an inductive quantity. The higher the coupling, the lower the shielding effectiveness, which can also be related to the area coverage via the number n of holes. On the right side in Figure 1.103 is the decoupling of the magnetic field, generated by the inner shield current H_s. The negative feedback inductance M between the inner conductor and the shield ensures the polarity reversal of the magnetic field between the field of the inner conductor H_i and the field of the outer conductor H_s. The reason that on the inside and on the outside of the shield the field runs in opposite directions is the skin effect. As a result, the current flows only on the surface of the conductor as the frequency increases (typically above 50 kHz). This is mainly the reason why coaxial cables have a lower cut-off frequency, but theoretically – neglecting losses – no upper cut-off frequency.

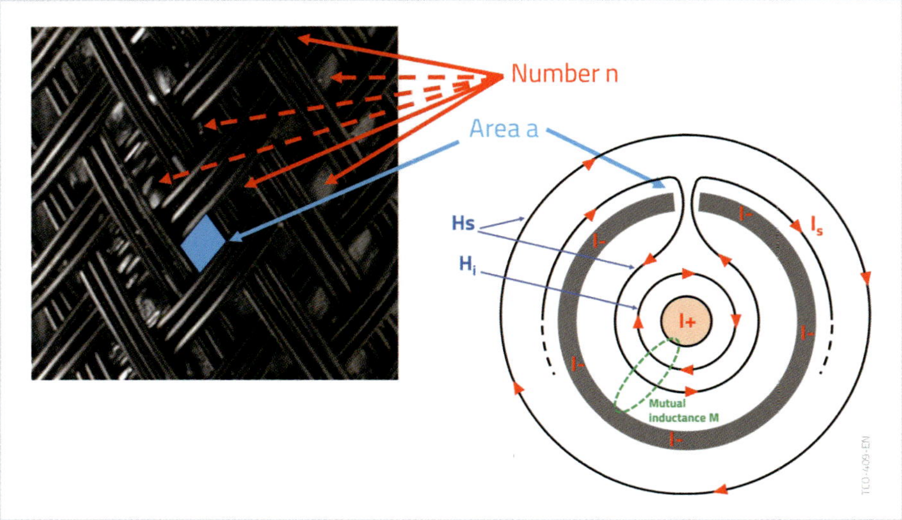

Fig. 1.103: Shielding braid with openings causing magnetic decoupling of the internal field.

The two procedures clearly show which parameters are important for good shielding effectiveness. It has already been mentioned that the cable shield must make all-round contact with the corresponding housing, so we will leave the topic of the "pigtail" aside at this point. However, there is another point that is essential for the contacting of the cable shield to the housing – the connector. This point will be described in detail in the next section.

Requirements for the connector when using shielded (coaxial) cables

When it comes to reducing EMC problems, it is often not the cable that is the cause, but frequently the connector. In the case of shielded or coaxial cables, the connector has to fulfil two essential tasks:

1. Connect the cable shield to the housing with as low impedance as possible.
2. Maintain the characteristic impedance of the cable.

Any deviation from the ideal state of the two points leads to problems in signal integrity and EMC. If no low-impedance galvanic connection can be made between the cable shield and the housing due to poorly connected cable shields, excessive contact resistances between connectors and housing, the use of a "better" cable has no effect at all on EMC and/or signal integrity. This is because the interference currents flow through the shielding of the cable. Typically, only about 3 μA of interference current is required through a 1 m cable shield to exceed the emission limit of the interference field strength according to CISPR 32! When the return current flows through the total impedance of the connector, cable shield terminations, and housing junctions, it creates a voltage drop, and this voltage between the housing and cable shield drives the common currents on the cable, resulting in an EMI disturbance. This is shown in Figure 1.104.

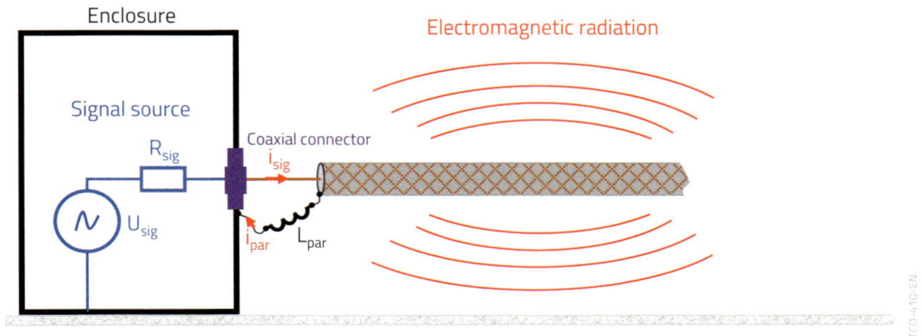

Fig. 1.104: Illustration of how a poor connection of the cable shield in a connector can cause common currents on a cable.

The parasitic inductance L_{par} causes the return current at the connector not to be perfectly symmetrical to the signal current. Any asymmetry in the coaxial cable means that the magnetic field lines of the signal current and the return current in the cable do not completely cancel each other out, resulting in a certain "leakage inductance" which leads to a "net signal current"; the part of the signal current which is radiated via the cable shield.

An approximate calculation of the radiation clarifies the problem:

- Signal current isig generated by the useful signal (1V/50 Ω cable): 20 mA
- Signal rise time (useful signal): 2 ns
- Parasitic inductance L_{par}, which arises due to connector transitions and cable shield connections: 0.2 nH

Thus, the voltage drop V_{par} at the parasitic inductance:

$$V_{par}=L\frac{dI}{dt}=0{,}2\ nH\cdot\frac{20\ mA}{2\ ns}=2\ mV \qquad \text{(EQ 16)}$$

To calculate the interference current at the cable shield, the total impedance, consisting of the parasitic impedance and the impedance of the cable as an antenna (radiation impedance) must be taken into account. Approximately, a value in the range between 100 and 200 Ω can be used, the "worst case" would be the impedance at λ/4 with approx. 40 Ω. For the calculation here, Ztotal = 150 Ω is selected.

$$I_{par}=\frac{1}{Z_{total}}\cdot L_{par}\cdot\frac{dI}{dt}=\frac{1}{150}\cdot 0{,}2\ nH\cdot\frac{20\ mA}{2\ ns}=13{,}3\ uA \qquad \text{(EQ 17)}$$

It was already mentioned above, that a cable shield current of approx. 3 µA can lead to failure of the field strength limits according to CISPR 32 – with 13.3 µA you are safe there! It is clear to see, connectors suffer from many of the same EMC problems as cables, after all they are just short pieces of conductor in a rigid housing. Thus, the overall design of the device is critical not only for passing EMC requirements, but also for good signal integrity and thus a good product. If one link in the overall chain does not function – in this case the connector – problems are to be expected.

Cable connectors are manufactured in many styles. In addition to application-related design features, a connector should have shielding characteristics that do not interfere with the shielding of the cable with which it is used. The most important points to consider in a connector are:

- The shielding of the connector must have a surface with high conductivity.
- The shielding must not have any interruptions/slots between the housing parts.
- The cable shield must be fixed to the housing in an all-round contacting, broad-surface manner.
- The connection between plug and socket must be all-round contacting, low impedance and mechanically stable.
- If cable impedances are defined, e.g. 50 Ω, these must be maintained by the connector system.
- The components of the connector must be corrosion-free.

The connector with its shielded cable thus represents an electromagnetically closed unit that must be contacted to the housing.

1.12.3 The connector and its connection to the housing

The best connector system is useless if it is installed in painted, anodized or bent housing surfaces and the RF-compatible connection between the two components cannot be established. The entire perimeter of the connector must be contacted with a good conductive, corrosion resistant surface without a gap; Why is this? The ground of the connector and thus also the cable shield require the direct ground reference to the useful signal; i.e. the ground of the useful signal and the connector must be at the same potential in terms of RF technology. A metal housing provides such a potential reference and also protects the electronics from direct field coupling from outside.

1.12.3.1 The housing and its shielding properties

Shielding involves placing a conductive shell around the electronics so that electromagnetic fields, both from the inside to the outside and vice versa, are attenuated through a combination of reflection and absorption. Furthermore, a metal enclosure provides a low impedance RF ground reference through the metal layer. This allows ground or earth references to be positioned at different locations without creating voltage potentials which then cause common mode interference currents to flow from one position to another, resulting in the radiation of interference via cables, as described above (see also Figure 1.100). For high shielding effectiveness, i.e. shielding efficiency, the enclosure must have a number of properties. For the sake of clarity, the most important points are briefly summarized here – the list is based on the ideal case in order to clearly illustrate the interrelationships in brevity.

- The ideal housing must surround the electronics seamlessly on all 6 sides, without openings.
- In order to be able to attenuate low-frequency magnetic fields (typ. < 500 kHz), the shielding material must have the highest possible magnetic permeability (iron), the minimum thickness of the material depends on the parameters frequency and field strength of the signal to be shielded.

- Electric and electromagnetic fields can also be shielded well with thin, non-ferromagnetic materials.
- Materials with high magnetic permeability generally have lower electrical conductivity and require greater sheet thickness to provide a good ground reference. In order to achieve both a high shielding effectiveness and an RF-compatible ground reference even with iron sheets, for example, the surface of the iron sheet should be provided with a highly conductive material (e.g. copper-nickel).

In practice, however, it is often not possible to realize that the shielding is continuous in all dimensions around the circuit – compromises have to be made. The metal structure should be as complete as possible between the interface connections, i.e. without openings, gaps or seams, since interruptions cause the impedance to the reference ground (e.g. bottom plate) to become higher and high-frequency leakage currents from filters or cable shields cannot be completely dissipated.

1.12.3.2 Openings in housing

The real effectiveness of the shielding of enclosures is not determined by the material properties, but by the necessary openings and interruptions in the shielding. Openings for ventilation, for controls, display, seams between sheet metal components reduce the shielding effectiveness from theoretical values around 100 dB to practical values around 30 to 40 dB. As mentioned above, the filter effect and the cable shielding are also significantly influenced by the "quality" or "stability" of the housing shielding attenuation. Enclose shielding attenuation is also directly related to the "ground", or "GND" of the electronic system; low enclosure shielding attenuation will also cause a weak ground, which will also cause low shielding attenuation of the connected peripheral cables. The worst aperture for shielding effectiveness is a long, thin slot; the best aperture is a round hole, but squares and hexagons are almost as good. Dividing a long slot into two or more shorter slots improves the shielding effect. If a slot is halved in length, the shielding effect increases by 6 dB. Slots in an area of cable connections should be avoided completely, especially because of the immunity to electrostatic discharges, since the potential differences at the openings cause massive field couplings into the system. Care should also be taken in interface areas that are assembled from several individual sheet metal parts. The junctions have neither a continuous contact nor a high conductivity and reduce the shielding effect of the shielded cables and considerably worsen the ground reference of the filters in this area. The problem is particularly serious with hinged front panels, doors and removable flaps that are part of a shielded enclosure. Conductive gaskets and female connectors can provide a remedy. Figure 1.105 illustrates the problem of openings, slots and peripheral connections.

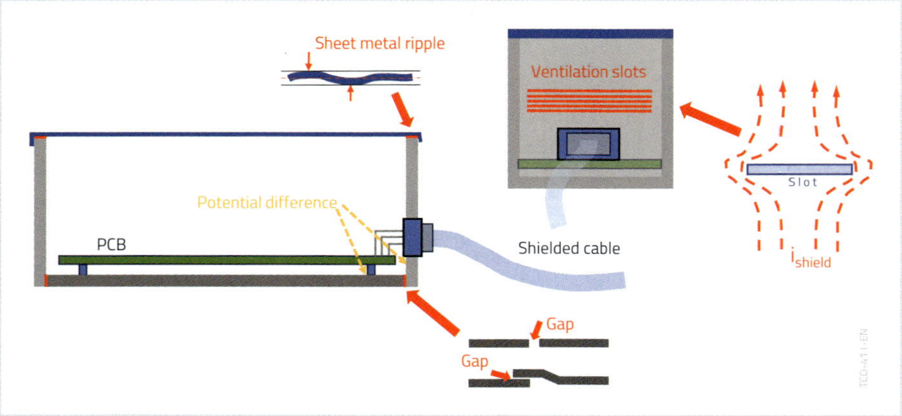

Fig. 1.105: Reduction of the shielding effectiveness of the housing, reduction of the cable shielding effectiveness and deterioration of the reference ground due to openings in the housing.

The effect of an opening in the housing is that the current in the shield must flow around the discontinuity, i.e. the opening. As the current is deflected, the local discontinuity creates a field coupling path. A long opening or a narrow seam has a greater effect on a current flowing at right angles to it than on a parallel current flow.

1.12.3.3 The ground system and ESD protection

Figure 1.105 shows that discontinuities between the rear panel and the base plate cause a potential difference which is "transferred" to the electronic assembly. This potential separation considerably reduces the shielding effectiveness of the housing and the shielding effectiveness of the cable. The immunity to electrostatic discharge on the housing is also significantly reduced in this way and, in the worst case, can lead to the destruction of the electronics in the event of discharge on the rear panel. This problem is illustrated in Figure 1.106.

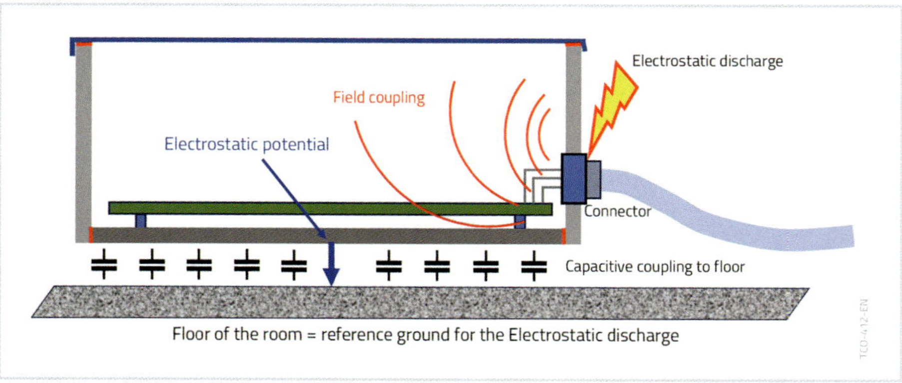

Fig. 1.106: Interrupted ground system leads to reduced immunity to ESD in the interface area.

The reference for the electrostatic discharge is the floor of the room. If a discharge takes place in the interface area, the back panel will capacitively couple the electrostatic potential to the interior of the device and from there to the bottom panel of the device. The discharge of the electrostatic energy occurs capacitively through the bottom of the device to the floor of the room. The shielding quality of the connector and the cable determine how much of the energy is coupled into the interface electronics. Potential differences generally lead to malfunctions of the device, but often also to the destruction of semiconductors (ICs). If the contact resistance between the connector and the backplane of the housing is too high, the energy is coupled directly via the cable shield into the interior of the cable, i.e. the signal wires, and is conducted from there into the interface electronics. Reliable protection against ESD can only be achieved by an enclosed housing on all sides with high-quality connectors and cable shielding connected in an HF-compatible manner. Any compromise leads to a reduction in immunity to ESD and also to increased radiation of RF interference via seams, openings and peripheral cables.

1.12.4 Laboratory experiment: comparison between good and bad connection of the connector ground to the housing.

A test generator was set up to illustrate the problem of an inadequate ground connection between the housing and the cable shield. The generator has two outputs, each with a 1 MHz square wave signal. The left BNC output socket is connected directly to the housing with its ground shield, the right output socket is connected only via the internal ground pin. The current clamp is used to measure the magnetic field emitted by the connected coaxial cable (LMR 400). The coaxial cable is terminated at its end with a 50 Ω load. The signal of the current clamp leads to a spectrum analyser. The whole set-up is illustrated in Figure 1.107.

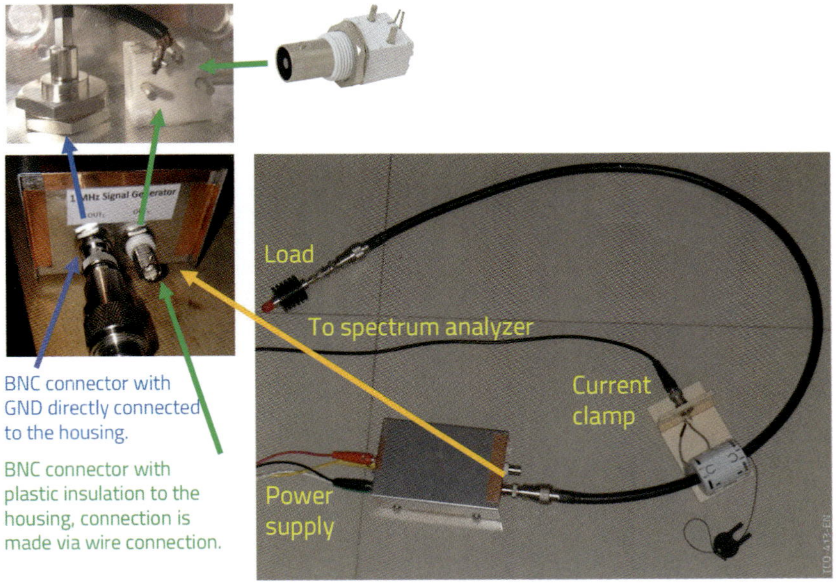

Fig. 1.107: Test setup, signal generator with two different BNC connectors, one connector with direct ground to the housing, the other with plastic collar.

The signal of the generator is shown in Figure 1.108, the frequency of the square wave signal is 1 MHz, the amplitude is 9 V. The signal has a rise time of about 1.3 ns, a fall time of 1.8 ns and contains prominent overshoots and undershoots which have a high amplitude in the frequency range up to 500 MHz.

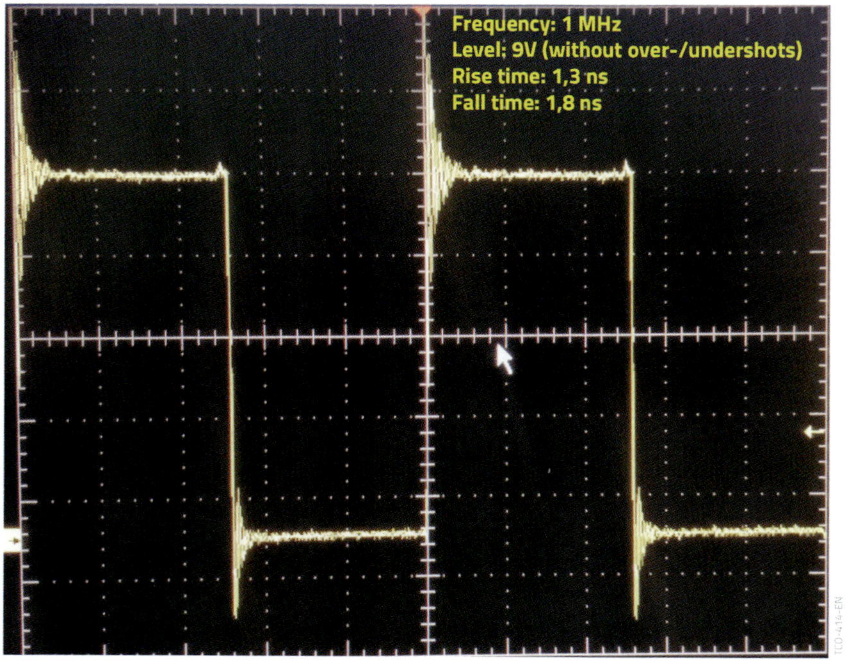

Fig. 1.108: Signal of the generator at 50 Ω

The comparison of the two types of connections is shown in Figure 1.109. The left diagram shows the amplitude spectrum from 1 MHz to 400 MHz for the case when the cable is connected to the socket with the plastic collar, in the right diagram the cable is connected to the socket with direct ground connection to the housing.

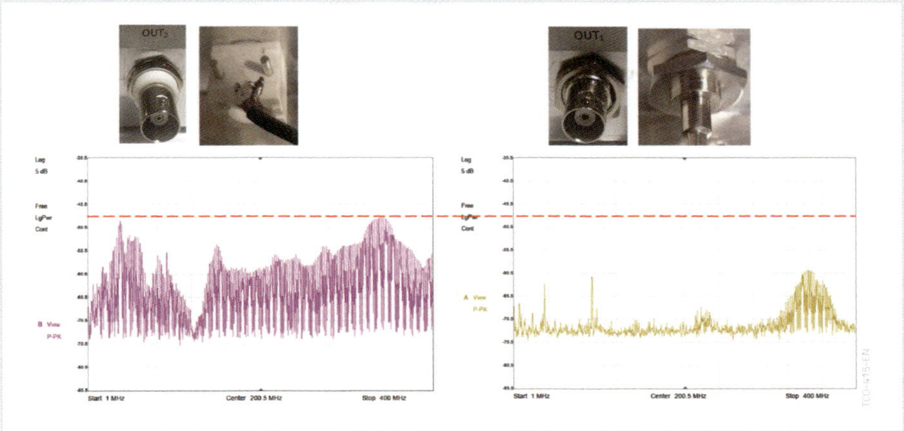

Fig. 1.109: Amplitude spectrum versus frequency, left at the socket with plastic housing, right with direct connection to the housing.

The result is clear. The level is up to 20 dB higher in the case of the plastic jack, and the increased interference signal is radiated directly via the cable. On the other hand, the cable is also 20 dB more "sensitive" to the coupling of interference into the device, which can lead to damage in the device in the case of transient interference due to discharge of static electricity or bursts.

1.12.5 Summary

Electromagnetic interference is caused by electronic switching operations on the printed circuit board. Reducing the interference at the point of origin is the best method, but not always possible. The overall concept of electronic assembly, housing with its openings, connector and cable determines the "EMC suitability". A common cause of EMC problems is an insufficient ground connection between the electronic assembly and the housing, which also has a strong influence on the shielding effectiveness of shielded peripheral cables. The quality of the connector, its low-impedance connection between the housing and the cable shield, plays a contributing role here. If the cable shield connection is inadequate, a cable can radiate interference energy as a monopole antenna; the cable's immunity to electromagnetic coupling is also drastically reduced and, if transient interference is coupled in, it can lead to interference with the device function.
A comparative estimation of EMC measures can be carried out by measurements with current clamp and spectrum analyzer.

1.13 Permanent Connections

In general, permanent connections are intended to be exactly that, permanent connections, once and done. Therefore, they are not limited by separability requirements which we have seen limit the deformation and forces that can be used in separable connections. Permanent connections can, and do, use much higher forces and larger deformations than separable connections. High forces result in better mechanical stability than that of separable connections. Larger deformations allow for the creation of larger contact areas, therefore, lower contact resistances than those of separable connections.

This chapter will focus primarily on mechanical permanent connections to wire and cable and to printed circuit boards. For wire and cable these include crimped and insulation displacement connections. For printed circuit boards the discussion will be limited to press-in connections to plated through holes in printed circuit boards.

Metallurgical permanent connections include brazing, welding, and soldering. Only soldered connections will be discussed in this chapter and only in terms of the requirements for solderability that influence the contact spring design and materials.

1.13.1 Permanent Connections to Printed Circuit Boards

The literature on Printed Circuit Boards (PCB) is extensive and details on the design and manufacture of PCBs are outside the scope of this book. A brief description of a multilayer PCB will be followed by a discussion of the components of a PCB that are most relevant to making permanent connections, mechanical and soldered, to a PCB and how those components impact on PCB and connector materials and design.

For mechanical permanent connections the focus will be on compliant press-in connection materials and design. With respect to the PCB this will require consideration of Plated Through Hole (PTH) manufacturing. For soldered connections the focus will be on PCB land/pad and connector contact solderability considerations, for both wave soldering and Surface Mount Technology (SMT). From the connector perspective, additional description about some of the key connector contact lead design issues as they apply to SMT applications will be necessary.

Printed Circuit Boards

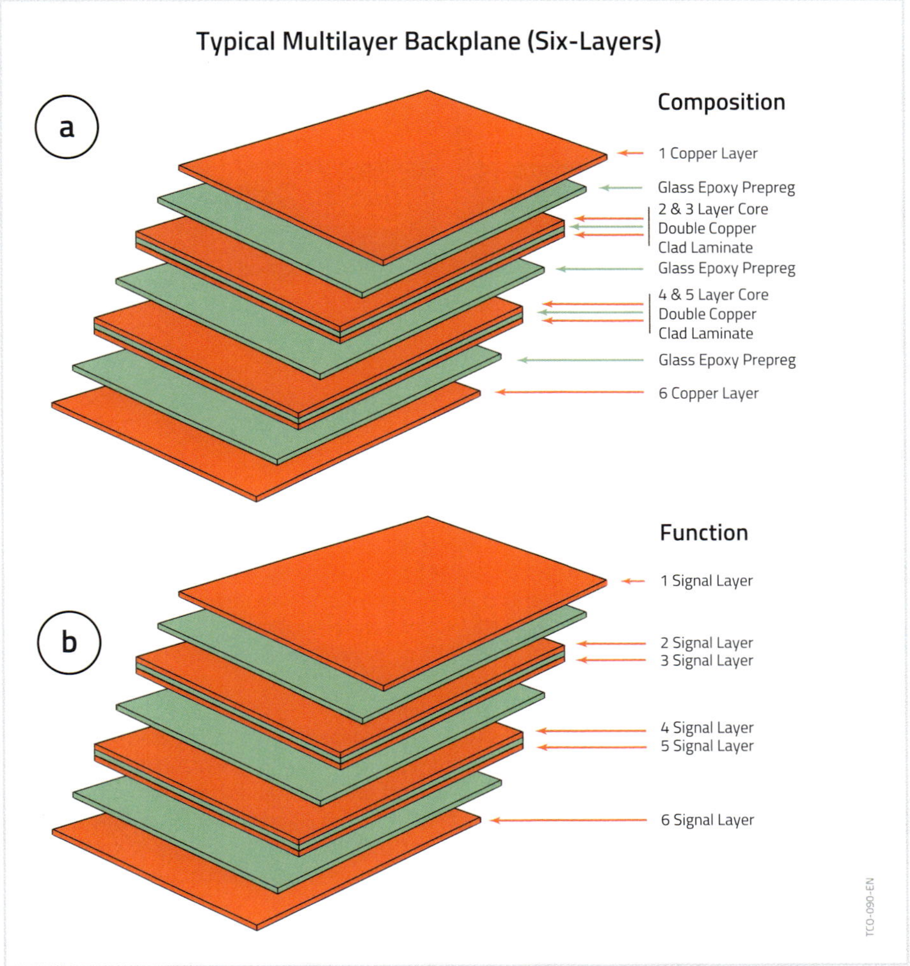

Fig. 1.110: Multilayer Printed Circuit Board

Figure 1.110 schematically illustrates the components of a multilayer PCB prior to laminating. This PCB will have six layers. Each layer consists of an FR4, fiberglass epoxy substrate with copper layers on both sides of the substrate. The thickness and distribution of the copper depends on whether the function of the layer is to carry signal currents, milliamps to a few amps, or power, tens of amps. A bonding medium, prepreg, is placed between each pair of layers. In this example, starting from the top, the first layer has signal traces on each side. The second layer has a power/ground plane on each side. The third and fourth layers also have signal traces on each side.

In the manufacturing process these four layers and the prepreg are stacked and laminated under heat and pressure to create a multilayer PCB.

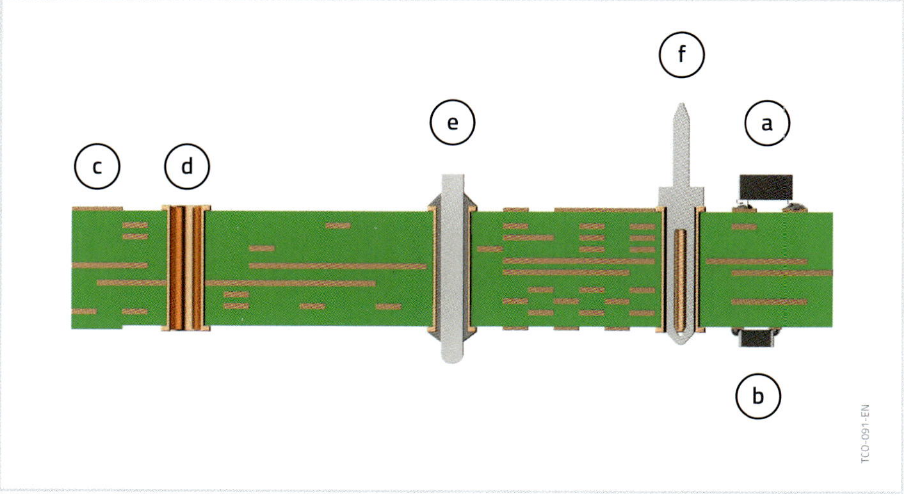

Fig. 1.111: Multilayer PCB cross section

Figure 1.111 illustrates a cross section of the laminated PCB described. The small rectangles shown in the first, third and fourth layers are signal traces that run between the many devices that are mounted on the PCB. The power and ground planes in the second layer are continuous with the exception of holes in the planes to allow inter-connections to be made between the first, third- and fourth layer circuitry using Plated Through Holes, PTH, three of which are shown in the drawing.

Consider now the variety of connections, mechanical and soldered, that can be made to this PCB. Note that some of the signal traces are on the surfaces of the PCB. Expanded areas along a surface trace to allow contacts to be made to the trace are called lands or pads. In addition to the copper circuitry, lands require an additional plating with the plating composition and thickness depending on the intended use of the pad, in particular whether mechanical or soldered connections will be made to the land. The plating considerations will be discussed with each application.

The top right component, Figure 1.111a, is soldered to surface pads on the PCB. For convenience, two different lead geometries, J-lead and Gull Wing are shown. In practice, of course, a component will use a single lead style. These geometries are two of the more commonly used methods for providing lead compliance to the soldered connection to minimize the effects of differential thermal expansion stresses on the solder joint as will be discussed in the section on SMT. Similar geometries are used as leads on connectors intended for SMT applications for the same reason. The PCB pads that are to be soldered will generally be treated so as to maintain or provide solder-ability of the base copper circuitry. Clean copper is readily solderable, but given the tendency of copper to corrode, the copper surface may receive an organic coating or a thin gold plating to prevent such corrosion and maintain the copper solderability. A second approach is to plate the copper with a solderable coating, tin depending on the environmental requirements to be met. The tin coating is generally screened on to the copper and reflowed. The coating thickness is typically of the order of 2 to 4 µm.

The bottom right component, Figure 1.111b, is a flip chip package which is soldered to the surface pads of the PCB using solder bumps or columns, a design sometimes called a leadless package. The plating requirement for this application is essentially the same as discussed for the previous component.

The pads on the left edge of the board, Figure 1.111c, are intended to mate to a card edge connector. The plating on the pad must be compatible with that on the card edge connector which is either gold over nickel or tin. The plating thicknesses on card-edge pads are typically in the range of 0.4–0.8 µm (15 to 30 micro-inches) for the gold and 1.25 to 2.5 µm (50 to 100 micro-inches) for the nickel. For tin coatings the thickness will be of the order of 2.5 µm (100 micro-inches). In addition to card edge connections, Central Processor Units, CPU are also connected to surface lands on multilayer boards. A Land Grid Array, LGA, socket contains spring contacts which are sandwiched between the lands on the PCB and the associated pads on the CPU package. LGA sockets use gold over nickel plating with the thicknesses noted above.

The three PTH connections are all different. As illustrated, Figure 1.111d shows a PTH that is used to make connections between layers within the board. For such applications a copper plating is generally used, the thickness depending on the current requirements to be supported. In this example the press-in solder contact, Figure 1.111e, makes a connection to one of the power/ground planes. A press-in solder pin is generally a solid, rigid pin and exerts only enough force to hold the pin in place until it is soldered, generally wave soldered. In contrast, the compliant press-in pin, Figure 1.111f, is intended to provide a robust mechanical permanent connection without the need for solder.

Two different permanent connection technologies, compliant press-in pins and soldering have been introduced in this introductory section. Each will now be discussed individually.

1.14 Soldering Technologies

As with PCB technology, the literature on solder and soldering technologies is extensive and beyond the scope of this volume. The focus of this chapter will be on how the requirements of soldering processes impact on connector design and materials options due to the time/temperature profiles of the soldering processes used. Two different processes, wave and reflow soldering will be briefly discussed.

1.14.1 Wave Soldering

Wave soldering, as the name indicates, is accomplished by passing the component to be soldered across the surface of a literal wave of solder as illustrated in Figure 1.112.

Wave soldering is used for Through Hole Technology, THT, as indicated by the pins in Figure 1.112. The board must be prepared prior to soldering in two ways. First, the board must have the appropriate coating at the points to be soldered. For THT this means the coating on the PTH which will, in most cases, be copper or copper plus an organic or tin coating, as noted previously. Second, flux, to facilitate solder wetting, will be applied to the bottom of the board.

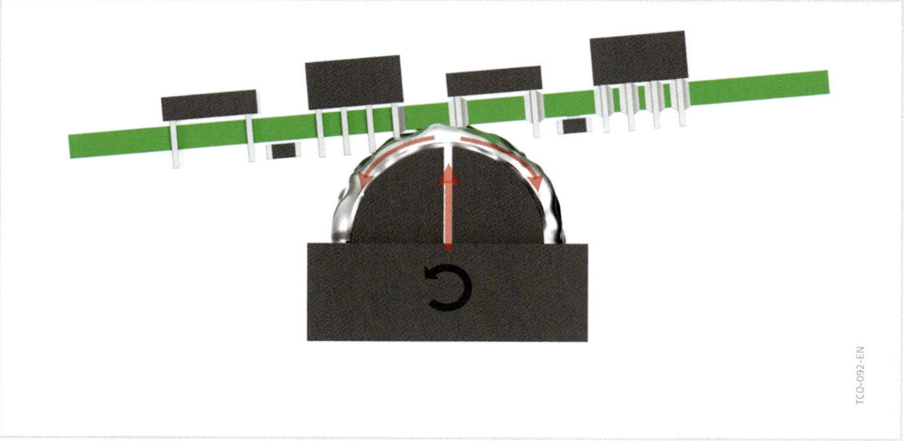

Fig. 1.112: Wave soldering process

A simplified description of the wave soldering process is in order. For this discussion, assume the solder wave is a fountain with solder flowing as indicated by the arrows. The board enters the wave from the left and is held with the bottom of the board below the surface of the undisturbed wave. Under these conditions the solder flows along the bottom of the board displacing the flux and soldering to the copper plating or reflowing the tin of the PTH annular ring and inside walls of the PTH. The solder wicks up into the PTH and onto the top surface producing a solder fillet on the top and bottom annular ring. This solder filled hole supports the pins mechanically as well as creating the desired electrical contact condition.

1.14.2 Reflow Soldering

Reflow soldering differs from wave soldering in that the solder must be added to the board and then reflowed through a secondary heating process, while the solder wave provides both the solder and the heat. The solder is applied to the PCB only at pads and lands where device leads are to be soldered to the board. A common means of applying solder is as a screened-on paste which contains both the solder and an appropriate flux to facilitate the soldering process.

Reflow soldering is used for Surface Mount Technology (SMT) and hybrid SMT/THT applications. SMT for components other than connectors is more common because of the smaller size of such components compared to connectors and the fact that they do not need to withstand mating/unmating forces. However, SMT compatible connectors have been seeing increasing use in recent years and are now common for pitches below 3 mm.

Several heating processes are used for SMT including Infrared (IR), IR/convection, and vapor phase. The temperatures and time/temperature profiles for these various processes are different and, therefore, have differing impacts on connector materials and design options. The process temperature is higher, and the time at temperature longer for SMT reflow soldering than they are for wave soldering. This fact places more demanding requirements on the connector housing as will be discussed.

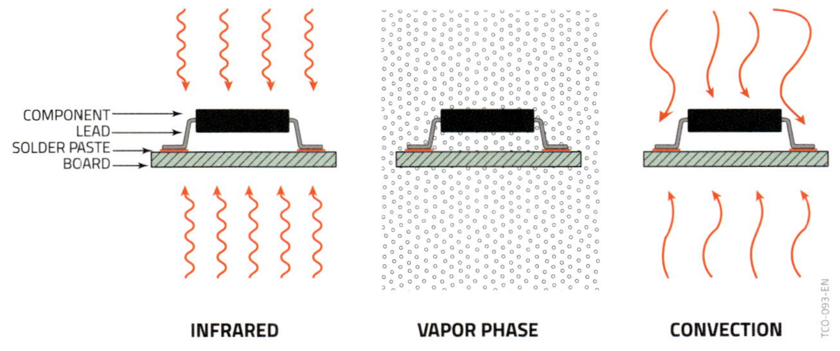

Fig. 1.113: Different reflow processes

JEDEC standards give important guideline to solder component in the right way. Consider now how these time/temperature profiles impact on the components of a connector.

Contact Finish

The primary function of the contact finish is to provide, and maintain, the solderability of the connector leads. For the most part this requirement leads to the use of tin coatings on the leads because such coatings are simply reflowed during the soldering process. If the tin is electroplated the quality of the plating process is important. Electroplated tin deposits directly on the surface of the contact whether it is copper alloy, a copper or nickel under plate. The quality of the electroplate becomes important when the tin is reflowed and attempts to form an Intermetallic Compound (IMC) with the contact surface. If the plating quality is inadequate the reflow process may not be able to form IMC uniformly over the contact. If this is the case, the soldering process may result in non-wetted or dewetted areas which can compromise the quality of the solder joint between the contact and the PCB pad or land. The potential for dewetting increases as the time at soldering temperature increases. There is a second IMC effect due to time at temperature, excessive IMC growth. IMC tend to be brittle and excessive IMC thickness can lead to a brittle zone between the solder and the pad or contact and compromise the mechanical performance of the solder joint. An example of an IMC related mechanical failure is shown in Figure 1.114. The lead design in this case is a J-lead. The use of a nickel underplate reduces the rate of IMC growth because nickel/ tin IMC grows more slowly than copper/tin IMC. Another benefit of nickel underplates comes into play since lead-free solders are required. Nickel underplates reduce the potential for tin whiskering, a concern in lots of applications. Use of contacts with reflowed tin coatings are not susceptible to these IMC and whiskering issues. On the soldering process side, these issues dictate that special attention be paid to minimizing the maximum temperature and time at temperature in reflow soldering.

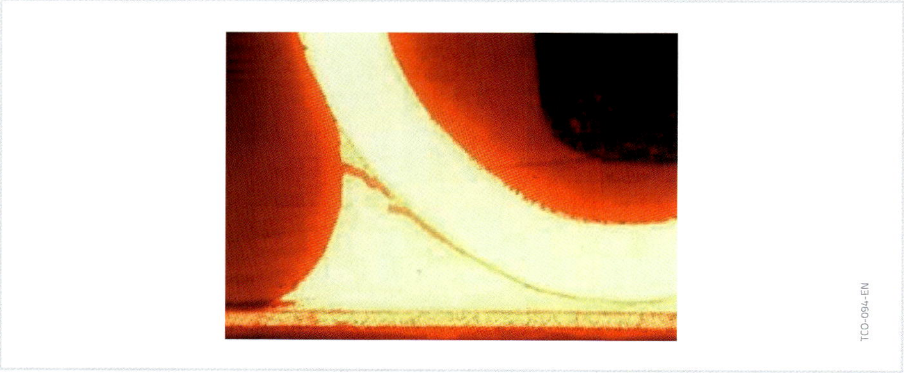

Fig. 1.114: Solder crack[7]

Contact Spring

Connectors are typically larger than most other SMT compatible components. This means that an SMT compatible connector generally must be attached to the PCB in some fashion to be able to tolerate handling during the reflow process. The design of the contact termination, the permanent connection end of the contact, for SMT applications focuses producing a solder fillet capable of satisfying the mechanical requirements of connector mating and unmating. There are three commonly used geometries, J-lead, gull wing and butt as shown in Figure 1.115. The gull-wing lead is the most suited for connectors because of the larger, and, therefore, stronger solder fillet.

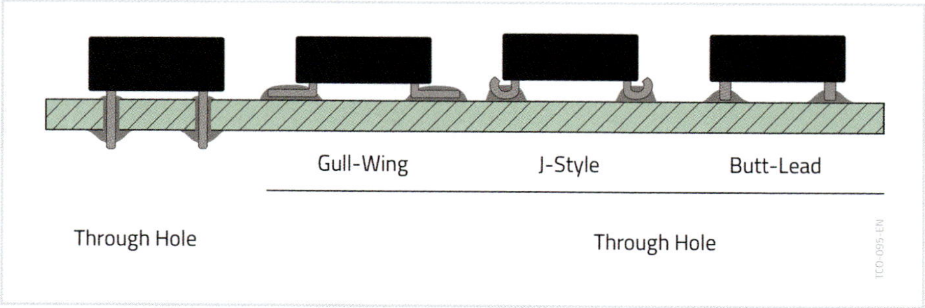

Fig. 1.115: Different lead designs

[7] courtesies of university of Bolton

The need to form the terminations of SMT compatible connectors to provide mechanical support introduces an additional issue, termination coplanarity. Coplanarity refers to a requirement that the bases of all the terminations lie on a common plane so that the solder joints of all terminations are acceptable. The solder fillet will depend on the contact between the termination and the solder pad.

There is an additional contact requirement that comes into play during the application lifetime of SMT compatible connectors. The termination end of such contacts must have some compliance to minimize the effects of differential thermal expansion during temperature cycling of the electronic equipment using the connectors. A common source of temperature cycling is the on-off cycling of the electronic equipment. The application ambient temperature will be of the order of 20 °C as will the original temperature of the board mounted connectors in the system. The system temperature will increase as the equipment is powered up, say to 55 °C. This leads to a differential temperature of 35 °C between on and off and, thus a 35 °C thermal cycling exposure.

TCO-096-EN

Fig. 1.116: Thermal expansion

Figure 1.116 shows the curvature introduced into a connector under such a cycle.

The Coefficient of Thermal Expansion (CTE) of the FR4 board is less than that of the polymers typically used in connector housings. Therefore, the connector would expand more than the PCB if not fastened to the PCB. Thus, the mounted connector will be under a compressive stress, the contacts will tend to bow towards the centre of the connector. The connector/PCB system will return to its original configuration when the equipment is turned off. This produces an opportunity for cyclic fatigue failures of the solder joint. In addition to the fatigue, however, the solder will creep to reduce the stress during each thermal cycle of the system. This process is the most significant degradation mechanism for SMT components, in particular for connectors because, as noted, connectors tend to be larger than other SMT components and the differential CTE effects increase with the length of the component. While the driving forces for CTE cannot be avoided, the stress at the solder joint can be moderated if some of the stress can be accommodated by the contact spring, that is, the compliance of the contact termination to the board can reduce the stresses at the solder joint. The length and thickness (in the direction of expansion) of the contact becomes a design consideration in such cases.

Connector Housing

There are two major effects of SMT time/temperature profiles with respect to the connector housing design/materials. The most significant effect comes from the peak temperature of the soldering process which limits the polymers that are suitable for SMT applications. The second effect is mechanical and arises from the need to fasten the connector to the PCB in some manner to ensure its retention during the handling of the board during the soldering process. Consider the peak temperature effects first.

Fig. 1.117: Connector degradation during reflow process

Simply stated, some of the polymers that are commonly used in connectors, NY66, PBT are not able to withstand the peak temperatures of the reflow soldering process without excessive degradation or distortion. Figure 1.117 dramatically illustrates this statement. The connector uses a polyester as a housing material and has been a reliable performer in the field for many years. After an exposure to a reflow soldering process, however, the housing has severely distorted. Polymers suitable for such soldering processes includes Liquid Crystal Polymers (LCP), Poly Phenylene Sulphide (PPS), Nylon 6T (NY6T), Nylon9T (NY9T). In addition to their high temperature properties, these materials have low moisture absorption and can be molded in thin wall sections to reduce the thermal mass of the connector and, therefore, the time at temperature during reflow. Nylon 46 is a bit particular and can be rated in between as even if it withstands high temperature, great care must be taken, because of its high moisture sensitivity (too much moisture creating bubbles on plastic surface during reflow).

The mechanical consideration, a retention mechanism to retain the connector on the board, is straight forward and dependent only on the ability to injection mold the desired retention feature components into the connector. One additional comment on the retention feature design is in order. Retention features can reduce the potential for problems due to coplanarity and differential thermal expansion. Coplanarity concerns can be moderated if the retention mechanism can preload the contacts onto the solder pad. Differential thermal expansions concerns can be moderated because the retention mechanism tends to hold the connector in place with the PCB expansion reducing the differential expansion and thus the stress on the solder joints.

1.14.3 THR (Through Hole Reflow) Technology

1.14.3.1 Introduction & History:

The through-hole technology dates back to the 50's and was primarily accomplished through wave soldering processes. Components, including connectors, were press-fit into Plated Through Holes (PTH) in a Printed Circuit Board (PCB) and then soldered, generally by wave soldering. The pioneer work in SMT started in the 60s and had its breakthrough in the 80s. This technology consists in soldering components to lands on the surface of a PCB using a number of different soldering processes. Typically, a solder paste was screened onto the lands on the PCB and the solder paste was subsequently reflowed by a variety of heating processes. SMT capability was originally limited to smaller components and devices but expanded to include larger components over time. SMT capability brought significant reductions in the PCB size due to reductions in the component size and in the board area required for routing signals. Figure 1.118 illustrates some stages of this evolution.

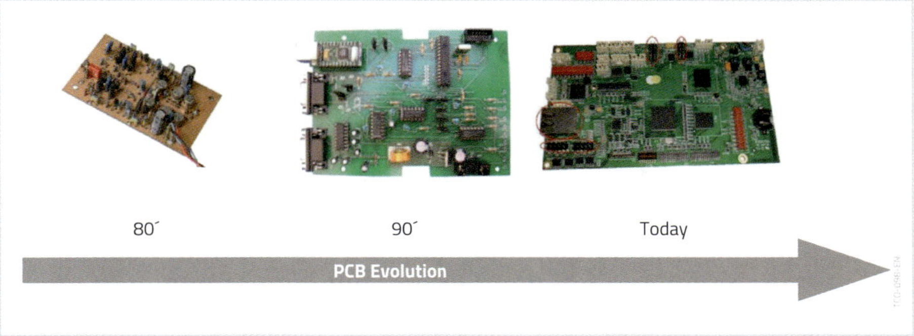

Fig. 1.118: PCB evolution

Initially, connectors were conspicuous by their absence from SMT processing. There were two major reasons for this absence: first, the size of connectors was large compared to that of other SMT components, and second, connectors are subject to mating and unmating forces that exceed the strength capability of the surface solder joints of SMT. This meant that connectors had to be added to a PCB using Through Hole Technology (THT) through compliant pin press-in technologies or through a secondary soldering process e.g. manual or wave soldering.

Eventually SMT versions of smaller connectors were developed. In recent years, however, Through Hole Reflow (THR) or Pin In Paste (PIP) processes have been developed that allow connectors to be reflow soldered in the same soldering process as the SMT components on the board. THR processing results in significant reductions in manufacturing costs as well as in additional savings in PCB board area.

1.14.3.2 THR Compatible Connectors

Producing a connector that is THR compatible requires changes in the design and materials of both, housings, and contacts. Two terminal block connector families, Würth Elektronik 691311500104 (THT version) and Würth Elektronik 691701510004B (THR version) will be used to illustrate these changes.

Housing Considerations

A schematic illustration of a four-position terminal block is shown in Figure 1.119.

Fig. 1.119: Four position terminal block connector

THR compatibility requires a housing material that is capable of withstanding the highest temperatures of reflow soldering (+245 °C for this product according to IPC/JEDEC J-STD-020D). There are two concerns relative to this temperature exposure. The first relates to the effect of soldering temperatures on the moisture content of the polymer. The absorbed moisture expands under the soldering temperature exposure thus producing blisters or bubbles in the housing, as illustrated in Figure 1.120.

TCO-100_EN

Fig. 1.120: Bubbles due to absorbed moisture exposure to reflow temperature

The second effect is the melting/distortion of the housing due to the temperature exposure as illustrated in Figure 1.121.

TCO-101_EN

Fig. 1.121: Melted/distorted housing due to reflow temperature exposure

To avoid these potential problems, Würth Elektronik has changed the material of THR terminal blocks to a Liquid Crystal Polymer (LCP). LCP polymers can meet SMT processing temperature requirements; in addition, they have lower thermal expansion coefficients, which is an important consideration when thermal expansion mismatch stresses come into play.

THR compatibility also requires changes in the housing design. In particular, standoffs in the housing are required to allow room for the solder paste and to promote air flow under the housing to facilitate the reflow soldering process. These features are illustrated in the following figures. Figure 1.122 shows a side view of a THR compatible terminal block housing with stand-off pegs.

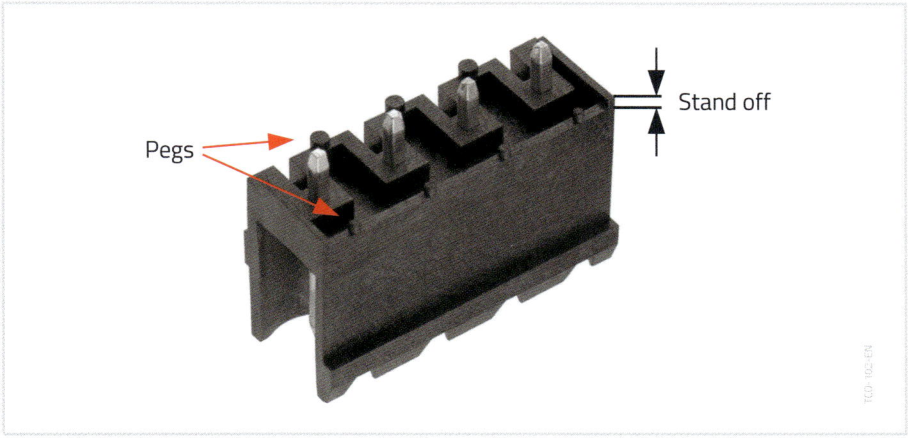

Fig. 1.122: Housing stand-off and stand-off pegs

Figure 1.123a and Figure 1.123b respectively show the planar base of a standard terminal block and the base of a THR compatible terminal block with stand-off features.

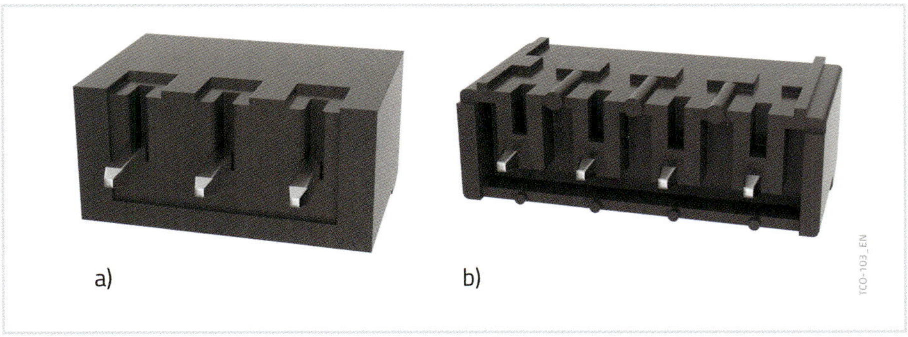

Fig. 1.123: a) Terminal block base/b) THR terminal block base showing stand-off features

The stand-off clearance between the connector base and the PCB should be at least 0.3 mm to prevent the base of the connector from touching the solder paste deposit. In addition, contact between the stand-off features and the solder deposit should be avoided to minimize the potential for solder balling during reflow.

Figure 1.124 shows the air flow paths through the THR compatible terminal block housing that are created by the stand-off features.

Fig. 1.124: Air flow paths around the stand-off features of a THR compatible terminal block

For manufacturing efficiency, it is also desirable that the housing design should be compatible with vacuum pick and place processes. The pick and place process may limit the size of the connector that is suitable for THR connectors.

Contact Considerations

Design and material change are also necessary for contacts, particularly the termination end of contacts. THR compatibility places additional requirements regarding the pin length. This is particularly true if the connector is intended to be both through hole and THR compatible. If the pin is too short, as per Figure 1.125, the stability of the pin in the through hole may be compromised with resulting handling problems during the reflow assembly process. In addition, the soldering of a short pin may not produce a solder joint that meets the requirements of the IPC-A610D Standard.

Fig. 1.125: "Short" pin termination

A pin that is too long may push some solder paste out of the PTH to accumulate on the end of the pin. If this occurs the solder paste on the end of the pin may ball up rather than wicking up into the through hole during reflow and compromise the solder volume and the soldered connection to the through hole plating.

Fig. 1.126: "Long" pin termination

To minimize the potential for these problems the pin length protruding from the PCB should be discernible in the solder and shall not exceed 1.5 mm. Würth Elektronik chose a 2.6 mm pin length for its THR contact pins to be compatible with PCB thickness from 1.6 mm to 2.3 mm. The pin centring in the PTH is also important to optimize the solder flow.

Solderability requirements for THR compatibility may dictate changes in the pin plating from bright tin to matte tin. Inadequate control of the brighteners may compromise the solderability of bright tin-plated pins due to surface film formation during storage of connectors.

Component Lead Terminations

In a lead-free world, the most common finish is tin because of its very good wettability properties. However, tin can be bright or matte. Matte tin is well adapted to high temperatures whereas bright tin tends to turn dark under high temperatures. If brass (CuZn) or a zinc-containing pin alloy is used as the base material, the pin must first be plated with nickel to a minimum of 1.25 µm before the tin/gold is applied, as explained in the first chapter of this book.

1.14.3.3 The THR Manufacturing Process

The basic requirement of a THR manufacturing process is to ensure compatibility with the existing SMT manufacturing process, typical stages of which are illustrated in Figure 1.127.

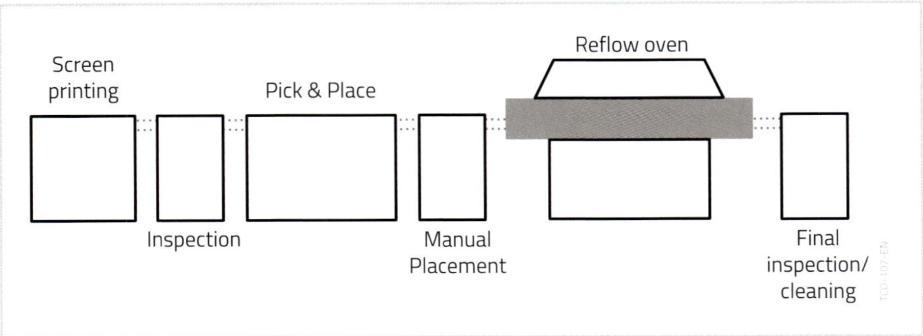

Fig. 1.127: Stages of a typical SMT/THR manufacturing line

Stage 1: Screen printing: the solder paste is applied to the PCB.

Stage 2: Inspection: ensures the dimensional control and alignment of the screen-printing process with the PCB.

Stage 3: Pick and place: puts the SMT, and any connectors which may also be pick and place compatible, onto the screened solder paste.

Stage 4: Manual placement: connectors are manually inserted into the PTH of the PCB.

Stage 5: Reflow oven: reflow soldering of the components onto/into the PCB via reflow soldering processes, typically vapor phase or convection reflow.

Stage 6: Final inspection: inspection using appropriate methods to ensure the proper placement of components and satisfactory soldered joints.

Stage 7: Cleaning: removes possible flux residues.

Figure 1.128 shows the status of the PCB with respect to the solder paste and soldering process through the manufacturing line. The first three illustrations show processes occurring in the screen-printing stage as the solder paste is applied to the PCB, the fourth illustration shows the PTH after the connector insertion with the solder paste distribution in the PTH, the last two illustrations show the reflow process and the configuration of the soldered component.

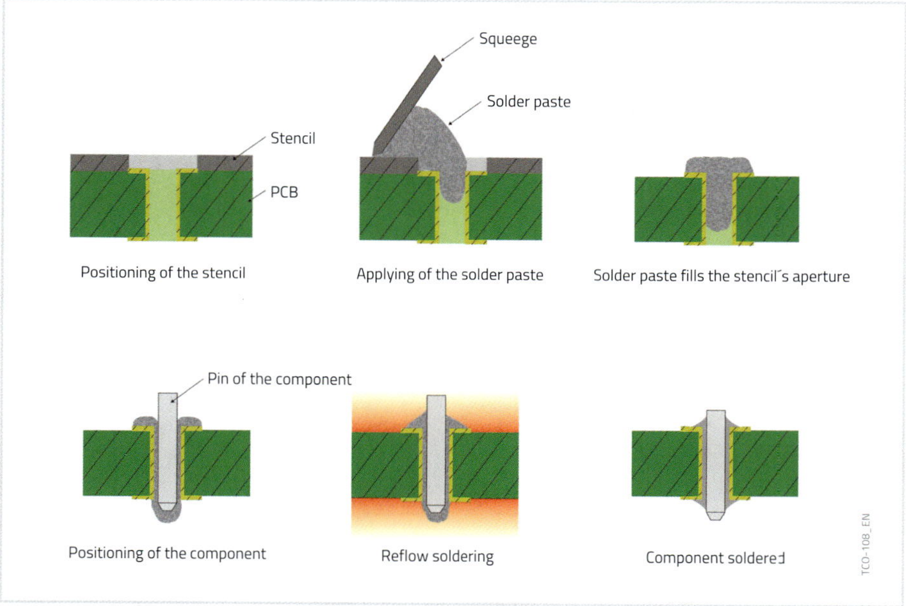

Fig. 1.128: Stages of printing and soldering THR processes

Screen Printing

The screen-printing process is particularly important because it determines the amount of solder paste that will be applied to the PCB and therefore, the amount of solder in the final soldered connection. The amount of solder, in turn, determines whether the applicable solder joint inspection criteria will be met. The design guideline for the amount of solder needed is that the solder shall be sufficient to fill the free volume of the PTH, that is (VPTH – VPIN). The solder paste is a combination of soldering flux and metal alloy, with the metal content being around 50 % of the volume of the solder paste. Therefore, it is necessary to apply twice as much paste than the free volume between the pin and the PCB hole. The flux components will either evaporate or be removed during the final cleaning process.

The following parameters have to be considered to maintain process control of the solder paste volume:

- Choice of the stencil thickness and design
- PCB design and connector layout
- Choice of the solder paste
- Speed, pressure and angle of the squeegee used to apply the solder paste to the PCB

The first two parameters determine how much solder paste will be retained in the stencil volume. The second two parameters will determine how well the solder paste will fill the PTH and stencil aperture.

Solder Paste Volume

The solder paste volume on the PCB surface will be determined by the thickness and area of the stencil opening. Stencil thickness is in the 100 to 200 μm range, and the area will depend on the PCB design. Typical stencil apertures are shown in Figure 1.129.

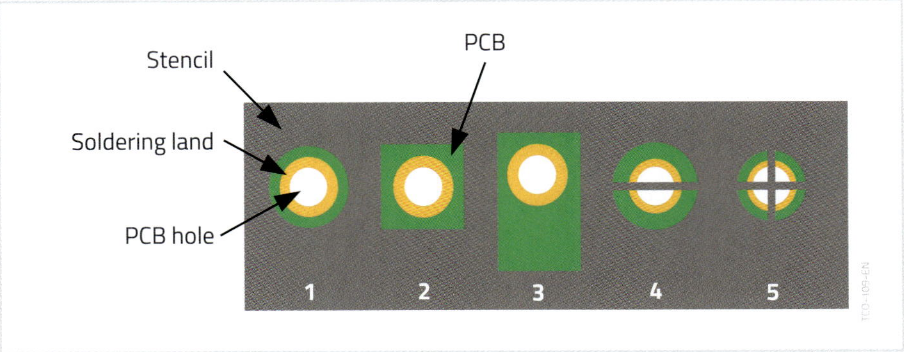

Fig. 1.129: Typical stencil apertures

Apertures 1 & 2 are typical for THR printing.

Aperture 3 is used when extra solder paste is needed.

Aperture 4 (paste reducer) is used for large PCB PTHs to prevent the solder paste from dropping through the PTH.

Aperture 5 is commonly used when double printing, top and bottom of the PCB, to add additional solder paste to meet the solder requirement.

Typical PCB pad dimensions are shown in Figure 1.130.

- T_b is the PCB thickness (generally 1.6 mm)
- D_d is the drilled hole diameter (not shown)
- D_h is the PCB PTH
- $D_h = D_p + 0.3$ mm, where D_p is the pin diameter (shown in Figure 1.131)
- D_{sp} is the solder pad diameter = D_h + 0.5 mm to 0.8 mm

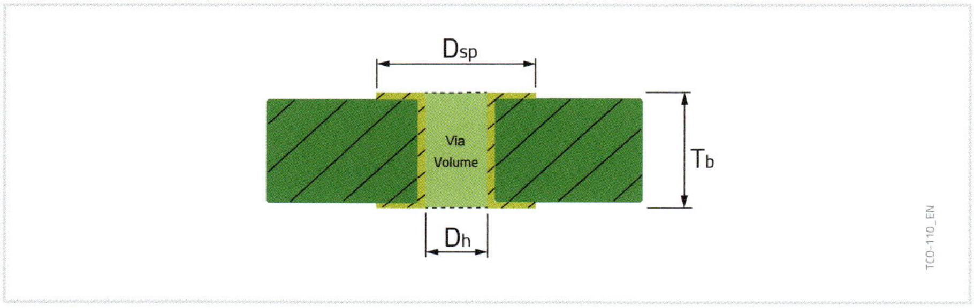

Fig. 1.130: PCB pad dimensions

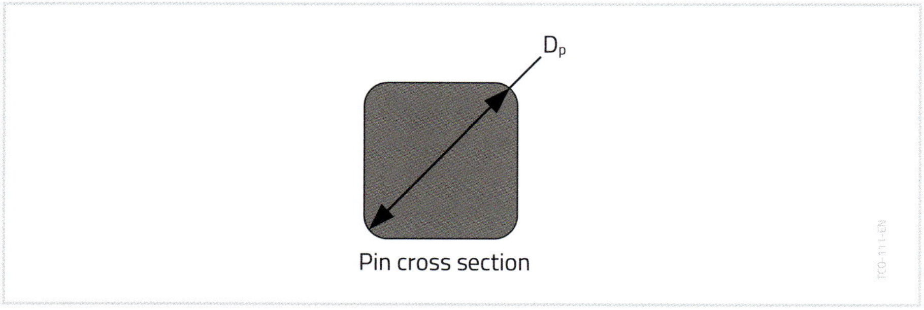

Fig. 1.131: Pin diameter

The stencil area, as shown in Figure 1.132, in combination with the stencil thickness (T_s), determines the solder paste volume on the surface of the PCB. Ideally, when an overprint is not necessary, the stencil diameter (D_s) = D_{sp} – 0.1 in order to guarantee a perfect paste deposition on the solder pad.

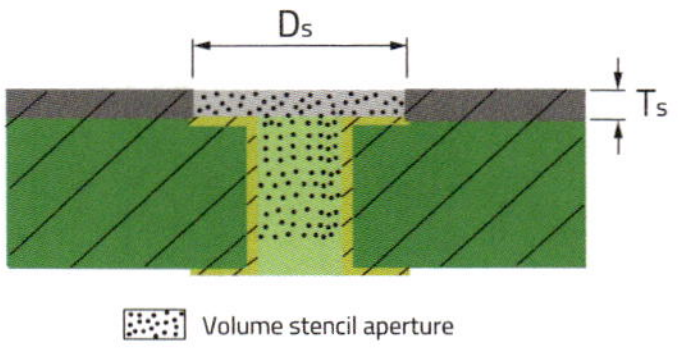

Fig. 1.132: Stencil dimensions

As noted previously, the solder paste consists of approximately 50 % of the volume of solder particles and 50 % of the soldering flux volume. The optimum solder particle size depends on the pitch of the PCB pads and on the thickness of the stencil. If the particle size is too large, as per Figure 1.133, the effective solder volume will be reduced.

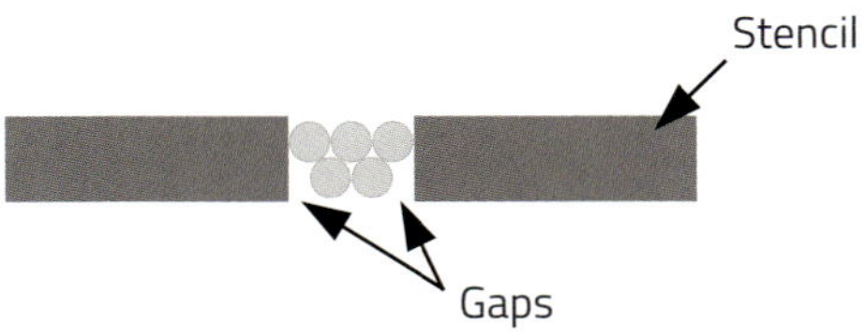

Fig. 1.133: Effect of the solder particle size on stencil fill

If the solder particle size is too small, viscosity issues may arise, thus affecting the retention of the solder paste in the stencil. The most commonly used solder particle size is in the range of 20 to 45 microns. JEDEC STD-005 covers soldering paste requirements. The pin geometry, square, round, or rectangular, may also impact the stencil fill and the solder volume.

Solder Paste Application

The solder paste is applied to the PCB and a subsequent squeegee pass is used to push the paste into the stencils' apertures, as indicated in Figure 1.134.

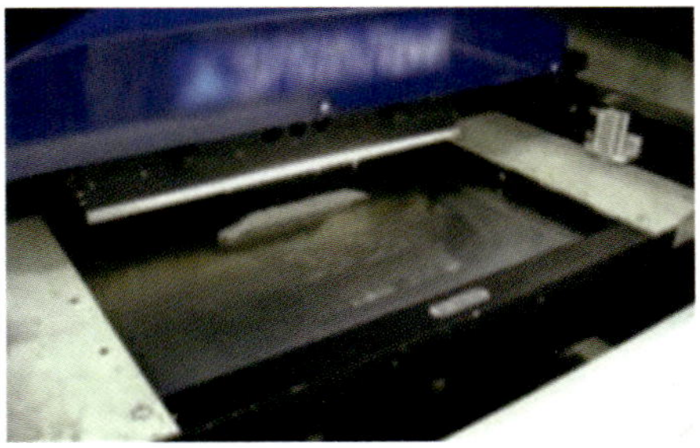

Fig. 1.134: Solder paste on PCB prior to squeegee pass

Three parameters are important when controlling the solder paste removal process, as indicated in Figure 1.135.

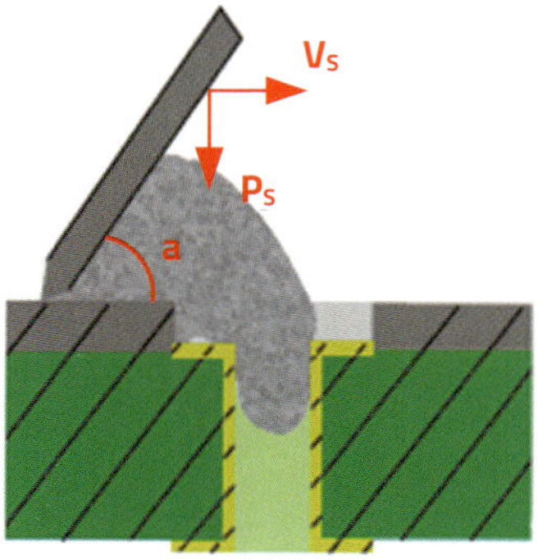

Fig. 1.135: Squeegee process parameters

V_s is the squeegee speed, P_s is the squeegee pressure against the PCB and a, is the angle of the squeegee blade with respect to the surface of the PCB. The squeegee angle a is set before printing, generally 45 degrees, and the remaining parameters are set depending on the stencil thickness and on the aperture configuration on the PCB. The fill of the stencil aperture and the penetration of the solder paste into the PTH of the PCB depend on the interaction of these three parameters.

Pin Geometry

Basically, pins can be round or any rectangular shape, as shown in Figure 1.136. The pin geometry plays an important role in the THR process. As explained previously, the pin volume has to be considered simultaneously with the stencil aperture definition and will vary with the pin geometry.

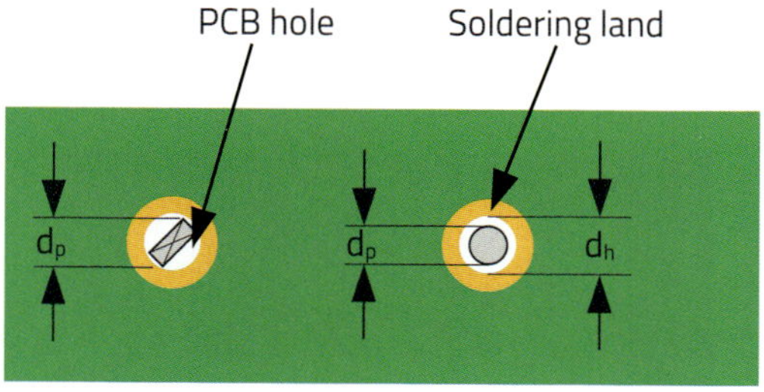

Fig. 1.136: Different pin shapes

Component Insertion

Whenever possible, it is desirable to use automatic pick and place technologies to insert the connectors into the PCB PTH.

Fig. 1.137: Typical pick and place insertion process

Figure 1.137 illustrates a typical pick and place insertion process. The size and configuration of the connector housing will impact the ability to use pick and place practices. Special features may need to be designed into the housing to enable the pick and place process. In our example of a WR-TBL vertical THR version, a plastic cap is positioned on the top of connectors (Figure 1.138) for pick and place machine handling.

For horizontal version, this cap is not needed as it can be handled directly from the top of the housing.

Fig. 1.138: Würth Elektronik Terminal block WR-TBL 691701510004B (THR version)

In addition, a suitable packaging has to be used in order to optimize this process for mass production. Commonly, the SMT and THR components are packed in tape and reel packaging, as illustrated in Figure 1.139, in order to be useable with automatic machines.

Fig. 1.139: Tape & reel packaging

Components and connectors intended to be mounted on a board are packaged in reels and placed in the pick and place feeder, as illustrated in Figure 1.140. The pick and place feeder then positions the correct component automatically on the board as previously shown in Figure 1.137.

Fig. 1.140: Pick and place feeder

Solder Reflow Practice

As noted previously, the THR reflow practice is basically identical to the SMT reflow practice. As such, it can be accomplished through infrared convection and vapor phase soldering. In either case, the temperature profile of the reflow soldering process is critical to a consistent and reliable reflow soldering. Figure 1.141 shows a representative temperature profile recommended for lead free infrared convection reflow soldering according to IPC/JEDEC J-STD 020D.

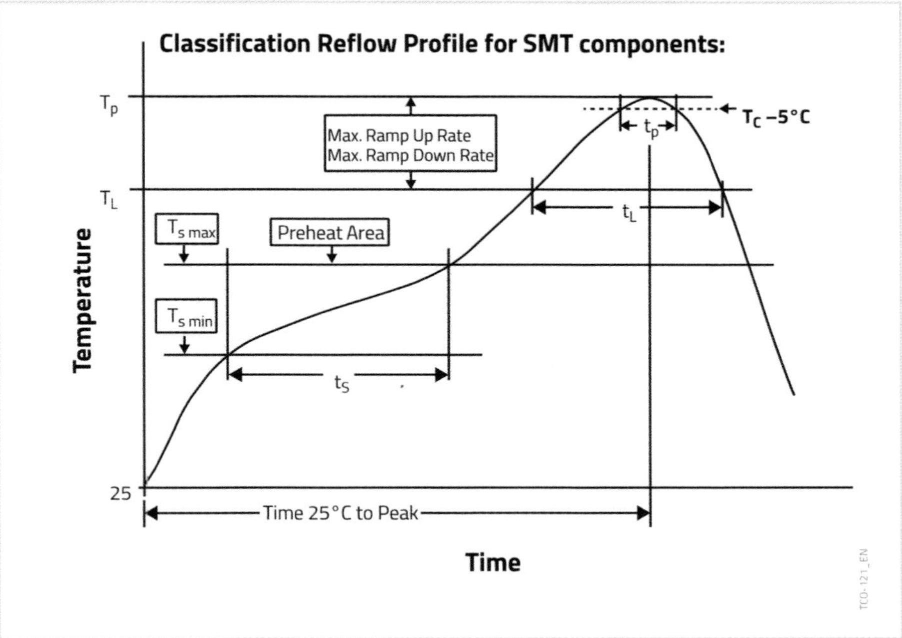

Fig. 1.141: Temperature profile for lead-free infra-red convection reflow soldering per IPC/JEDEC

Process details are provided in Table. 1.2. Peak temperature recommendations are provided in Table. 1.3

Profile Feature	Pb-Free Assembly
Average ramp-up rate (Ts_{max} to T_p)	3 °C/second max.
Preheat **± Temperature min (Ts_{min})** **± Temperature max (Ts_{max})** **± Time (ts_{min} to ts_{max})**	 150 °C 200 °C 60–180 seconds
Time maintained above: **± Temperature (T_L)** **± Time (t_L)**	 217 °C 60–150 seconds
Peak/classification temperature (T_p)	See table below
Time within 5 °C of actual peak temperature (T_p)	20–40 seconds
Ramp-down rate	6 °C/second max.
Time 25 °C to peak temperature	8 minutes max.

Tab. 1.2: Reflow profile details

Package Thickness	Volume mm^3 <350	Volume mm^3 350–2000	Volume mm^3 >2000
<1.6 mm	260 °C	260 °C	260 °C
1.6 mm–2.5 mm	260 °C	250 °C	245 °C
2.5 mm	250 °C	245 °C	245 °C

Tab. 1.3: Connector size dependent parameters

1.14.3.4 Example with Würth Elektronik WR-TBL 691701510004B Connector

In accordance with the previous discussion, typical process parameters for Würth Elektronik connector WR-TBL 691701510004B are defined as follows:

PCB thickness (T_b) = **1.6 mm**
Pin diagonal (d_p) = **1.2 mm**
Pin length = **2.6 mm**
Pin width (W_p) = **1 mm**
PTH hole diameter (D_h) = d_p + 0.3 = **1.5 mm**
Pin volume (V_p) = W_p x W_p x T_b = 1 x 1 x 1.6 = **1.6 mm³**
PCB PTH volume (V_h) = π x $(D_h/2)^2$ x T_b = π x $(1.5/2)^2$ x 1.6 = **2.8 mm³**

In consequence, the needed solder paste volume (Vsp) is:

$$V_{sp}=2 \times (V_h - V_p)=(2.8-1.6)\times 2=2.4mm^3 \quad \text{(EQ 18)}$$

Considering that the solder paste will fill about 80 % of the PTH at the time of the printing, the aperture of the stencil has to be accordingly smaller.

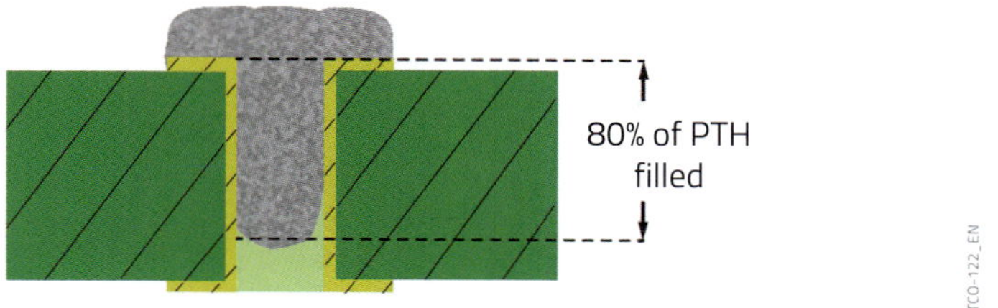

Fig. 1.142: Solder paste fills 80 % of the PTH

Typically, with a stencil thickness (T_s) of 150 µm (namely 0.15 mm) the stencil aperture volume (V_{sa}) should be calculated as follows:

By experience, we recommend to apply a coefficient of 1.5 to V_{sa} in order to have a proper soldering joint that matches the standard in soldering quality, the IPC-A-610D. After adding this coefficient, V_{sa} becomes **0.24 mm³**

Therefore, the theoretical diameter (D_s) of the stencil aperture in case of a round aperture should be:

$$V_{sa}=\pi \cdot \left(\frac{D_s}{2}\right)^2 \cdot T_s$$

$$0.24= \pi \cdot \left(\frac{D_s}{2}\right)^2 \cdot 0.15 \qquad \text{(EQ 19)}$$

$$D_s=\mathbf{1.43\ mm}$$

In this particular example D_s is smaller than D_{sp}, the solder pad diameter that we advise at 2.3 mm (D_h + 0.8). Therefore, we recommend using the theoretical formula mentioned previously (Pad diameter – 0.1), i.e. **2.2 mm**.

According to the previous extract from IPC/JEDEC J-STD 020D, Würth Elektronik recommends the following profile, shown in Figure 1.143, for both WR-TBL THR connectors series 7013 and 7015.

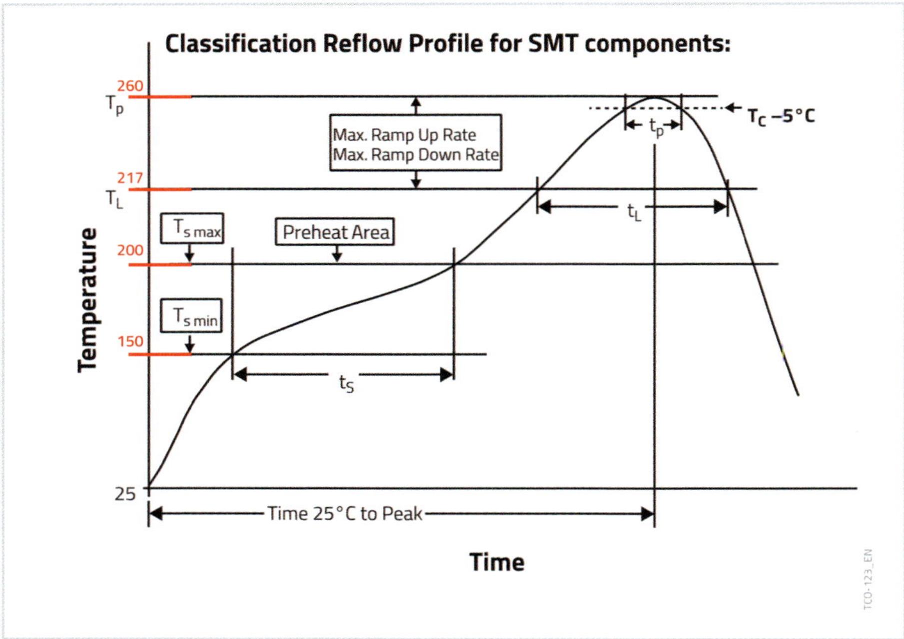

Fig. 1.143: Recommended soldering profile for WR-TBL THR

1.14.3.5 Soldering Quality

As per the IPC-A-610D, a compliant solder joint must fill at least 75% of the free volume between the pin and the PCB hole and there must be less than 25% of air cavity. Figure 1.144 illustrates the soldering joint acceptability.

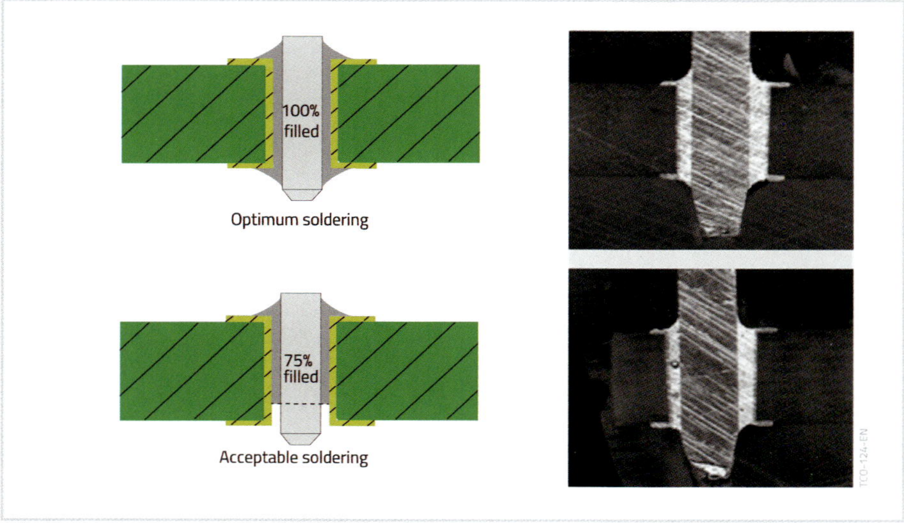

Fig. 1.144: Acceptable soldering joints

If the requirement for pick and place and solder paste application process are not complied with, then the quality might be affected, and it might produce defects due to insufficient solder paste resulting from:

a) incorrect solder paste calculation,
b) improper PTH/pin volume or
c) inappropriate screen-printing machine settings.

Figure 1.145 shows an example of insufficient solder paste.

Fig. 1.145: Unacceptable soldering joints: poor solder joint

Defects may also be caused by an excess of solder paste as per Figure 1.156.

Fig. 1.146: Unacceptable soldering joints: solder paste excess

1.14.3.6 Conclusion

The manufacturing of THR compatible connector products requires to pay attention to material and design changes directed towards the optimization of the components solderability with respect to typical SMT soldering processes. Housing considerations include the selection of a polymer having appropriate temperature capability and the redesign of the housing to provide stand-off and air flow capacity. The contact pin length and plating must be optimized to ensure solderability and pin geometry must be considered in solder volume calculations. The process plays an important role in this technology: therefore, it has to be carefully set and adapted to the components intended to be mounted. As for SMT components, the reflow profile must be calculated considering several parameters that are specific to each different board and to the reflow technology being used.

1.15 Press-in Technololgy

1.15.1 Compliant Press-in Technololgy – flexible Press-Fit

Compliant press-in technology was developed primarily by, and for, the telecommunications industry in the '70s to replace soldered press-in pins. A major driving force was to eliminate the soldering process to avoid the thermal exposures of soldering processes and problems associated with those exposures. The issue became more important as PCB took on higher functionality leading to increased number of layers and board thickness and mass. In addition, compliant press-in connections are readily repairable in comparison to the difficulties of soldered pin repair.

Compliant pin application methods range from mass termination of board-to-board connectors containing up to 1000 positions in a single cycle in telecommunications applications to insertion of individual compliant pins into a PCB and pressing connectors onto the inserted pin field. Tooling for mass insertion ranges from single "flat rock" presses to automatic equipment that monitors insertion forces continuously to control the insertion process. The individual pin insertion process is very flexible and allows for different options to be customized on a PCB board family. Over the years compliant press-in technology has spread too many other industries including automotive and aerospace due to this manufacturing flexibility and excellent field experience.

Compliance is the key to this field experience. Consider a solid pin being inserted into a PTH in a PCB. A solid pin is essentially rigid and will experience little elastic deformation upon insertion. The PCB is also essentially rigid in the plane of the board, the plane in which deformation must occur. This rigidity results from the woven fibre glass matrix of the FR4 layers making up the PCB. The end result is a low residual force to maintain the integrity of the contact interface and a potential for damage to the PTH. That is why rigid pins are only intended to retain their position until soldered. Compliant pins have a significantly different deformation kinetics and, therefore, the ability to provide a much higher residual force against the walls of the PTH.

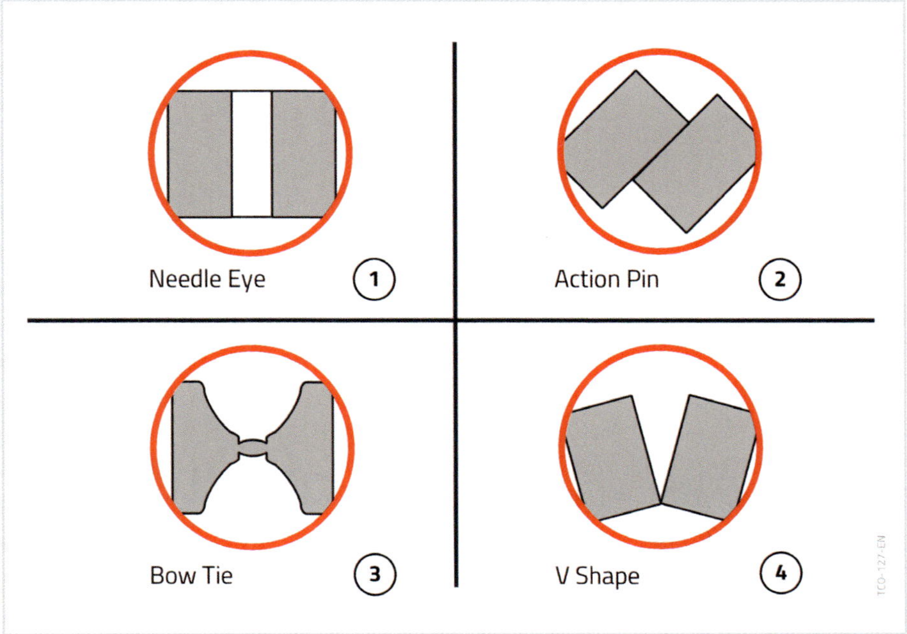

Fig. 1.147: Press fit section design

There are many different compliant section designs in use today. The inserted cross sections of four examples are shown in Figure 1.147. "Eye and Needle" (a), "Action Pin" (b), "Bow Tie" (c) and "V Shape" (d) are the most popular ones. Note that there are several different compliance mechanisms in these designs. The most used is called an "Eye of the Needle" (EN) for reasons apparent in Figure 1.148. The EN compliance is essentially that of a double supported cantilever beam. Figure 1.148a shows the original compliant section and Figure 1.148c the deflected compliant section of an EN compliant pin fully inserted in a PTH. Given the board rigidity, noted above, the deformation on insertion is primarily in the compliant section of the pin. The relationship between the compliant section dimensions, both width and length, and the diameter of the PTH control the amount of deflection and, therefore, the residual force between the compliant section of the pin and the PTH wall.

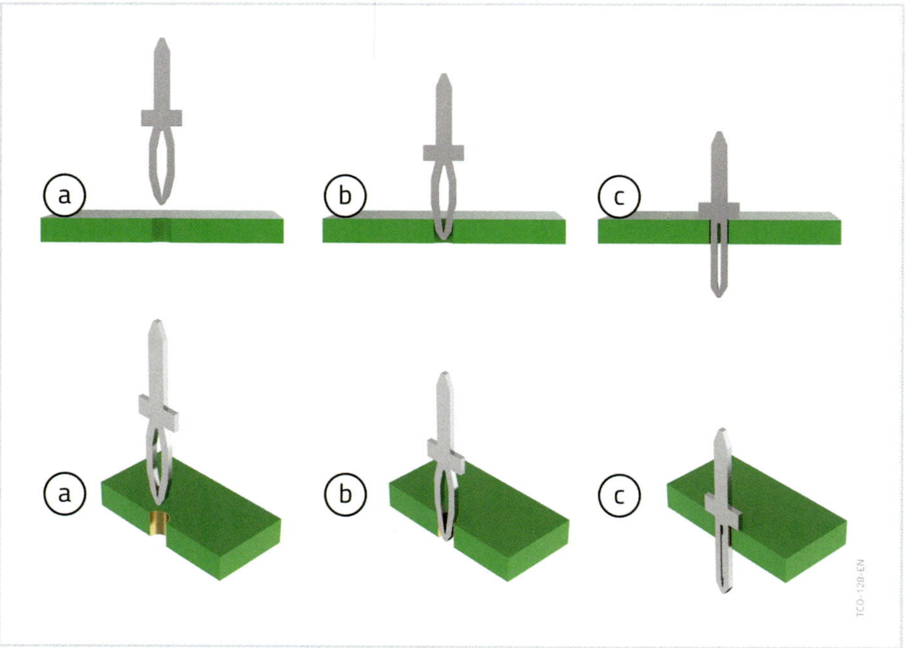

Fig. 1.148: Press fit process

Recall that the objective of a connector is to create and maintain an appropriate magnitude and distribution of metal-to-metal contact areas at all interfaces. Consider the contact areas in a compliant pin connection. Figure 1.148 shows the linear contact areas between the length of the compliant section and the wall of the PTH in a longitudinal cross section. There are at least two longitudinal contact interfaces, one at each beam of the compliant section. The transverse cross sections in Figure 1.147 show contact areas along the width of the compliant section. Again, there will be at least two transverse contact areas. The sum of these areas determines the total contact area between the PTH and the compliant section. The area so generated will be much larger than that of typical separable interfaces and, thus, the resistance of a compliant pin connection will be significantly lower. With this analysis of contact area creation in hand, attention turns to the generation of contact force to maintain the integrity of the contact area.

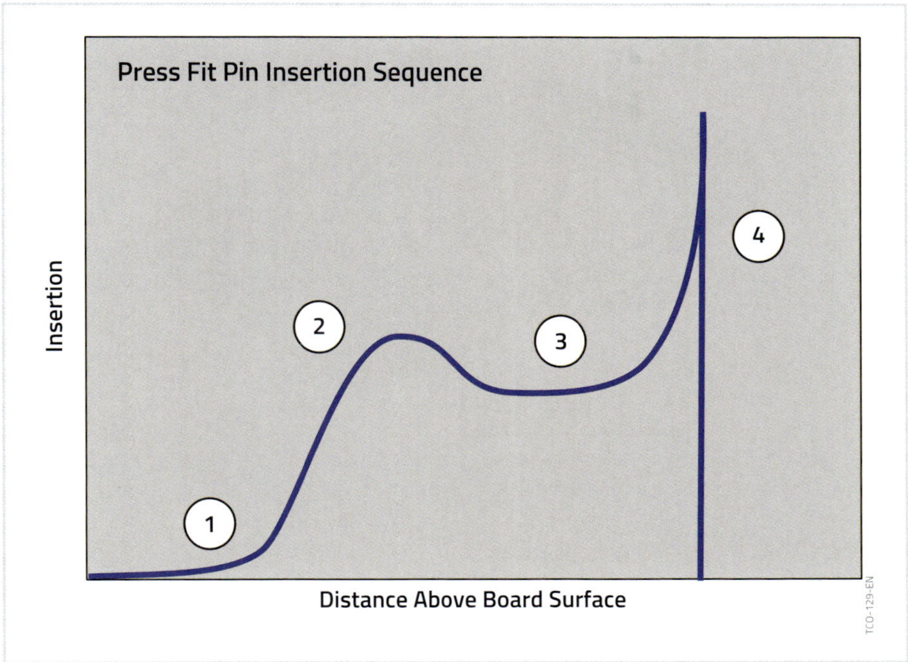

Fig. 1.149: Press fit insertion forces

There are similarities between the insertion of a compliant pin into a PTH and the mating of a pin contact into a receptacle socket (discussed in section 1.10.1 Separable Interface Requirements) but the deflection is in the pin rather than in the beams of the socket contact. The same lead-in frictional issues exist as with pins and sockets.

In this case the lead-in geometry of the compliant section of the pin determines the effective entry angle because the geometry of the PTH is constant. Figure 1.149 schematically plots insertion force versus the insertion depth of the compliant section of the pin into the PTH. Note the similarity to the insertion force curve for a separable connector in Figure 1.64.

The initial increase in insertion force comes from the deflection of the compliant section as the lead-in makes contact to the PTH wall, Figure 2.67b. Friction forces change in magnitude and direction as the lead-in entry angle varies until the uniform portion of the compliant section slides against the walls of the PTH when the insertion force plateaus, as it does in the separable connection when the receptacle beams are fully deflected. The final increase in force occurs when the tooling bottoms against the PCB. There is a significant difference, however, in the magnitude of the forces. Compliant pin insertion forces are in the range of several kilograms while separable connection insertion forces are typically a few hundred grams. The residual contact forces that create and maintain the integrity of the metal-to-metal contact interfaces between the compliant pin and the PTH are also of the order of kilograms, compared to the hundred or so grams of a separable connection. So, in addition to a lower resistance because of the larger contact area, the integrity of the contact interface is also superior because of the higher residual contact force.

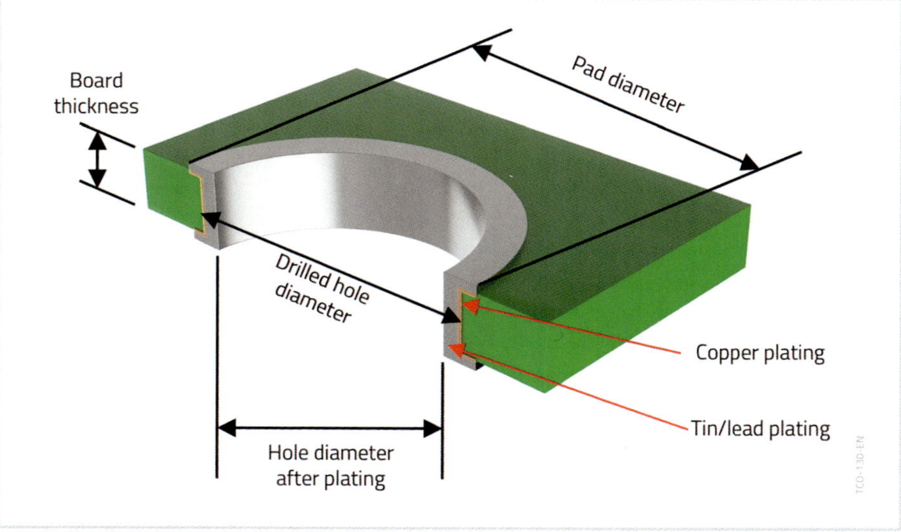

Fig. 1.150: Plated through hole cross section

Consider now the dimensional relationships that determine the deflection mechanics and, therefore, the performance of compliant pin permanent connections. Refer back to Figure 1.148 and note the original width, W_o, of the compliant section as compared to the diameter of the PTH. Clearly the PTH diameter must be less than Wo for the compliant section beams to be deflected on insertion. Figure 1.150 shows a cross section of a PTH. Note that there are several "diameters" in the figure. The basic diameter is that of the drilled hole in the PTH. This is the critical dimension as it defines the deflecting surfaces, the rigid PTH wall that determines the deflection of the compliant section. Despite the importance of the drilled hole diameter, the PTH dimension is usually specified by the finished hole diameter, the diameter after plating, because that is the diameter that is readily measured. All PTH will have a copper plate, with a typical thickness in the range of 25 to 75 µm (1000 to 3000 micro-inches). The secondary plating for compliant pin applications, if any, is generally tin/tin-lead in the range of 5 to 15 µm (200 to 600 micro-inches). The tin plating thickness influences the insertion force, due to the higher co-efficient of friction of tin as compared to copper but has little effect on the residual contact force.

As noted previously, the residual forces in compliant pins are significantly higher than those of separable connections. Recall that high normal forces in separable connections affect the wear, the mating durability, of the contact finish. In compliant pins the concern about the effect of high forces is on the magnitude of the pin insertion force and the potential for damage of the plating in the PTH. For reference, the insertion forces for a compliant pin in a nominal 1.0 mm finished hole are in the range of 50 to 250 Newton per pin. A minimum retention force of 30 Newton is also specified for this hole size.

The PTH damage due to such high insertion forces can include lifting of the annular ring around the PTH (Figure 1.151a), radial hole distortion and fracture of the bond between an inner layer trace with the PTH plating (Figure 1.151b). In addition to these force effects, damage to the PTH is dependent on the compliance of the pin and the contour of the edges of the pin that make contact to the PTH plating.

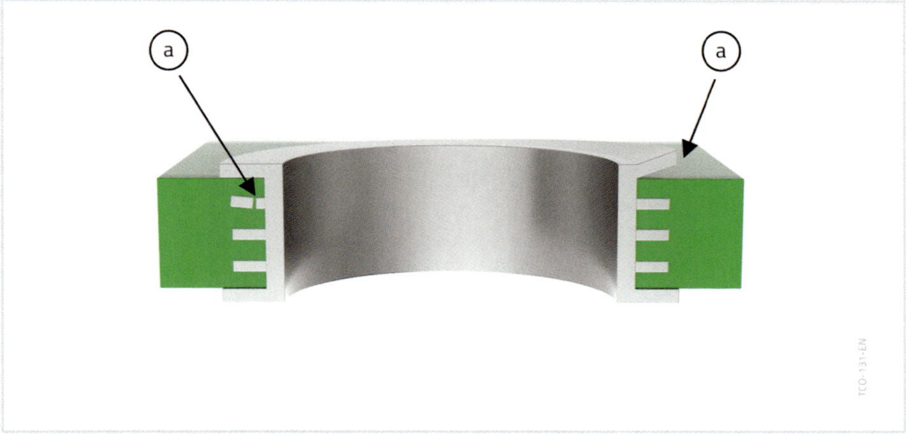

Fig. 1.151: Plated through hole damage

The finished hole diameters for compliant pin applications have decreased dramatically since their introduction in the 70's. At that time the finished hole diameter was nominally 1.5 mm (0.06 in). In the 80's and 90's the finished hole diameters decreased to nominal 1.0 and 0.6 mm (0.04 and 0.024 in) respectively. Recently holes in the range of 0.2 mm (0.05 in) have been introduced. As an example of hole tolerances, for a nominal finished hole of 1.0 mm the tolerances are +0.09 and −0.06 mm. The corresponding drilled hole dimensions are 1.15 mm ±0.025 mm. Naturally, these tolerances vary with hole size. The reduction in PTH dimensions were necessary to open new markets for compliant pins as electronic equipment increased in functionality and decreased in size.

1.15.2 REDCUBE – Solid Press-Fit

REDCUBE Terminals are the most reliable high-power contacts on PCB level with a high current rating. With the same ampacity, the components show the lowest heat development compared to other parts that supply power for PCBs.

Pressing the pins into the PCB, a high friction between pin and plated through-hole generates a homogenous cold-welding between the pin and the copper plated via in the PCB. This results in a gastight, strong mechanical connection with a contact resistance less than 200 µΩ. No other technology transmits currents up to 500 A with such low self-heating.

The via for a press-fit system is essentially made in the same way as the holes for accepting components for THT soldering. Thus, there are no changes required in the manufacturing process of printed circuit boards.

Regarding long-term reliability, **REDCUBE** PRESS-FIT is an improvement since it has the lowest **FIT** rate (Failure in Time) of the overall system. It is up to 30 times better than the FIT rate of an SMT solder joint. A single solid press pin has a typical **extraction force of 100 N** for a 1.6 mm PCB. Such a small component with 8 pins could withstand a weight load of an average person without extraction out of the PCB. Therefore, **REDCUBE** PRESS-FIT Terminals are perfect to provide not only an electrical but also a mechanical connection solution for electrical components.

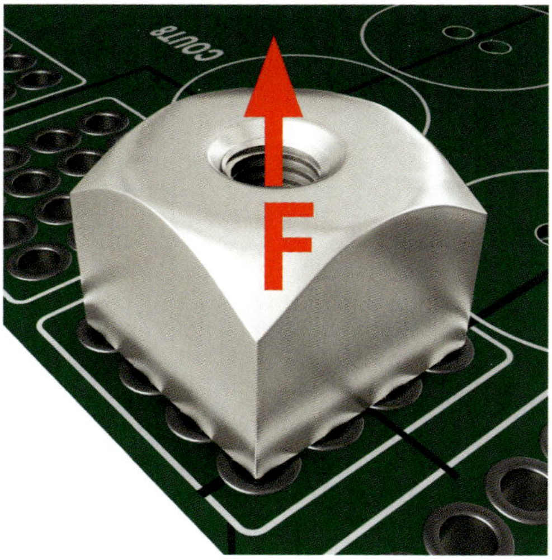

Fig. 1.152: A single solid press pin has a typical extraction force of 100 N for a 1.6 mm PCB.

If after a press-fit process a solid press pin in a 2.4 mm thick printed circuit board fits on each corner with more than 3° against the sleeve, the press connection zone has a lower electrical resistance than the brass pin itself and thus does not pose an electrical or thermal bottleneck. The connection surface angle is normally much greater, which provides a generous safety buffer for the electrical connection.

Advantages of REDCUBE PRESS-FIT

- **Very high ampacity**
- **Ideally suited for continuous and peak currents**
- **Extremely high environmental stability**
- **Suitable for rough environmental conditions**
- **Low-resistance connection (<200 µOhm)**
- **No cold solder joints**
- **Double sided mounting**
- **Space saving design**

Press-Fit versus Soldering

Press-fit technology provides several advantages in comparison to solder technology. Very thick circuit boards with high copper plating can be processed easily. Furthermore, **two-sided mounting** of circuit boards is possible without any problems, which usually enable a very compact design of modules. As a result, current paths are shortened which is thermally very beneficial for the processing of high currents.

There is reliable contact between pin and copper layer over the complete case length of a press-fit zone. It is not guaranteed for soldering that the solder rises the complete length of the via whereby many higher transition resistances are produced. Therefore, long-term reliability and mechanical stability are also not as high as with **REDCUBE** PRESS-FIT.

The processing of **REDCUBE** PRESS-FIT Terminals integrates seamlessly in the production process of the systems and is thus very cost-effective. In contrast to soldering the circuit board is not loaded thermally during the press-fit process.

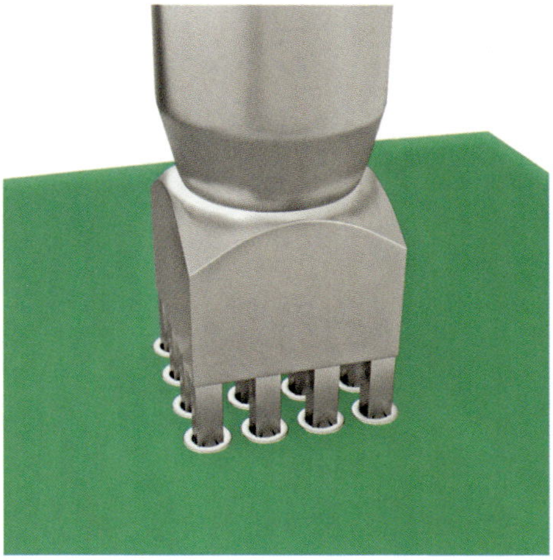

Fig. 1.153: Processing of REDCUBE PRESS-FIT Terminals.

Materials and tolerances

REDCUBE PRESS-FIT from Würth Elektronik are manufactured from the material CuZn39Pb3 and are therefore RoHS-compliant according to the RoHS stipulation concerning copper alloys.

The circuit board thickness should ideally be between 1.6 mm and 3.2 mm. Tested surfaces are chemical tin, HAL and ENIG. The **immersion tin** coating process is recommended. Using this process usually guarantees that the tin is evenly distributed in the case whereby the tolerances can be complied with more easily and thus chip formation can be prevented. Due to the uneven distribution of the tin in the case for the HAL process, WE recommend the immersion tin process for circuit board thickness of 2.4 mm and greater. ENIG can be used but not recommended for Press-Fit technology.

Unless otherwise noted in the corresponding drawing, Würth Elektronik **REDCUBE** PRESS-FIT have quadratically designed press-fit pins. The through-hole plating in the PCB must therefore have the characteristics shown in Figure 1.154:

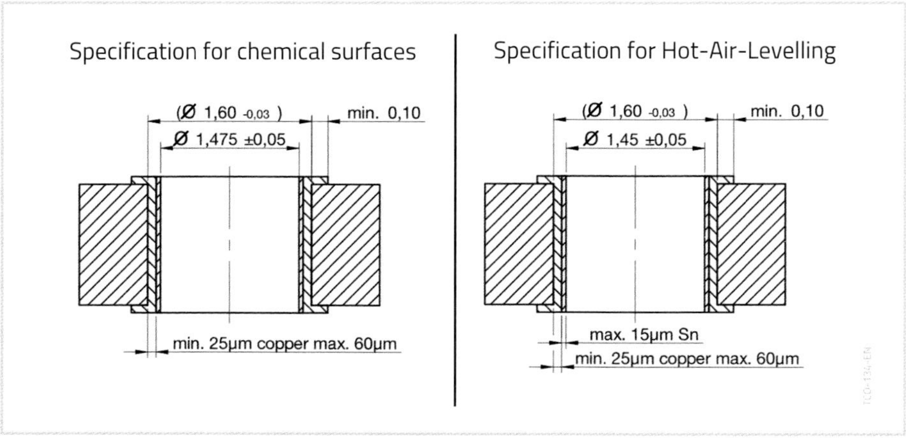

Fig. 1.154: Characteristics for Through-Hole plating in the PCB.

Current load

With **REDCUBE** PRESS-FIT Terminals from Würth Elektronik, currents of more than **500 A** can be carried on the circuit board. The current-carrying capacity of **REDCUBE** PRESS-FIT must always be considered in the context of the complete system. Many factors such as conductor path thickness, conductor path width, cable cross section, ambient temperature and heat distribution should be taken into account for the selection of the individual **REDCUBE** PRESS-FIT.

In comparison with a solder connection (R = 300 to 400 µΩ), the press-fit zone itself with 100 to 200 µΩ has extremely low resistance so that the **limiting factor** can usually be found in the layout of the connected **conductor paths** or the **connection of external feed lines** to a press-fitted component.

Further information at: **www.we-online.com/redcube**

1.16 Permanent Connections to Wires and Cables

There are two mechanical permanent connection technologies for wire and cable, crimped and insulation displacement connections. Crimped connections require the stripping of the insulation of the wire to expose the conductors and are made to individual wires. Insulation Displacement Connections, IDC, in contrast, do not require stripping of the insulation, the insulation is displaced during the connection process, and are generally made to multi-wire cables. Crimped connections date back to the 1940s and IDC to the 1960's.

Before discussing crimped and insulation displacement connection details an introduction to wire and cable construction is in order.

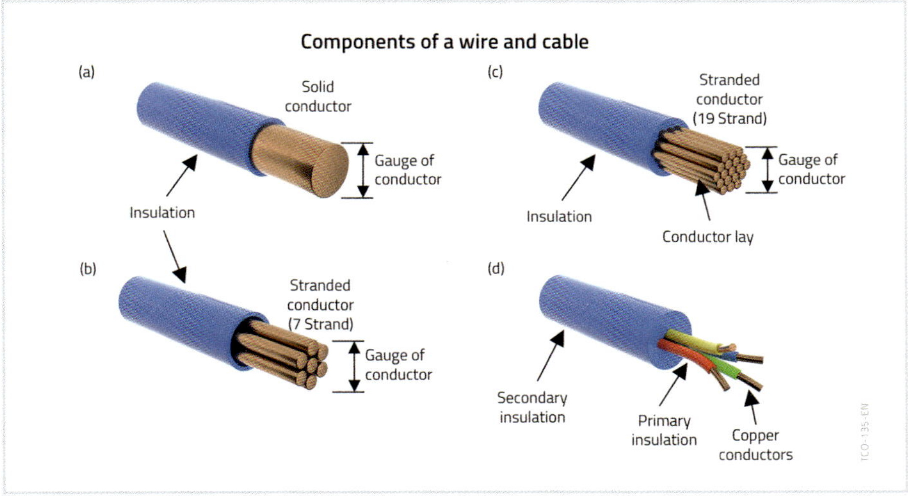

Fig. 1.155: Components of a wire and cable

In general, a wire is understood to consist of a conductor and its insulation. A single conductor is generally referred to as a "bare wire". In the simplest terms a cable is a collection of wires held together by a variety of means and mechanisms. Figure 1.155 schematically illustrates several examples of wires and a jacketed cable. Figure 1.155a illustrates a wire with a solid conductor. Figure 1.155b illustrates wire with a stranded conductor, in this example seven strands. Figure 1.155c illustrates a wire with a 19 connector strand. Multistrand conductors can be arranged in several ways as will be discussed. Figure 1.155d illustrates a jacketed cable with four wires within the cable jacket. The cable jacket both constrains and protects the wires within it. Consider now the individual components of wire and cable.

Conductors

The most common conductor material is annealed copper because of its high electrical conductivity ease of manufacturability and cost. Higher strength copper alloys are used when wire strength is an important consideration. Such alloys always compromise the conductivity, however.

Aluminium conductors are sometimes used to take advantage of its weight/conductivity ratio. Aluminium is not as conductive as copper, but it is also much lighter and for a given conductivity will be lighter than a copper conductor. When aluminium conductors are used, aerospace applications for example, special attention must be paid to permanent connection techniques due to the intrinsic aluminium oxide on the surface and the tendency for aluminium to flow under load.

Copper conductors may be plated. The most common plating's are tin, silver and nickel. Tin plating is used to provide corrosion protection and to enhance solderability shelf life and mechanical termination performance. Silver plating is used to optimize surface conductivity in high frequency applications. Nickel is used for high temperature applications but with a compromise in solderability and mechanical termination ease.

As noted, conductors may be solid or stranded. Stranded conductors are the most common due to their greater flexibility compared to solid conductors. Solid conductors have a higher conductivity than stranded conductors of the same diameter or gauge. Stranded conductors are preferred for crimped connections and solid conductors for IDC applications though both can be successfully used in either technology.

There are two types of stranded wires, bunch and concentric. Bunch stranded wires contain any number of strands and there is no attempt to control the relationship among strands. Concentric wires contain a central conductor with one or more layers of additional conductors wound around the central core. Concentric wires contain specific numbers of conductors because of the packing of the wires in space. The most common are 7 and 19 strands because six strands can be close packed around the center conductor and 12 more can be close packed around the seven strand conductor bundle as illustrated in Figure 1.156.

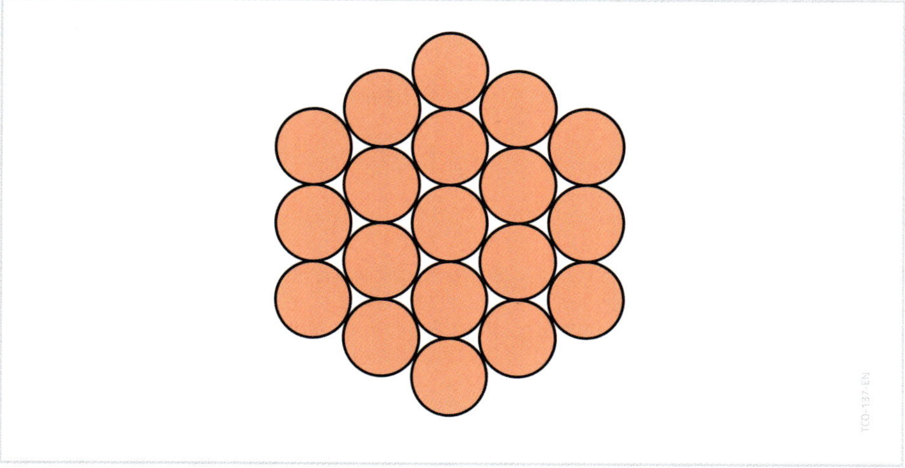

Fig. 1.156: Strand packing of conductors

Stranded wires may be twisted. The length of a complete twist of the conductor strands is referred to as the lay of the wire. Concentric wires are always twisted while bunch stranded wires are generally not. The purpose of the lay is to provide and maintain some relationship among the strands and some control over the diameter of the bundle. Thus, twisting reduces the flexibility of the stranded conductor as well as increasing the cost. The lay and lay direction define three different types of concentric wire constructions. Consider Figure 1.157, in this case the lay of the conductors is in the same direction and with the same pitch (length of lay). This is a unilay construction. If the direction of the lay is changed but the pitch retained it is an equilay construction. If both are changed it is called a true concentric construction because the change in both pitch and direction results in a tighter and more stable conductor bundle diameter. Bunch stranded wire without a lay is commonly used in the white goods industry for both cost and flexibility reasons. Crimped connections are made primarily to bunch stranded wire with copper conductors. Of course, the insulation must be stripped from the wire to facilitate the mechanical crimped connection. The disadvantage with bunch stranded wires is that the strands can easier spread after the cable is stripped. This could cause a problem when processing the crimping. At AWG cables, which are twisted, this is not the case. Concentric constructions are preferred for IDC applications due to the diameter control and its impact on contact force as will be discussed.

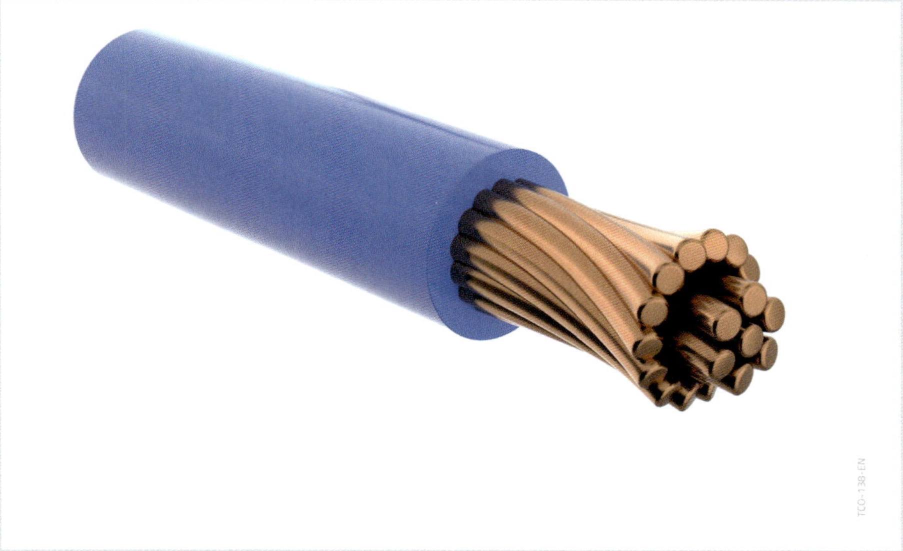

Fig. 1.157: Concentric constructions/conductor lay

Insulation

The wire insulation provides both mechanical and electrical functions with the relative importance of each depending on the application. The mechanical functions are dependent on whether the application requires wire or cable. Figure 1.158 illustrates three examples of cables.

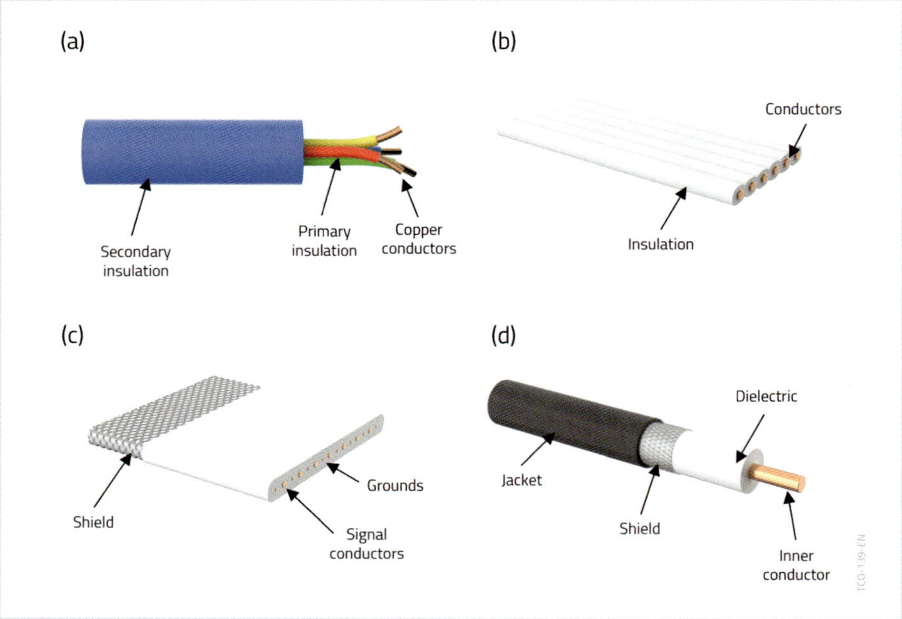

Fig. 1.158: Ribbon, transmission and coax cables

Figure 1.158a shows a jacketed cable, in this case there are four wires within the organizing and protective jacket. Note the reference to primary and secondary insulations. The primary insulation is the insulation on the wires themselves and the secondary insulation is the cable jacket. The primary insulation function is primarily electrical. There are two electrical functions, insulation and impedance control with insulation being common to all applications. Impedance control is relevant in high frequency applications such as coaxial cables and transmission lines. The secondary insulation performs primarily protective functions, such as abrasion and chemical resistance.

Ribbon cables, Figure 1.158b, are analogous to jacketed cables in the sense that their function is to establish and maintain dimensional control of the conductors, or wires, within the cable. The ribbon cable shown contains conductors while other constructions may consist of a collection of individual wires. Ribbon cable is widely used in IDC applications as will be discussed. In such cases dimensional control is required to ensure the proper engagement of the cable with the insulation displacement contacts within the connector housing during the termination process. Jacketed and ribbon cables are generally used in low frequency applications.

Transmission cables, Figure 1.158c are intended for high frequency applications and consist of a collection of signal and ground conductors in a controlled distribution and spacing to provide impedance control. The insulation material is selected for impedance control as well. Transmission cables are often shielded, as indicated. Transmission cables also use IDC technology.

Coaxial cables, Figure 1.158d are also intended for high frequency applications. The term "cable" applies because a shield and an outer jacket surround the "wire", the insulation (dielectric) and inner conductor. Given this construction, different termination technologies are used with coaxial cable.

The discussion of insulation materials will be limited to the most commonly used materials. Many specialized applications exist and are outside the scope of this book.

Primary Insulations

Primary insulation materials are selected for their electrical and mechanical characteristics. For low frequency applications insulation resistance is sufficient. For high frequency applications where controlled impedance and signal transmission characteristics are important, the dielectric constant is important as is the potential for foaming the insulation to further reduce the effective dielectric constant. Mechanical requirements are application dependent as well with flexibility and abrasion resistance among the more important characteristics.

PolyVinylChloride

PVC is the dominant low frequency insulation material due to a combination of cost and manufacturability of a variety formulations to optimize mechanical performance for a range of applications. High frequency use is limited by dielectric constant and associated capacitive losses.

Polyolefins

The properties of polyethylene, PE, insulations are dependent on the processing of the polymer. Both low and high density formulations are used. Low Density PE, LDPE, has good mechanical properties and a low dielectric constant making it suitable for high frequency applications, particularly in foamed versions. High density PE, HDPE, has better mechanical characteristics and a slightly higher temperature range than LDPE.

Polypropylene, PP, has similar properties to PE but with better abrasion resistance and a higher temperature capability, 105 degrees C compared to 80/90 for LDPE/HDPE.

Fluorinated Hydrocarbons

The fluorinated hydrocarbons, polytetrafluoroethylene, PTFE, polyfluoroethylenepropylene, PFEP, and polyvinyledinefluoride, PVDF, have superior electrical, mechanical and temperature capabilities than PVC and the polyolefins, but at a cost premium that limits their use to the most demanding application

Material	Volume resistivity (ASTM D257) (Ω-cm)	Volume resistivity (ASTM D150) (nominal)	Voltage breakdown (ASTM D149) (V/mil)	Hardness (Shore)	Abrasion resistance	Temp. Range (°C)
PVC	10^{11}	7.0	500	D85/90	fair	−20/+80
PE	10^{18}	2.5	600	D45/60	good	−60/+80
PP	10^{15}	2.2	650	D50	excellent	−40/+105
PTFE	10^{18}	2.1	600	D60	excellent	−70/+250
PFEP	10^{18}	2.1	600	D60	excellent	−70/+250
PVDF	10^{15}	5.5	250	D70/80	excellent	−40/+150

Tab. 1.4: Properties of insulation material (source – IICIT)

Secondary Insulations

As noted earlier, cable secondary insulations are primarily intended to provide mechanical and chemical protection to the cables during routing and manufacturing of equipment. Several secondary insulation materials are natural and synthetic rubber compounds.

Polychloroprene Rubber

Like PVC, PCR, is compounded in a variety of formulations to support different application requirements. PCR has a good balance of mechanical properties and chemical resistance.

Ethylenepropylene Rubber

EPR is a polyolefin copolymer with good chemical and solvent resistance as well as good mechanical and electrical properties.

Polyurethanes

Thermoplastic and elastomeric polyurethanes have good mechanical properties but limited electrical performance.

Polyamides

Several polyamide formulations are available with good chemical and solvent resistance and low coefficients of friction. Some varieties, however, have high moisture absorption and limited flexibility.

With this overview attention now turns to mechanical permanent connections to wire and cable, beginning with crimped connections.

1.16.1 Crimped Connections

From its beginnings in the 1940s, crimped connections have been the dominant permanent connection technology for the white goods and industrial markets. This dominance has been realized through the successful field history of crimped connections. A history that is dependent on the consistent use of the crimping system. The crimping system consists of:

- Wire
- Terminal
- Tooling

The system begins with the selection of an appropriate wire gauge for the application. A typical wire selection criterion is the intended current capacity of the connector in its intended application.

The selected wire gauge then determines the crimp terminals that are suitable for that wire gauge. A crimp terminal may be capable of accepting more than one wire gauge. In this example the terminal, shown in Figure 1.159 can accept AWG 24 up to AWG 18 wires.

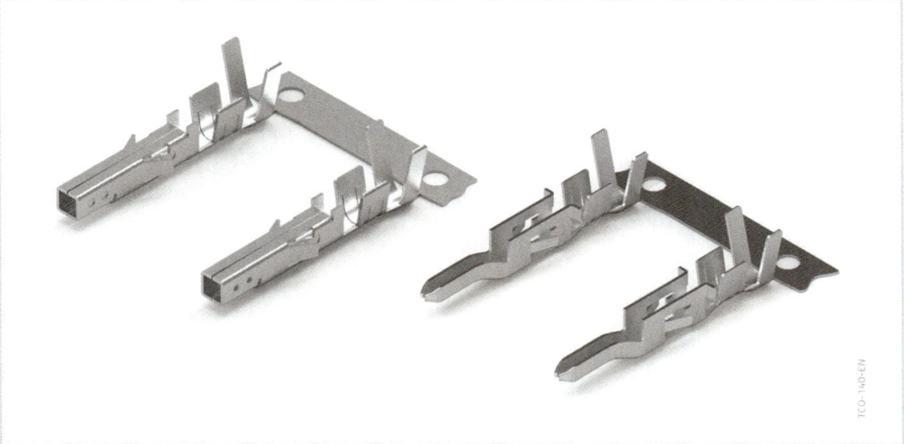

Fig. 1.159: Male and female crimp terminals (WR-MPC4 series)

Finally, the combination of wire and terminal determines the appropriate crimp tooling and crimping process specifications. A set of tooling may be capable of accepting a range of wire/terminal combinations by varying the crimping process specifications. The tooling, Figure 1.160 used for crimping this wire/terminal, combination can crimp the following wire/terminal: AWG 24 to 18. Details of the crimping process and process specifications will be discussed later in this chapter. A cross-section of a crimp connection using this crimping system is shown in Figure 1.161. This configuration, commonly called a "B" crimp because of its geometry, is the dominant open barrel crimped connection geometry.

Fig. 1.160: Crimping tool

Fig. 1.161: Crimp cross section

There are several important features of interest in the crimp cross-section of Figure 1.161. First note the deformation of the terminal and tooling. All of the conductor strands have been deformed, either against other strands or against the sides of the crimp terminal. The bundle of crimped strands approaches a solid mass of copper conductor. There are multiple contact areas created between individual strands and between strands and terminal wall. These result in a large total contact area and a highly redundant contact area. Note also the deformation of the wings of the crimp barrel as they penetrate the conductor bundle and come together, this wing contact is an important contribution to the performance of crimped connections.

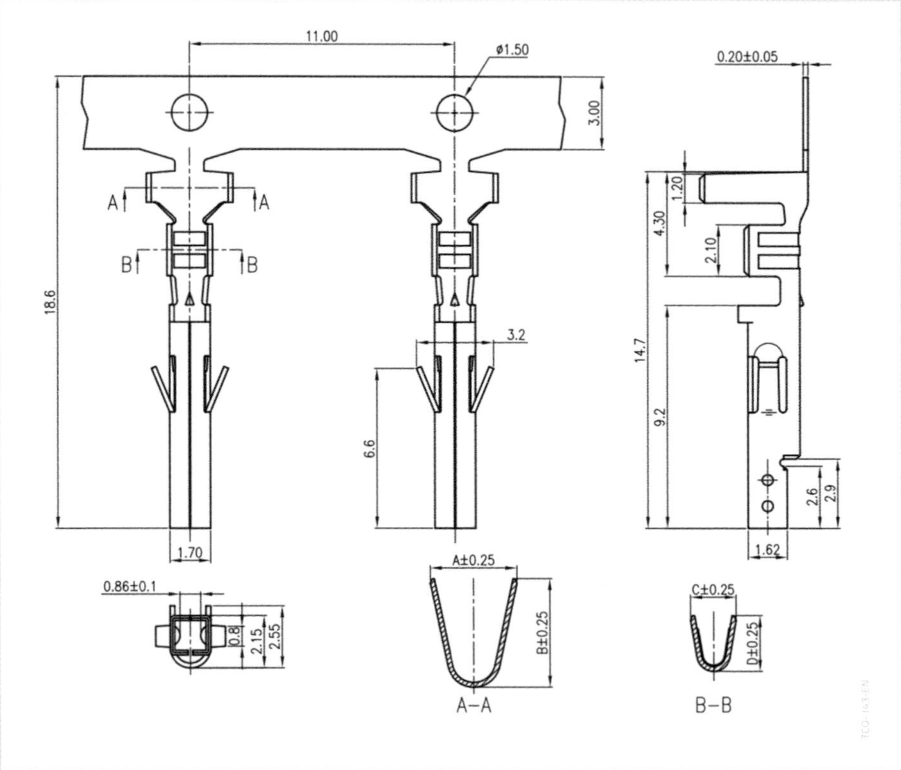

Fig. 1.162: WR-MPC4 female crimp contact

Figure 1.162 shows the original cross section of the crimp barrel. A comparison of the crimped connection cross-section with the original cross-sectional areas of the conductor, the wire gauge, and the cross-sectional area of the terminal in the crimp zone will show that the crimped connection cross-section has a smaller area than the original combined area. This reduction in area indicates that the conductors and crimp barrel have been extruded along the axis of the crimp terminal during crimping. This extrusion creates additional contact area and opportunities for stabilization of the multiple contact areas.

The necessity to control the relative deformation of the conductors and crimp terminal is the reason for the crimping system. Control of the original area, by selection of conductor gauge and crimp terminal, and control of the final crimped cross-section area by the envelope of the crimp tooling as it creates the deformation allows for a consistent and operator independent deformation mechanism. This capability for consistency is the reason for the excellent performance history of crimped connections in the field as noted earlier.

With this overview as a context attention turns to details of the crimping process.

Crimp Terminals

There are two basic types of crimp terminals, closed barrel and open barrel as illustrated in Figure 1.163 The most dramatic difference between the two cases is the presentation of the stripped wire to the terminal. (The strip length of the conductor is an important process control as will be discussed in the discussion of crimped connection inspection.) For closed barrel terminals the conductor must be presented axially, Figure 1.163a. The target area for insertion is the cross-sectional area of the inside of the crimp barrel. In contrast for an open barrel terminal the conductor can be presented from above, Figure 1.163b, and the width of the "U" shape of the crimp barrel rings provides a significantly larger target area. This simple geometric consideration explains why open barrel terminals are used in high volume automatic crimping equipment. Closed barrel terminals are used in semi-automatic equipment, often with hand presentation of the conductor and with hand tooling in lower volume applications.

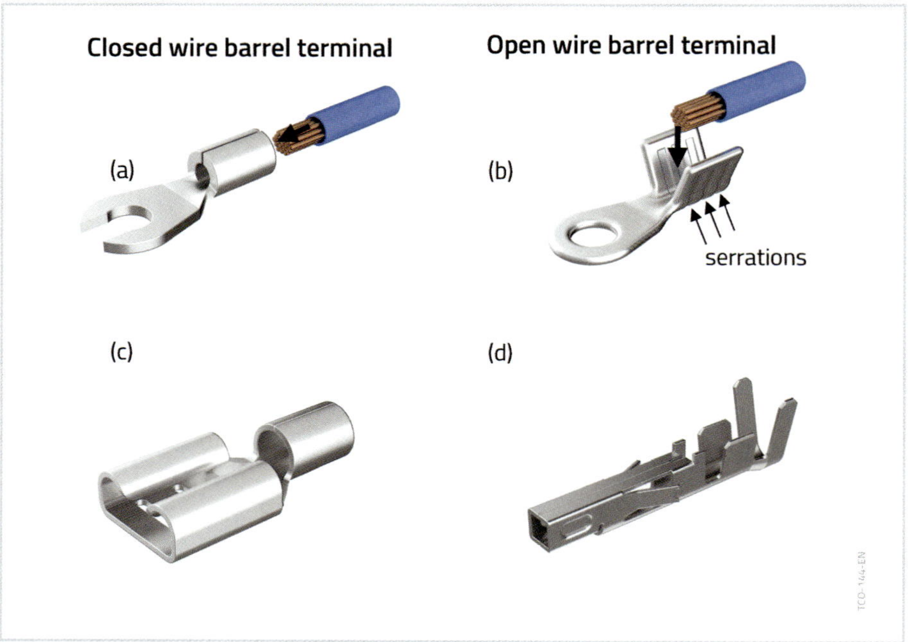

Fig. 1.163: Open/closed crimp barrel terminal

Figure 1.163c illustrates a pre-insulated closed barrel terminal. Pre-insulated terminals provide protection against shorting between contact positions in terminal blocks as used in white goods and industrial applications. The plastic pre-insulating sleeves may be colour coded to indicate the wire size. Some pre-insulated terminals include an internal copper sleeve to facilitate insulation support crimping in addition to the insulating plastic sleeve.

Figure 1.163d shows an open barrel crimp terminal that also includes an insulation support barrel. All crimp terminals will have a crimp barrel where the electrical permanent connection is made. The crimp barrel shown also contains serrations, shallow grooves in the crimp barrel. Serrations serve to enhance the integrity of crimped connections as will be discussed later in this chapter. The purpose of the insulation support barrel is to provide a strain relief to reduce the stresses transmitted through the conductors to the crimp barrel during vibration or mechanical disturbances in the field. Insulation support crimps may be either compression, where the insulation is lightly crimped, or overlapping, where the wings of the insulation barrel wrap around the insulation of the wire.

The Crimping Process

As noted, open barrel terminals are generally crimped using high volume automatic crimping equipment while closed barrel terminals are generally crimped using semiautomatic or hand tooling. The kinetics of the crimping process with respect to the deformation of the conductors and crimp barrel, however, are very similar for both terminal styles, and both take advantage of the consistency of the crimping system of wire, terminal and tooling. The following discussion will use open barrel crimping as an example.

Figure 1.164 shows three steps in the crimping process. The first figure shows the containment of the conductor strands and the engagement of the wings of the crimp barrel with the lead-in section of the crimping tooling as the crimping process begins. In the second figure, compression of the strands begins as the wings fold over in the forming section of the tooling. Forming of the wings around and into the strand bundle is a critical part of the crimping process as will be discussed. The final figure shows the total deformation of the strands and the crimp barrel in the completed crimp.

Control of this final stage of the crimping process provides the desired control of the overall crimping deformation and geometry. Control of these parameters is critical to the integrity of the crimped connection as will be explained in the following discussion.

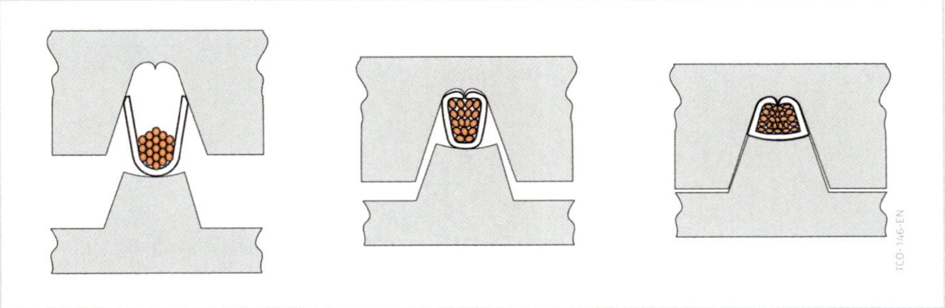

Fig. 1.164: The crimping process

Crimping Deformation and Crimped Connection Performance

As discussed in some detail in section 1.6.2 The Mechanical Interface: Friction and Wear/Durability, the deformation of two surfaces as they come in contact with each other determines both the electrical and mechanical characteristics of the interface between them. During the crimping deformation multiple interfaces are created as the conductor strands deform against one another and against the walls of the crimp barrel. Consider first how the crimping deformation affects the mechanical strength, the pull strength, of the crimped connection.

Figure 1.165 schematically illustrates the pull strength as a function of deformation during crimping. The initial pull strength is essentially zero as the conductors are simply sitting in the crimp barrel. As deformation begins friction forces are created between the strands and the strands and the crimp barrel and the pull strength begin to increase. As the conductor strands deform, and as they wipe against the walls of the crimp barrel, surface films and contaminants are displaced and metal-to-metal contact interfaces are generated as previously discussed. As deformation continues the number of interfaces and the average size of the interfaces increases. The friction forces, work hardening and cold welding at the interfaces also increase. Eventually the bonding between the strands and the crimp barrel increases to the point where the strands and crimp barrel approximate a solid mass and the pull strength levels off. The decrease in pull strength occurs when the cross-sectional area of the strands and crimp barrel becomes smaller than the original combined area of conductors and crimp barrel.

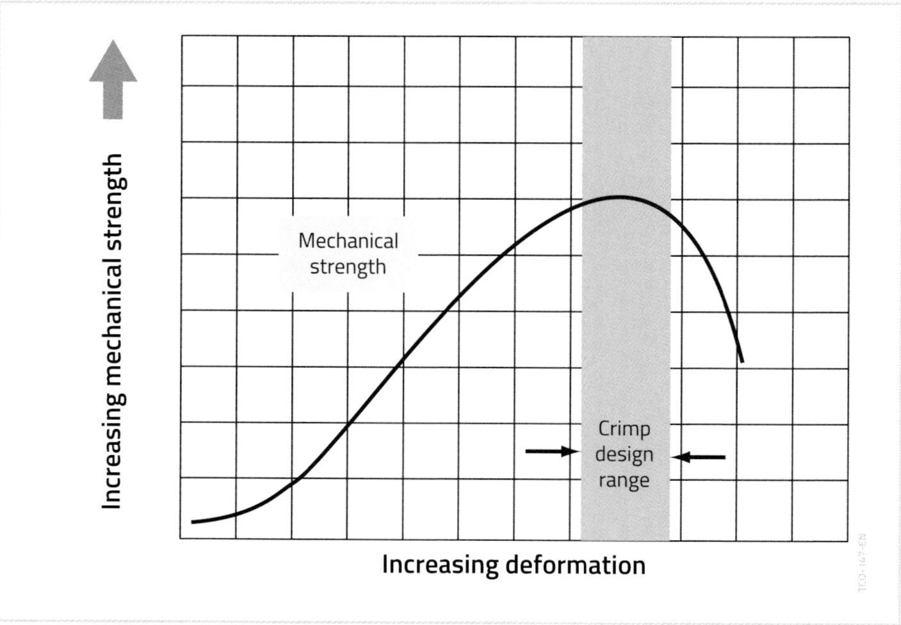

Fig. 1.165: Mechanical performance vs. deformation

Obviously, the same description of interface formation applies to the electrical contact interfaces. Figure 1.166 schematically illustrates the change in crimp conductivity with crimping deformation. The increase in conductivity is due to the increasing contact area. Once again, at some point in the deformation the multiple redundant contact interfaces areas reach a cumulative area where these contact areas are essentially equivalent to the full cross-sectional area of the crimp barrel and conductor system.

This is analogous to the "full area contact" of a separable contact interface as discussed in section 1.6.2. The Mechanical Interface: Friction and Wear/Durability. This is the reason for the plateau in the conductivity versus deformation curve. Conductivity increases incrementally beyond this "full area contact" region. The decrease in conductivity comes when the crimped cross-sectional area drops below that of the original conductor plus crimp barrel area.

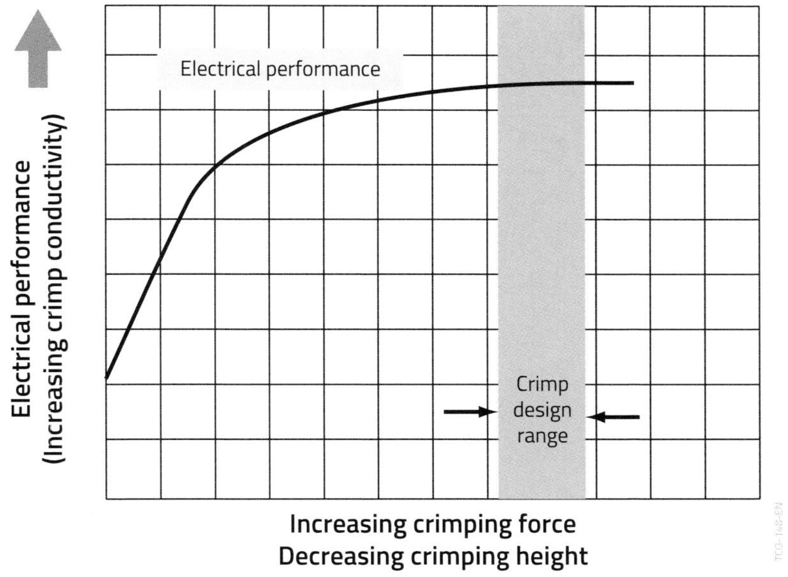

Fig. 1.166: Electrical performance vs. deformation

Figure 1.167 plots both pull strength and conductivity as a function of deformation on the same graph. In this figure deformation is characterized by crimp height, basically the parameter that controls the overall deformation during the crimping process. Crimp height will be discussed in detail during the discussion of crimp inspection methodologies. Note that the very broad plateau in conductivity overlaps the much narrow mechanical strength curve. Thus, crimp deformation can be controlled to optimize mechanical or electrical performance. In the early days of crimping technology, the tendency was to optimize mechanical performance, in particular to ensure that connections were not over crimped, that is on the falling side of the mechanical strength versus deformation curve. In part this was to allay concerns that the deformation during crimping was not "cutting" the strands. The failure mode for crimped connections on the upside of the curve tends to be by pull-out of the strands. On the downside the failure mode is conductor breaks, inside or outside the crimp barrel. Pull-out failures clearly indicated that the conductor strands were not being "cut". As field history of successful crimped connections was accumulated, optimization of electrical performance tended to be emphasized and crimping deformations were increased. A commonly used gauge for deformation is the Area Index, the ratio of the cross-sectional area of the crimped connection to the original cross-sectional area of the conductors and the crimp barrel. Area index recommendations fall in the range of 70 to 85 % with the lower ranges applying to automated crimping processes and the upper to hand tool applications.

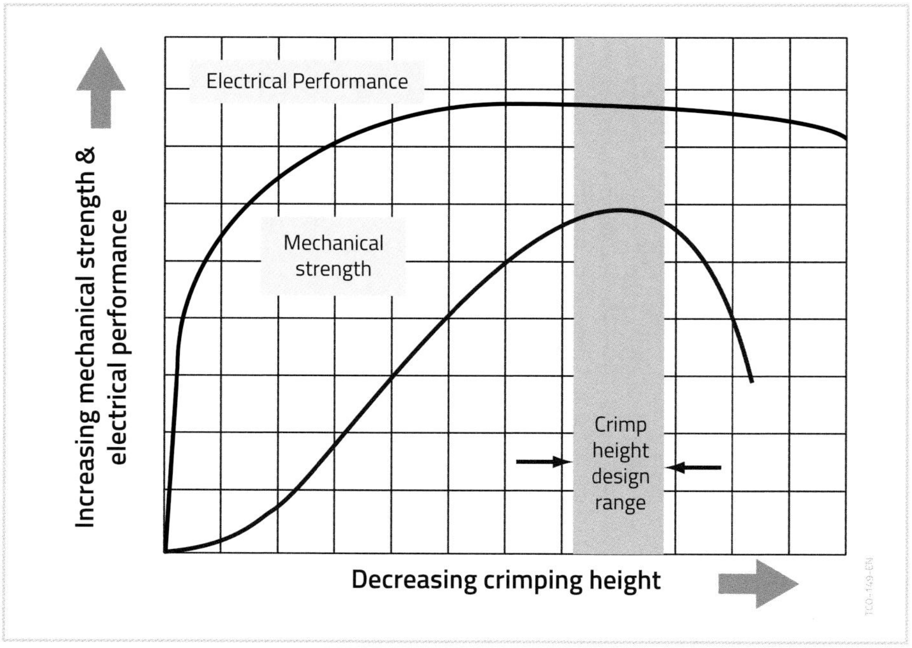

Fig. 1.167: Mechanical & electrical performance vs. deformation

Crimp Integrity

The discussion in the previous section has highlighted how the multiple redundant contact interfaces between conductor strand and between conductor strands and the crimp barrel are created. These interfaces, however, are small and of low mechanical strength individually. To provide consistent field performance these multiple contact areas must be protected so as to function collectively. The excellent field performance history of crimped connections testifies to the fact that this condition has been realized. There are several crimp connection design parameters that provide this protection on a consistent basis. These include residual forces in the crimp barrel, mechanical confinement by the crimp barrel and mechanical confinement/interlocking due to serrations in the crimp barrel. Each will be considered separately.

Residual Forces

It is clear that both the conductor strands and the crimp barrel experience significant plastic deformation during the crimping process. Despite this plastic deformation, there are always residual elastic stresses in the deformed configuration that tend to spring back towards the original pre-deformation configuration of the materials. In the crimping process, this elastic spring back occurs when the crimp tooling is removed. At this point the crimp barrel and the strands tend to return to their original geometries. The crimp barrel will tend to return towards its original open configuration and length and the conductors will tend to return towards their original cross sections and lengths. The relative spring back of the crimp barrel and conductors will depend on material parameters such as Young's modulus and yield strength and the geometries of the components. The desirable response is that the crimp barrel tends to respond less than the conductor strands, both radially and longitudinally resulting in a radial compressive stress and a tensile longitudinal stress on the strands. This compressive stress would, in turn, provide mechanical stability to the individual contact areas within the crimped connection.

Mechanical Confinement

If the residual force systems described in the previous section are realized, they will provide a confinement mechanism to retain the integrity of the contact interfaces. A second confinement system can also be created. Note in the crimped cross-section of Figure 1.161, that the wings of the crimp barrel come and close against each other to form the crimped connection. This in itself provides a confinement system for the conductor strands. The geometry of the crimp tooling to form the wings into the crimped connection configuration as well as the material and dimensional characteristics of the crimp terminal will determine how successfully this confinement mechanism can be realized in practice.

Serrations

Crimp barrel serrations provide an independent mechanism for mechanical stability of crimp contact areas. Figure 1.168a, show four serrations in a crimp barrel terminal. Serrations provide two benefits. First, as the conductor strands deform into the serrations they wipe along the edges of the serrations. This wiping action displaces surface films and enhances contact area formation. Second, the conductors are deformed into and interlocked between the serrations, as shown in Figure 1.168b. This interlocking mechanism provides additional mechanical stabilization in the crimped connection to preserve the integrity of the multiple metal-to-metal contact interfaces in the crimped connection.

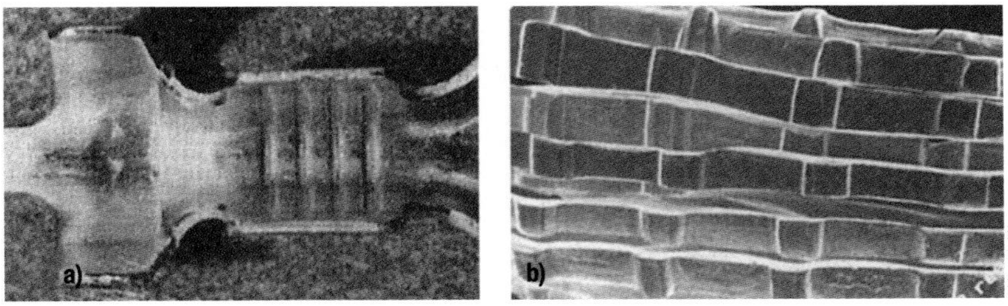

TCO-150-EN

Fig. 1.168: a) Serrations in a crimp barrel / b) Conductor deformation into crimp barrel serrations

To enhance the potential for all of these mechanical stabilization possibilities the crimping process and deformation must be tightly controlled. This is, of course, the intent of the crimping system of wire, terminal and tooling. The wire and terminal combination determines the cross sectional area going into the tooling and the tooling design determines the final cross sectional area of the crimped connection, that is controls the amount and geometry of crimping deformation.

Crimped connection geometries

As would be expected, crimped connection geometries are different for open and closed barrel terminals. Figure 1.169 schematically illustrates some common closed barrel crimped connection geometries. Recall that the "b", Figure 1.169a, crimp is the most common geometry and therefore most standard for open barrel terminals and dates back to the 1950's.

Common closed barrel crimps are the "indentor" crimp (Figure 1.169b), "confined" crimp (Figure 1.169c) or "4 indent" (Figure 1.169d), which is used in military or aerospace applications.

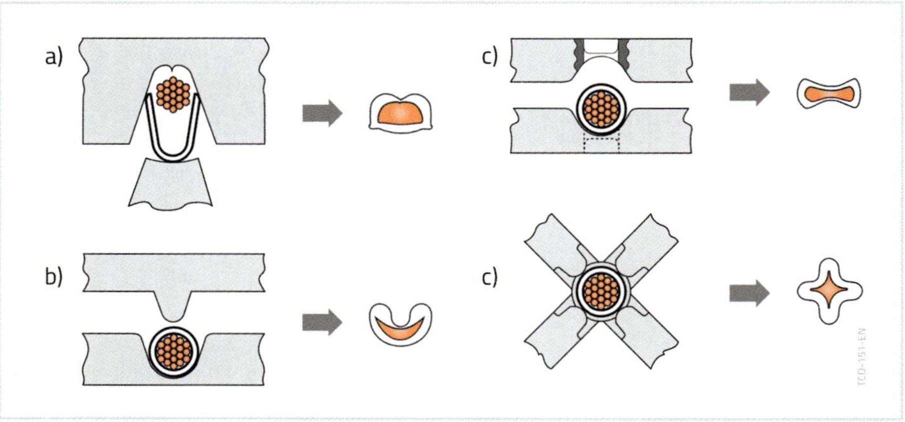

Fig. 1.169: Open barrel "B" crimp and types of closed barrel crimp cross sections

The W crimp dates back to the 1940, and is still commonly used in larger wire sizes and for high current applications. The indent crimp geometries are more recent and are used in military and aerospace applications.

Consider now the crimping system as a whole. Recall that the fundamental purpose of the crimping system is to control the overall deformation of the conductors and the crimp barrel. The following discussion covers the approach to determining an appropriate crimp height, the critical process control parameter for crimping.

Figure 1.170 illustrates the basic factors in the control of deformation during crimping.

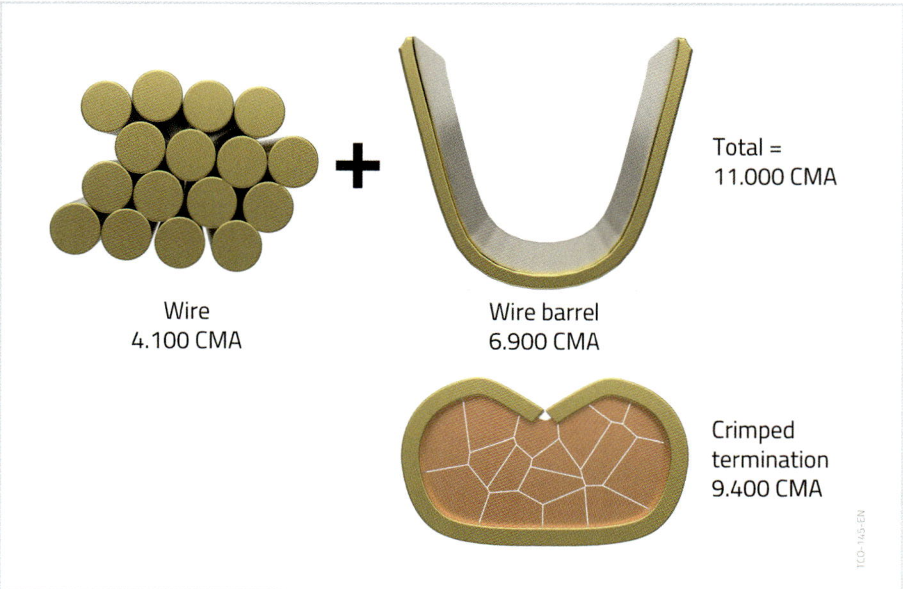

Fig. 1.170: Crimped connection area factors

In the early days of crimp technology, the cross sectional areas of a crimped connection were calculated in terms of Circular Mil Area (CMA) which is calculated by squaring the diameter of the conductor. For example, the CMA (1 CMA, Circular Mil Area is a unit of area equal to that a circle whose diameter is one Mil) of the 16-strand wire shown in Figure 2.48 is given as 4100 CMA, so the CMA of an individual strand is 256 CM yielding a strand diameter of 16 mils (26AWG strands/22AWG wire). The CMA of the cross-section of the crimp barrel, given as 6900 CMA must be calculated from the dimensions of the strip stock and the crimp barrel cross section. This calculation is generally accomplished by using the crimp barrel dimensions noted and the conversion factors:

1 Circular mil **= 0.7854 x 10^{-6} Square Inch**

= 5.066 x 10–10 m^2 = 5.066 x 10–6 cm^2

Figure 1.170, the total CMA of conductor and crimp barrel is 11000 CMA. The final crimped cross-section area is given as 9400 CMA, yielding an Area Index ($CMA_{wire\text{-}terminal}/CMA_{crimped\ connection}$) of 85 % for this example. Recall from the main text that an 85 % area index is at the high end of the design zone.

Component Selection

Wire

The majority of crimped connection applications use stranded annealed copper conductors. Within a given wire gauge the number and diameter of the individual strands can vary. For example, a solid 16AWG conductor has a cross-sectional area of 2,03 x 10^{-3} in^2 while the range of cross-sectional areas for stranded 16AWG conductors is (1.91–2.22) x 10^{-3} in^2. As mentioned earlier, wire gauge selection is primarily based on the current capacity for the series of wire gauges.

AWG	Conductor diameter (solid wire)		Area
	(in)	(mm)	(mm^2)
0000 (4/0)	0.4600	11.684	107
000 (3/0)	0.4096	10.405	85.0
00 (2/0)	0.3648	9.266	67.4
0 (1/0)	0.3249	8.251	53.5
1	0.2893	7.348	42.4
2	0.2576	6.544	33.6
3	0.2294	5.827	26.7
4	0.2043	5.189	21.2
5	0.1819	4.621	16.8
6	0.1620	4.115	13.3
7	0.1443	3.665	10.5
8	0.1285	3.264	8.37
9	0.1144	2.906	6.63
10	0.1019	2.588	5.26
11	0.0907	2.305	4.17
12	0.0808	2.053	3.31

AWG	Conductor diameter (solid wire)		Area
	(in)	(mm)	(mm^2)
13	0.0720	1.828	2.62
14	0.0641	1.628	2.08
15	0.0571	1.450	1.65
16	0.0508	1.291	1.31
17	0.0453	1.150	1.04
18	0.0403	1.024	0.823
19	0.0359	0.912	0.653
20	0.0320	0.812	0.518
21	0.0285	0.723	0.410
22	0.0253	0.644	0.326
23	0.0226	0.573	0.258
24	0.0201	0.511	0.205
25	0.0179	0.455	0.162
26	0.0159	0.405	0.129
27	0.0142	0.361	0.102
28	0.0126	0.321	0.0810
29	0.0113	0.286	0.0642
30	0.0100	0.255	0.0509
31	0.00893	0.227	0.0404
32	0.00795	0.202	0.0320
33	0.00708	0.180	0.0254
34	0.00630	0.160	0.0201
35	0.00561	0.143	0.0160
36	0.00500	0.127	0.0127
37	0.00445	0.113	0.0100
38	0.00397	0.101	0.00797
39	0.00353	0.0897	0.00632
40	0.00314	0.0799	0.00501

Tab. 1.5: AWG

Conductors may be bare (un-plated) or plated, generally with tin, depending on the application. The performance of bare and tin-plated conductors is comparable with respect to crimping. For completeness it should be noted that the insulation type and thickness is also variable across a given AWG. The insulation does not affect the crimping system because it is stripped prior to crimping, but it can influence the design of crimp terminal strain relief features. Insulation stripping practice varies from hand stripping of individual wires to semi-automatic stripping to high volume automatic stripping within an overall crimping system. While wire stripping is simple in principle, care must be exercised to ensure that the insulation is cleanly cut to the appropriate strip and that the individual conductors are not damaged during insulation stripping. The two most important examples of conductor damage are cutting of individual strands and displacement of individual strands from the conductor bundle as will be discussed later. The effect of cut or displaced strands is to reduce the amount of conductor area that is inserted into the crimp barrel which, in turn, will change the deformation of the conductor/crimp barrel during the crimping process. The stripping length has to respect the specifications given for each individual terminal.

Terminal

A WR-WTB plug crimp terminal is shown in Figure 1.171.

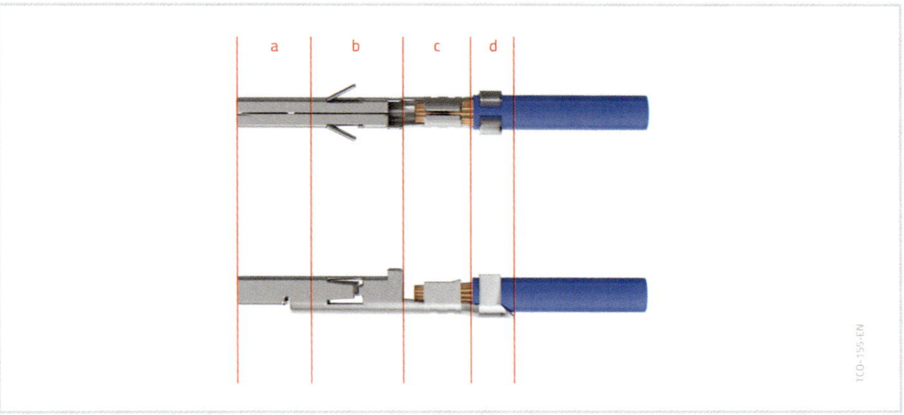

Fig. 1.171: WR-MPC4 crimp terminal

This terminal has a separable interface beam (a), a transition zone (b), a crimp barrel (c) and an insulation grip (d). The separable interface contact design, (a), is application dependent and must not be compromised during the crimping process. The function of the transition zone, (b), is to ensure this independence and, in some designs, to provide latching features to secure the terminal in the housing. The terminal shown in Figure 1.171 has an open crimp barrel, (c), so that the conductors can be placed in the crimp barrel from above. This mode of entry facilitates mechanization of the conductor insertion process and allows for higher crimping rates. There are several designs of insulation grips, (d), including a crimp and a variety of wrap around grips. Insulation grip designs are common to many terminals and are independent of the design of the crimp barrels and separable contact beams.

Tooling

As noted previously, tooling varies from hand tools to fully automatic high speed presses for high volume applications. A variety of Würth Elektronik hand tools is shown in Figure 1.172.

Fig. 1.172: Examples of Würth Elektronik hand crimping tools

Hand tools are tailored to a crimping system by exchanging the crimping dies in the hand tool body. Some hand crimping tools offer adjustable crimping force and include a stepped ratchet system that ensures that the crimping process is carried to completion. If it is not completed, the tool will not open.

For higher volume applications a semi-automatic or automatic crimping machine with interchangeable dies may be used.

As per the Figure 1.173, an automatic crimping machine needs to have a mini applicator that is made up of a base tool (generally common to all machines) and a die set (dedicated to a terminal).

Fig. 1.173: Mini applicator

Crimped Connection Inspection Practices

Before beginning the discussion of crimp inspection practices a discussion of wire stripping practices is necessary. There are three commonly used methods for wire stripping, mechanical, thermal and chemical. While the practice of each of these methods is significantly different, the criteria with respect to the resulting stripped conductors are common. As noted earlier, two of the most important aspects of wire stripping practice are control of the insulation stripping and strip length and avoidance of conductor damage or displacement from the conductor bundle. The insulation strip length is specified for reasons to be discussed. Limits on conductor damage or displacement are also specified. Conductor damage includes nicking or cutting of individual strands as shown in Figure 1.174.

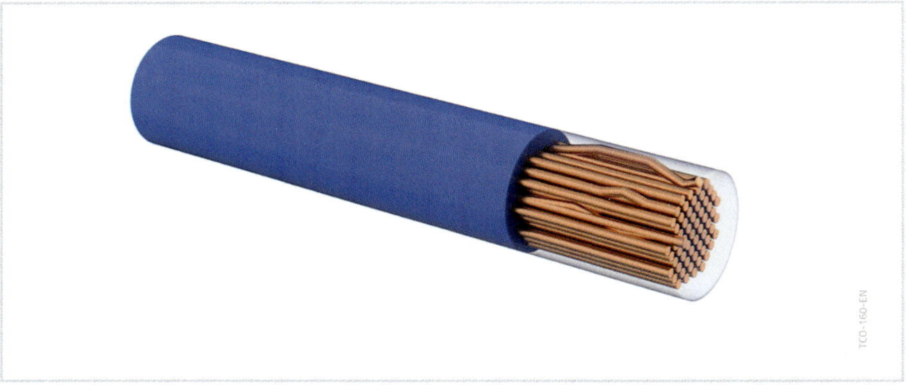

Fig. 1.174: Nicking and cutting strands

Nicking of strands makes the conductor more susceptible to failure due to mechanical stresses applied to the crimped connection during its application lifetime, in particular, vibration related stresses. Cutting of strands reduces the amount of conductor material inserted into the crimp barrel and, therefore, the amount of deformation produced during crimping which, in turn, can compromise the crimped connection performance in the field. Displacement of conductor strands from the conductor bundle, as shown in Figure 1.175, has a similar effect in reducing the amount of conductor material inserted into the crimp barrel.

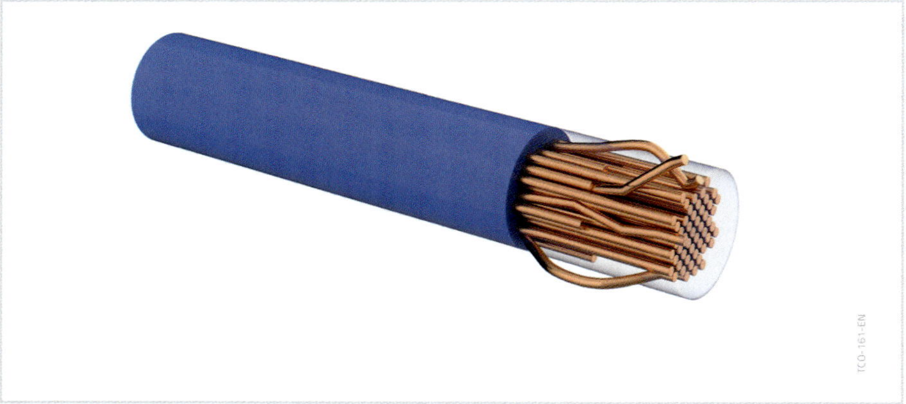

Fig. 1.175: Displaced strands

Inspection of crimped connections performs two functions. First, it validates that the crimping process was properly performed. Second, the data obtained may be used to monitor any trends and/or deviations in the crimping process over time. There are three different crimp inspection methodologies, visual, to ensure the crimping process was followed, crimp height, to ensure that the crimping deformation was properly controlled and pull strength, to validate the consistency of mechanical performance. Consider each separately.

Visual Inspection

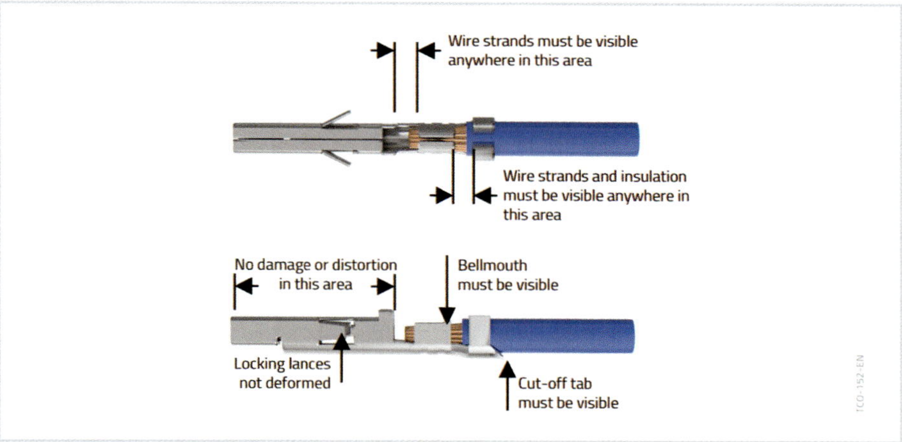

Fig. 1.176: Visual crimp inspection

The comments in Figure 1.176a) evaluate the stripping of the insulation from the wire and the placement of the stripped wire into the crimp barrel. The requirement that wire strands and insulation be visible between the insulation barrel and the crimp barrel is to ensure that the insulation has been properly stripped and the wire properly inserted into the barrel. If no wire strands are visible the strip length may have been too short and insulation may have been included in the wire barrel on insertion. If this is the case, the insulation will be extruded throughout the crimp barrel during crimping and will compromise the mechanical and electrical performance of the crimped connection. If no insulation is visible, the wire strip length may have been too long so that the insulation does not fill the insulation crimp barrel and the strain relief function of the insulation crimp is compromised.

In addition to these comments, there should be no visible cut or nicked strands and no individual strands visible outside the crimp barrel. Defects of this type should have been identified earlier in the process after wire stripping.

The requirement that conductor strands be visible between the crimp barrel and the contact section of the terminal has a similar function. If no conductor strands are visible the strip length may have been too short. The conductor strands should also not extend across the transition zone into the contact area which would indicate the strip length was too long.

The comments in Figure 1.176b) are intended to ensure that the terminal was not damaged during the crimping process. No damage or distortion of the contact area is allowed because the insertion of the pin contact and, therefore, the contact force may be compromised. Damage to the locking lance, if any, may affect the mechanical stability of the contact in the housing and, subsequently, the mechanical stability of the contact interface due to vibration and shock in the application environment. The check for the cut-off tab is to verify that the insulation crimp barrel has not been damaged with potential effects on its strain relief function.

The comment "bellmouth must be visible" is a check to ensure the proper location of the crimp zone in the crimp barrel as shown on Figure 1.177.

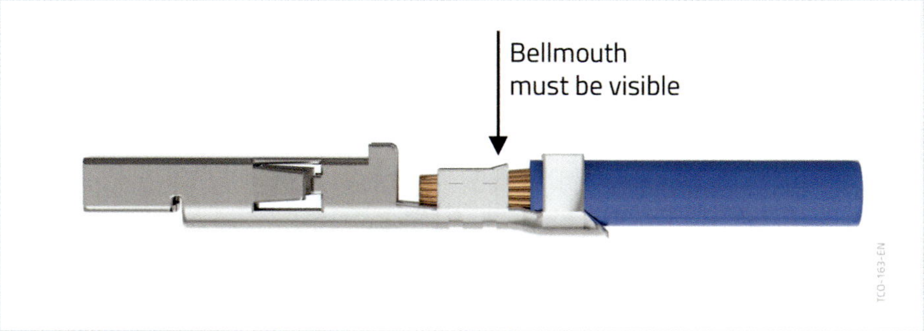

Fig. 1.177: A proper bellmouth

The bellmouth also has a functional role in that it reduces the potential for the crimp tooling to cause nicking or cutting of strands during the crimping process. A bellmouth is generally required at the conductor end of the crimp barrel with a bellmouth width of the order of one or two times the metal thickness used in the crimped barrel.

A final visual inspection criterion is that the crimped connection does not have a "banana" shape. Excessive longitudinal extrusion of the crimp barrel during crimping, resulting in the "banana", could compromise the mechanical properties of the crimp in addition to making insertion of the contacts into the housing problematic. An additional extrusion effect is "flash" at the base of the crimp barrel due to extrusion of the crimp barrel material between the walls of the crimping die and the anvil. Excessive flash is an indication of worn tooling.

Crimp Height

Crimp height is arguably the most important crimped connection inspection practice. As noted previously control of the crimp deformation is critical to crimped connection performance and crimp height provides a means to both control and monitor that deformation. Figure 1.178 schematically illustrates the methodology of crimp height measurement. The crimp height should be measured at the highest point on the base of the crimp terminal. Crimp height tolerances depend on wire size.

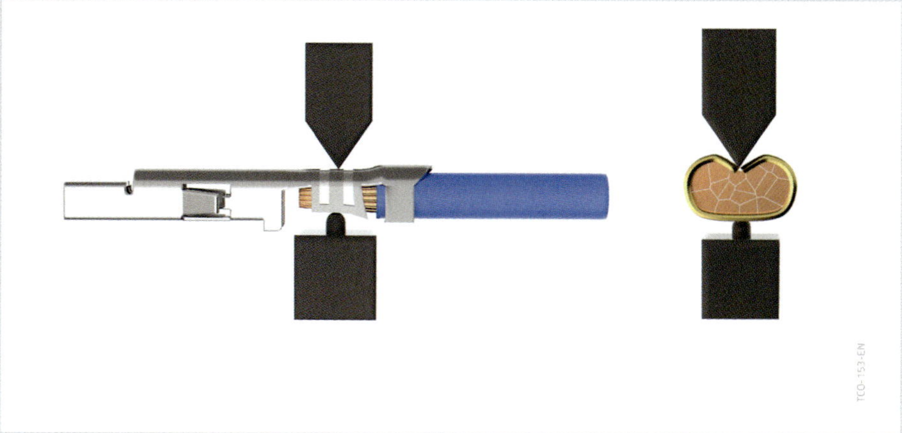

Fig. 1.178: Crimp inspection – crimp height

From Figure 1.178 it can be seen that the two key measurement requirements are the alignment of the crimp barrel wings against the micrometre flat and the positioning of the micrometre tip at the high point on the base of the crimp barrel. Because crimp height tolerances can be as tight as one or two mils, precise measurements are required.

Crimp height has been emphasized because it is an important process control parameter. Crimp width, the distance across the base of the crimp barrel, is also important in determining the crimped connection cross section and, therefore, amount of deformation that occurs during crimping. Crimp width, however, is not a variable in the sense that it is controlled by the design of the crimp tooling and the crimp height.

It should be noted that crimp height measurements are made continuously during high volume automatic crimping production. Deviations from crimp height specification cause the machine to shut down. In some equipment crimp force is similarly monitored.

Pull Force Testing

There are two approaches to pull force testing of crimped connections, pull to failure and pull to specified force value. Pull to failure testing is obviously destructive and may be done periodically to confirm conformance to pull test requirements. Pull force criteria are wire size and application dependent. In either case, care must be taken to ensure that the applied load is uniformly distributed among the conductors in the crimped connection. A commonly used practice to achieve this distribution is to remove or open the insulation crimp and wrap the wire around a mandrel prior to applying the load. Specialized gripping means have also been developed for this purpose.

If the pull force testing is done to failure, the failure criteria will be dependent on the wire gauge. A crimped connection failure typically approaches the wire breaking strength. As noted in the text, in samples pulled to failure the failure mechanism depends on the crimp height used for the crimping process and whether the deformation is on the increasing or decreasing side of the pull strength versus deformation curve. In some cases, the failure mechanism is the basis for acceptance of the crimping process.

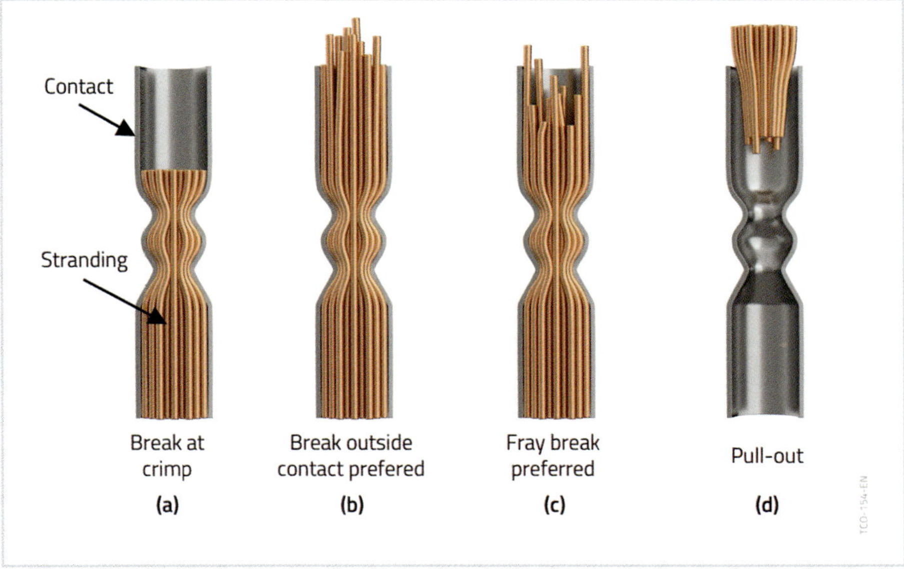

Fig. 1.179: Crimping Pull out strength

Figure 1.179 shows four different failure mechanisms. In a) the conductors break in/at the crimp zone. Such failures may indicate excessive deformation, and work hardening, of the conductors during crimping. In b) and c), the conductors break away from the crimp zone indicating that the breaking strength of the conductors was realized. In d), the conductors pull out of the crimp barrel indicating insufficient deformation during crimping. Pull testing values may also be indicative of cut or nicked strands during insulation stripping. Table. 1.6 shows the tensile strength.

Conductor cross section			Tensile strength
AWG	mm	mm^2	(N)
30	0.25	0.05	6.7
28	0.32	0.08	13.4
26	0.40	0.13	22.3
24	0.51	0.20	35.6
22	0.64	0.32	53.4
20	0.81	0.52	89
18	1.02	0.82	142
16	1.29	1.31	222.3
14	1.63	2.08	311.5
12	2.05	3.31	489.5
10	2.59	5.26	667.5
8	3.26	8.37	978.6

Tab. 1.6: Tensile strength/pull out force (Value per IPC/WHMA-A-620A)

Conclusion

Crimped connections remain a consistent and cost-effective approach to terminating wire and cable when the crimping system of:

Wire: selected primarily for current carrying capacity
Terminal: appropriate to selected wire gauge and
Tooling: selected for the wire/terminal combination is followed.

Once the crimping system is in place the primary process control parameter is the crimp height which ensures the proper amount of deformation of the system to establish and maintain the intended electrical and mechanical performance.

The key point of this discussion of crimped connection technology is to highlight the importance of the crimping system of wire/terminal/ tooling as a means to ensure consistent control of the deformation during the crimping process. Controlled deformation, in turn, ensures the creation and integrity of the multiple metal-to-metal contact areas created between the conductor strands and between the strands and the crimp barrel during crimping.

1.16.2 Insulation Displacement Connections

Insulation Displacement Connectors (IDC) are used primarily on cables, in particular ribbon cables, in cable-to-cable or cable-to-board applications. Prior to IDC technology contacts were individually loaded into multi-position connectors by a process known as crimp-snap. For, say a 60 position connector, the crimp-snap process required handling 60 individual wires, each intended for a specific position in the housing, stripping the insulation, crimping a contact onto the conductor and inserting the contact into the appropriate cavity in the connector housing.

In contrast using IDC technology a 60 conductor ribbon cable is oriented and inserted into a 60 position housing preloaded with IDC contacts and all of the conductors are terminated with a single stroke of the IDC tooling. IDC production rates are much higher given that a single cable is handled rather than 60 individual wires; no insulation stripping is required as compared to 60 individual insulation stripping operations; all conductors are terminated simultaneously rather than 60 individual crimping operations and the terminated IDC contacts are simultaneously inserted into the pre-loaded housings as compared to insertion of 60 individual crimped contacts into the appropriate cavity in the connector housing.

The improvement in production rates is obvious from Figure 1.180. But there is more. In the crimp-snap process the individual wires must be selected and inserted into the proper cavity in the housing – 60 opportunities to make a wiring error. In contrast, the ribbon cable needs only to be placed "right side up" into the preloaded connector housing to eliminate wiring errors. It is easy to understand why IDC technology dominates in cable-to-cable and cable-to-board applications.

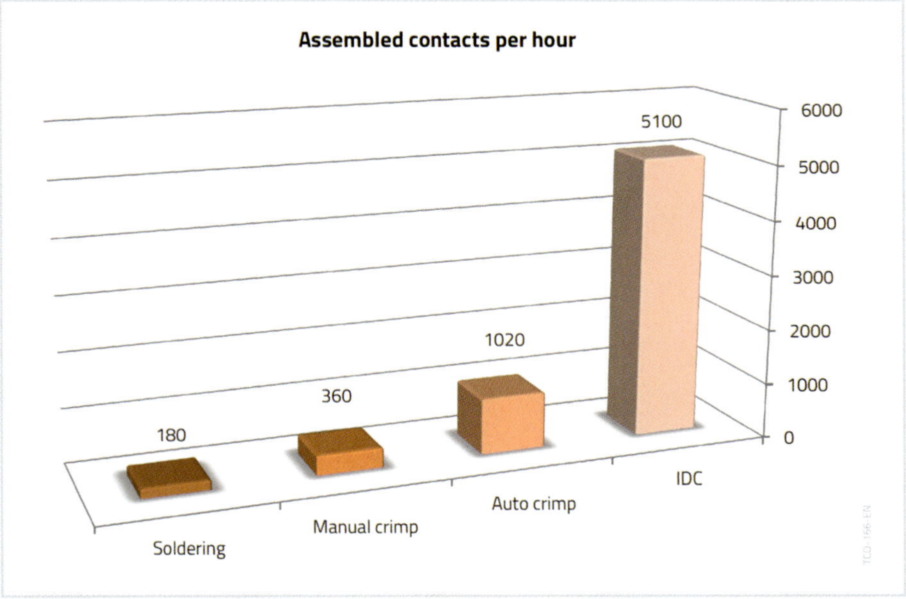

Fig. 1.180: Process efficiency comparison

The Insulation Displacement Connection Process

IDC connections must meet the same requirements for permanent connections as crimped connections an acceptable and stable value of connection resistance. In other words, an appropriate area of metal-to-metal; contact and a contact force system to ensure the integrity of the connection interface under the mechanical and thermal stresses of the application environment. The system for creating and maintaining the metal-to-metal contact interfaces is quite different from that of crimping.

Figure 1.181 schematically illustrates the insulation displacement process and the creation of the metal-to-metal contact interfaces in an IDC contact that has both an insulation displacement slot and an electrical contact slot. This design was chosen for ease of discussion and many IDC contacts have a single slot with a more gradual transition between the insulation displacement and electrical termination functions.

Figure 1.181a shows the wire as it is being inserted into the insulation displacement portion of the IDC contact. In Figure 1.181b , the insulation has been displaced and the conductor is moving downward into the electrical contact portion of the IDC contact.

In Figure 1.181c the conductor is at its final position in the contact slot. Figure 1.181d shows a cross section of the conductor at its final position in the contact slot. Consider the processes taking place at each position.

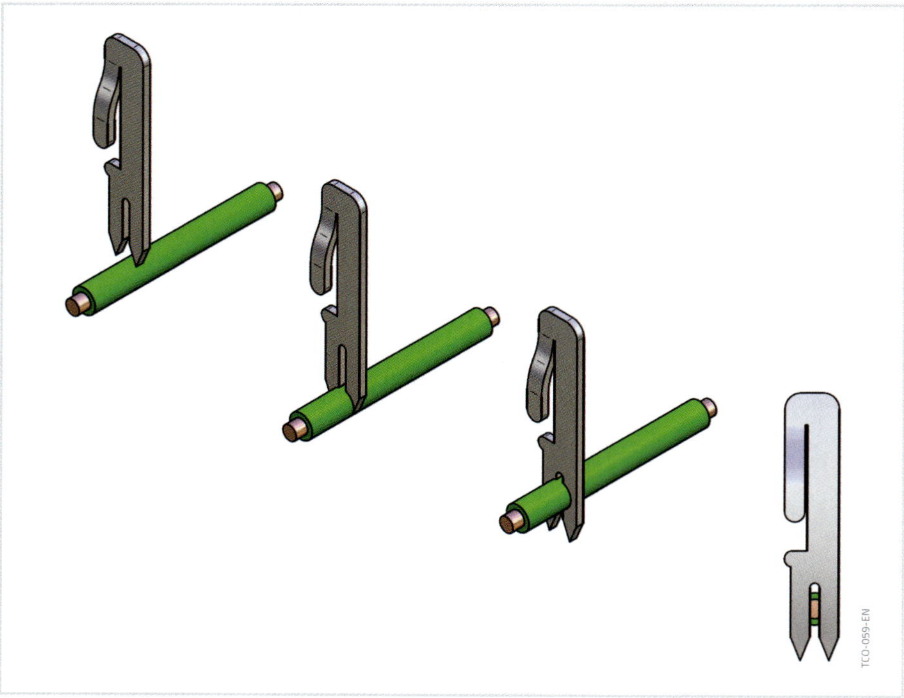

Fig. 1.181: IDC process

Figure 1.182a provides a more detailed schematic view of the insulation displacement process. The width of the gap in the insulation displacement slot is wire size dependent and is intended to compress the insulation and create friction forces between the insulation and the sides of the slot. The ramp angle and roughness of the slot can significantly influence these friction forces. In addition, some slot designs coin edges, along the ramp, or create edges at the ends of the ramp to initiate the separation of the insulation from the conductor. There are two reasons why complete displacement of the insulation is important. First, electrical contact, as noted, requires metal-to-metal interfaces. Second, if insulation is carried into the electrical portion of the IDC slot it can compromise the contact force as will be discussed.

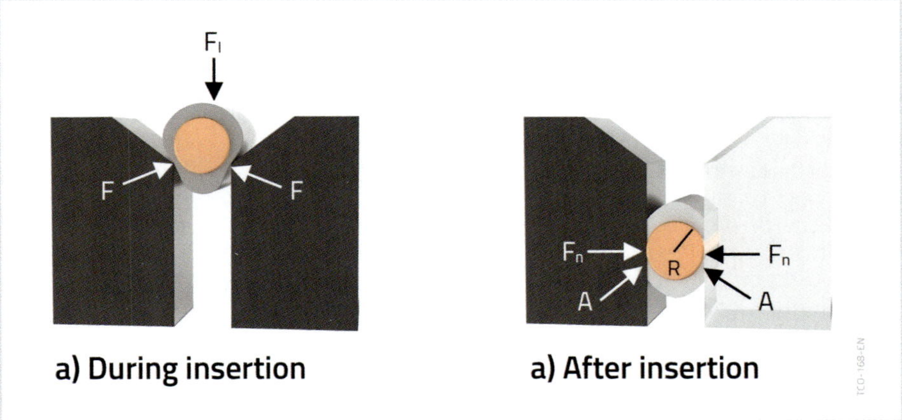

Fig. 1.182: IDC insertion kinetics

Consider now Figure 1.182b as the conductor moves downward into the electrical contact portion of the slot. The slot design shown consists of two cantilever beams. The conductor will displace both beams creating a force between the conductor and the beams. This force will lead to deformation of the conductor and a change in conductor cross-section as it is inserted into the slot. This deformation creates two contact areas, one against each beam. The contact areas, A, in Figure 1.182 are determined by the deformation of the conductor and the stock thickness of the material from which the contact is stamped. A guideline for an appropriate contact area is that it approximates the cross-sectional area of the conductor being terminated. The wiping action during the insertion of the conductor into the electrical slot enhances the metal-to-metal area in addition to the effects of increasing contact force. The magnitude of IDC contact forces will be discussed in the next section.

IDC Terminal Designs

Figure 1.183 illustrates three variations of an IDC slot design. Recall that for a cantilever beam the contact force due to beam deflection is given by:

$$F=\left(\frac{1}{4}\right) E\, D\, w \left(\frac{t}{l}\right)^3 \qquad \text{(EQ 19)}$$

F: contact force [dimension]
E: elastic modulus of the base metal of the contact [dimension]
D: deflection of the contact beam [dimension]
w: width of the beam
l: length of the beam
t: thickness of the beam

In case of the IDC design in Figure 1.183a, l is the length of the contact from the base of the slot to the point on the beam in contact with the conductor, w is the thickness of the contact beam and t is the dimension of the contact geometry at the base of the IDC slot.

Note that for IDC beam deflections there is an interaction between the beam deflection and the position of the conductor in the slot, which is l, the effective length of the beam. The beam deflection will be given by the reduced diameter of the conductor due to the deformation that occurs as the conductor slides down into the slot.

The parameter t, the effective thickness of the cantilever beam is also more complicated. Rather than deflecting at the attachment point of a cantilever beam, the IDC beam deflection varies with the curvature of the base of the IDC slot. Figure 1.183 shows some variations in the geometry of the base of an IDC slot. Figure 1.183a shows a straight slot. The beam deflection, and the stresses in the beam, will be distributed primarily around the base of the slot. Figure 1.183b shows a straight slot with a circular opening added at the base of the slot. This configuration is sometimes called a "keyhole" slot. The effective thickness of the beam is decreased by the addition of the circular opening and the deflection and associated stresses are distributed over a longer length of the beam yielding a more compliant structure. Figure 1.183c introduces a varying thickness along the length of the IDC beam to further spread out the deflection and stresses. This geometry stores the most elastic energy of the three designs and makes the most effective use of the beam material.

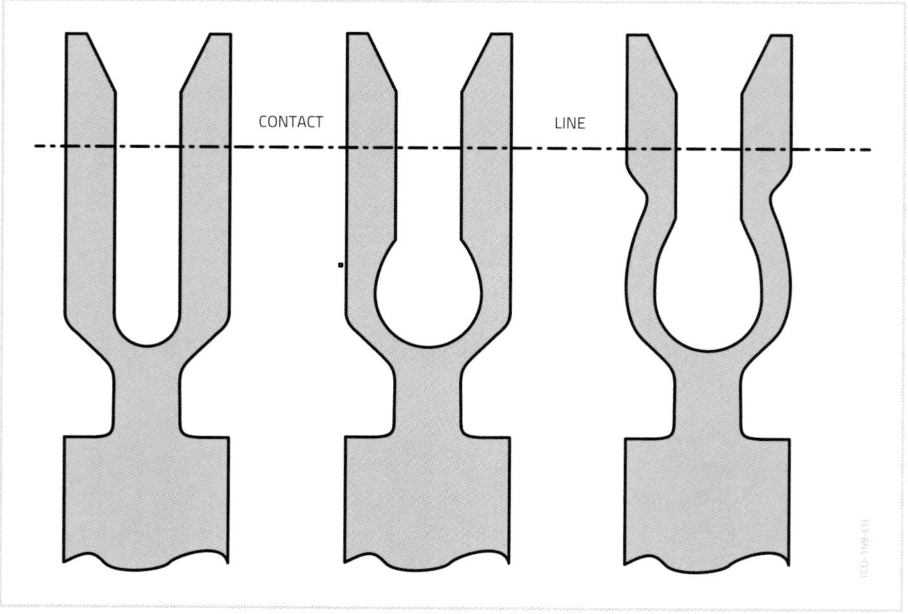

Fig. 1.183: Beam deflection depending on contact shape

While the mathematics will be different, each of these designs will have a relationship between the beam deflection and the contact force exerted by the deflected beam on the deformed conductor positioned in the slot. Figure 1.184 shows an example of such a relationship for a straight slot IDC. The curve for beam deflection is shown as linear which would be the case if the deformation of the beams was elastic. The band around the beam deflection curve represents the tolerance in control of the slot width. The curve for conductor deformation shows considerable curvature to represent the plastic deformation of the conductor as it is pushed down into the slot and deforms as it deflects the IDC beams. Once again, the band around the conductor deformation curve represents the tolerance in conductor diameter during manufacturing. The "window" where the two tolerance bands overlap shows the expected range in contact force and conductor deformation. The values for contact force and deformation depend on the conductor size and IDC slot/beam geometries. The conductor deformation is typically somewhat larger than for crimping, nominally 25 % reduction in area. Contact forces are typically much higher than in separable connections, 500 to a few thousand grams compared to 50 to a few hundred for separable interfaces. These forces are sufficient to provide the desired mechanical stability.

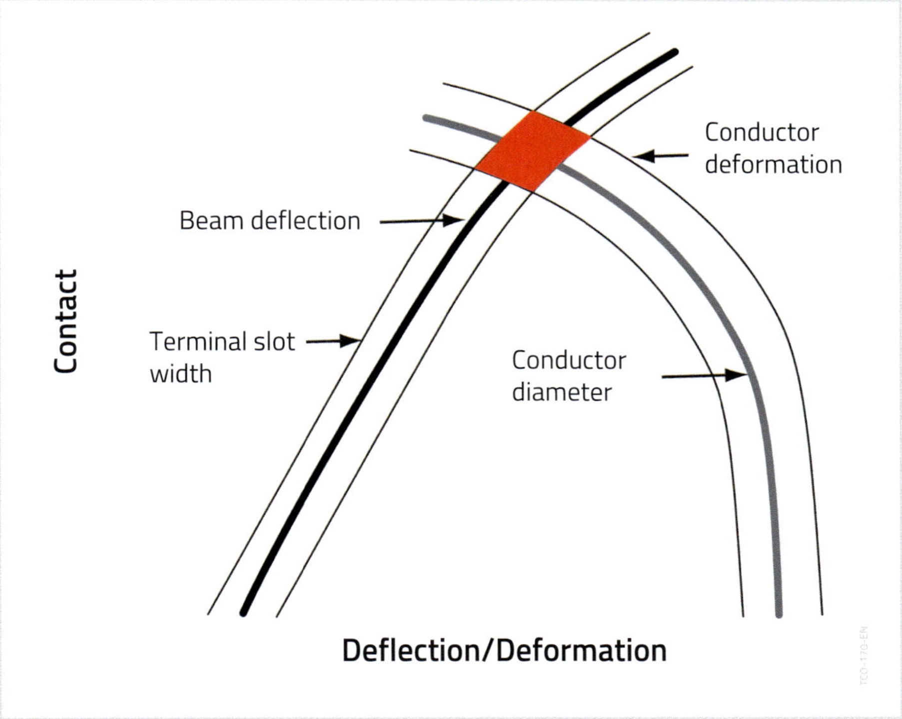

Fig. 1.184: Contact force vs. contact deflection/conductor deformation

While many IDC slots are flat stampings there are also stamped-and-formed dual slot options. Figure 1.185 shows examples of two dual slot contacts. In these designs both of the slots may have an electrical function, or the outside slot, the slot away from the separable contact interface, may be a strain relief slot. The dual electrical slot option will provide enhanced electrical performance due to parallel and redundant contact interfaces. The strain relief option enhances mechanical stability.

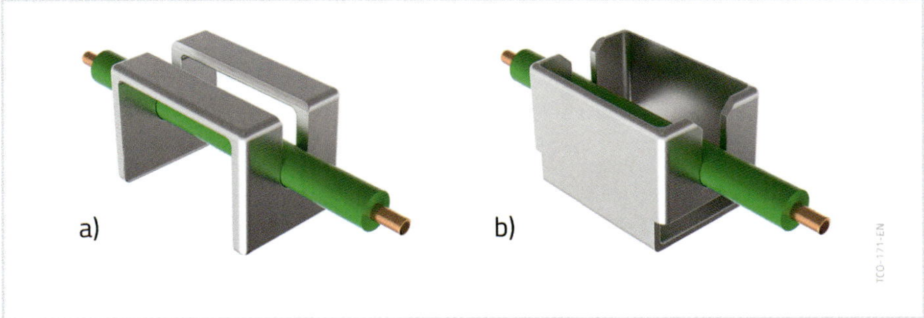

Fig. 1.185: IDC dual slot contacts

The two designs shown in Figure 1.185 are quite different in deformation mechanics. The slots in Figure 1.185a will show similar performance to the stamped slots previously discussed with the exception of contributions due to the coupling link between the slots. The slotted box of Figure 1.185b has a smaller degree of coupling between the two IDC beams than the design of Figure 1.185a. The deflection of the beams takes place along the base of the side walls of the box and will include additional effects due to flexure of the side walls.

Solid and Stranded Conductors

The figures used in this discussion have all shown solid conductors for ease of visualization. In practice, however, stranded conductors, primarily in ribbon cables, outnumber solid in IDC applications and generally meet the same specifications. The difference between solid and stranded conductors in IDC applications is obvious.

The deformation of solid conductors is more predictable and controllable than that of stranded conductors because the stranded conductors can move relative to each other as they are inserted into the slot. Such movements change the effective diameter of the stranded conductor bundle, therefore, the deflection of the beams and, in turn, the contact force. Figure 1.186 schematically illustrates the deformation of solid and stranded (seven strand) conductors, of the same wire size, terminated in the same IDC slot design. The tendency towards such conductor reorientation is reduced by the concentric wire construction and the lay of the outer layer of conductors as they wrap around the central conductor. In addition, the residual wire insulation and insertion tooling design, to grip the wire circumferentially or at two points during wire insertion also reduces the potential for conductor reorientation during insertion.

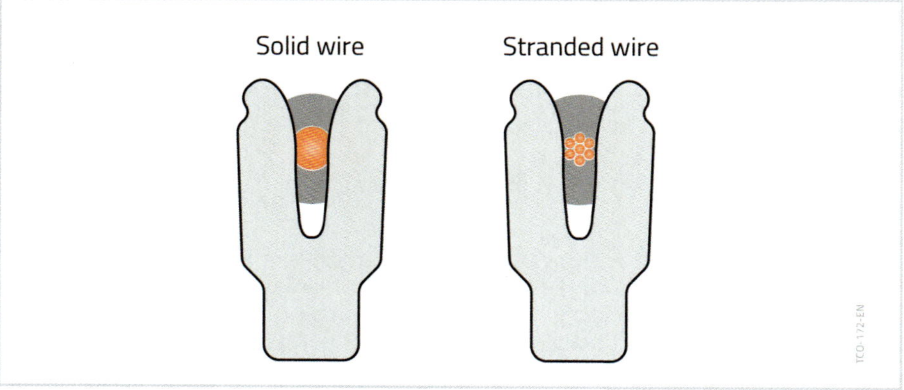

Fig. 1.186: IDC cross section: Conductors deformation

The beam deflection caused by the solid conductor is entirely due to the deformation of the conductor creating two contact areas with the IDC terminal. In contrast the stranded conductor has changed its shape from the original close packed structure, with the six outer conductors abutting the centre conductor, to a more open ovalized configuration. Note that four of the seven strands have been deformed against the IDC beams with little strand-to-strand deformation. The contact area produced in a stranded IDC connection appears to be less than that for a solid conductor. It should also be noted that stranded conductors are more prone to cut or broken strands during insertion. The most significant impact of such defects will be a possible reduction in contact force.

Despite these concerns, stranded conductor IDC connectors are able to meet contact area and contact force requirements as evidenced by the inclusion of both solid and stranded conductors as options in many IDC connector product specifications and qualification programs.

Plated and Unplated Conductors

Conductors for IDC applications are generally annealed copper with an elongation of greater than 10 % to ensure ductile deformation of the conductors on insertion. As noted, wire sizes are application dependent, but solid conductors of 0.25 to 3.6 mm nominal diameter and seven strand concentric constructions of 0.05 to 10 mm^2 cross-sectional area are preferred, although higher strand counts have been used.

AWG Gauge	Diameter (mm)	Section (mm^2)
9	2.906	6.634
10	2.588	5.264
11	2.305	4.172
12	2.053	3.309
13	1.828	2.624
14	1.628	2.081
15	1.450	1.650
16	1.291	1.309
17	1.150	1.038
18	1.024	0.823
19	0.912	0.653
20	0.812	0.518
21	0.723	0.410
22	0.644	0.326
23	0.573	0.258
24	0.511	0.205
25	0.455	0.162
26	0.405	0.129
27	0.361	0.102
28	0.321	0.081
29	0.286	0.064
30	0.255	0.051
31	0.227	0.040

Tab. 1.7: AWG – mm conversion

Solid conductors can be used without plating, but plated stranded wires are preferred. Platings include tin and silver, for general and high frequency applications respectively.

Strain Relief

Despite the high contact forces of IDC connections, external forces applied parallel to the IDC slot axis can disturb the conductor. Protection against such forces can be important. One of the most popular type is the external strain relief used for 2.54 mm box header (WR-BHD) to avoid breaking IDC connection while operator is pulling directly the flat cable for easier handling (instead of pulling connector).

Fig. 1.187: Strain relief example

Summary

To summarize, how does the IDC process address the fundamental permanent connection requirement to create and maintain the integrity of metal-to-metal contact interfaces? Two, or more, contact areas are created as the conductors wipe against the sides of the IDC slot under the force created by the deflection of the slot beams. The contact area is given by the combination of the deformation of the conductors and the thickness of the IDC slot. The contact force that maintains the integrity of the contact interfaces is the force due to the deflection of the IDC slot beams by the conductor as it deforms and slides down the slot. A guide for the contact area is that it should approximate the cross sectional area of the conductor. The contact forces in an IDC are an order of magnitude higher than those of separable connections to provide the needed mechanical stability. Strain relief features are necessary to resist forces applied parallel to the IDC slot. Both solid and stranded conductors are suitable with plating being recommended for stranded conductor applications.

1.17 Separable Interface with PCB

1.17.1 SKEDD Technology

1.17.1.1 Introduction

SKEDD is a technology, developed by Würth Elektronik, with which connectors can be connected to the circuit board directly without soldering. This type of contact offers considerable advantages over soldered connectors.

The REDFIT IDC (insulation displacement connector) is the first series of connectors from Würth Elektronik which comes with SKEDD technology. In the meantime, renowned companies worldwide trust the advantages of SKEDD technology. In this chapter you will learn about the advantages of SKEDD and suitable applications for using SKEDD.

1.17.1.2 Functionality

At the heart of SKEDD technology are innovatively designed stamped contacts, which provide an electrically stable connection directly in the through-hole plating (via) of the PCB. The prestressing arising during contacting ensures a stable electrical connection even in case of intense vibration.

The use of high-performance alloys as the contact material enables a high contact force, which safeguards the electrical connection given high mechanical requirements. SKEDD technology thus provides a stable electrical connection (Figure 1.188)

Fig. 1.188: SKEDD stamped contacts

The difference to press-fit technique with solid pins

In a solid press-fit connection, the high forces during connection create a gas-tight connection between the contact members. The through-hole plating of the circuit board adapts to the contact through the press-fit process.

SKEDD technology means there is no permanent deformation of the through-hole plating. Only the SKEDD forks are pre-stressed and spring back to their original position once the connection is released. This means that SKEDD is utilizes the elastically section of the copper material (Figure 1.189).

The widest possible elastic range is essential for compensating hole and offset tolerances, whereby an appropriate contact force must be guaranteed. This range is influenced by the selection of suitable materials and a design that forms the ideal compromise between rigidity and elasticity.

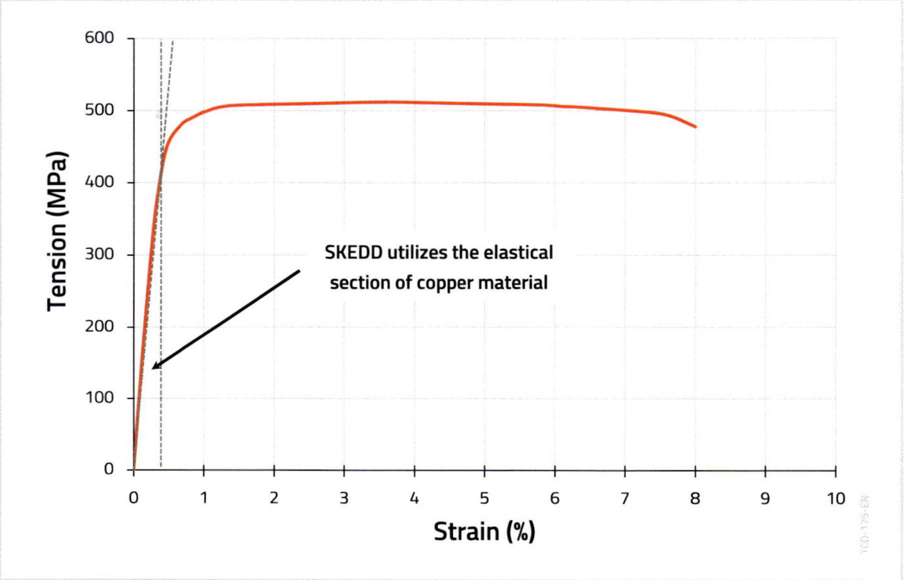

Fig. 1.189: Tension and elongation behaviour of copper material

Vibration resistance

Thanks to the low insertion forces (compared to conventional press-fit contacts), SKEDD connectors can be manually inserted into the circuit board. Nevertheless, the spring force of the contacts is sufficient to ensure a very stable electrical connection, even during intense vibration. This is ensured with vibration tests, which test the mechanical stability of the connector on the circuit board as well as the quality of the electrical connection (Figure 1.190).

Fig. 1.190: Test setup for vibration test

For example, the REDFIT IDC is tested for 7.5 hours at accelerations of up to 10 g (maximum accelerations of up to 5 g occurred during the launch of a Space Shuttle). It is ensured that no contact interruptions >1 µs occur even under high mechanical load.

You can find more details about reliability at **www.we-online.com/redfit**

1.18 Connector Testing

1.18.1 Introduction

The purpose of connector testing is to evaluate the performance of a connector in its intended application. Three questions must be answered in order to meet that objective:

- What are the application requirements/conditions?
- What are the connector requirements?
- What are the performance criteria to meet those requirements?

Application Requirements

The connector application environment must be defined in terms of parameters such as the operating temperature range; the severity of the environment with respect to connector degradation mechanisms such as corrosion and mechanical disturbances; and special conditions such as water immersion and high altitude. Other application requirements include the current and voltage, and safety and standards requirements.

Connector Requirements

Connector requirements include, as a minimum, requirements on contact resistance magnitude and stability, current and voltage, mating and durability and environmental stability (thermal, mechanical, and chemical).

Performance Requirements

Performance requirements are obviously application dependent. Consider connector resistance as an example. An industrial application with current requirements of the order of amperes may require an overall connector resistance of several to a few tens of milliohms with a contact resistance stability of the order of ten milliohms. A power application, tens to hundreds of amperes, in contrast, will require an overall connector resistance below a milliohm with a contact resistance stability of a milliohm or less.

Once answers, or, at least, guidelines, for these questions are in hand a testing program can be developed. But first, let's provide some definitions.

1.18.2 Definitions

What is a test?

A test is a combination of an exposure and a measurement.

What is an exposure?

An exposure subjects a connector to a test environment intended to simulate an application environment for a time intended to simulate an application "lifetime". Tests are usually referenced by the exposure. For example, a corrosion test subjects a connector to a laboratory environment intended to drive a particular corrosion degradation mechanism. Example environments include steady state or cyclic humidity and Mixed Flowing Gas (MFG). Humidity exposures are used for tin plated connectors and MFG for noble metal plated connectors.

The two major concerns about exposures are the accuracy of simulation of the intended environment and the relationship between the duration of the exposure and lifetime in the field as will be discussed.

What is a measurement?

A measurement is a quantitative evaluation of a performance parameter. For example, electrical measurements include contact resistance and dielectric withstanding voltage, and mechanical measurements include mating force and retention force. A resistance measurement is made prior to and after any exposure(s) and changes in the resistance, commonly referenced as Δ_R (R), are taken as indicative of performance.

Consider now the following simple example of a test sequence:

- Contact resistance, R_1 (measurement)
- Unmate/mate (x cycles, conditioning) MFG (exposure)
- Contact resistance, R_2 (measurement)

This sequence is a corrosion test, MFG exposure, and change in contact resistance is the measurement. The initial resistance measurement R_1 is the base measurement. The unmate/mate cycles are a conditioning process because durability cycling can lead to wear through of the plating increasing the susceptibility of the contact interface to corrosion. The number of unmating/mating cycles specified depends on the application requirements. The conditioned samples are now exposed to the appropriate application dependent MFG environment to drive the corrosion degradation mechanism. The final step is the second contact resistance measurement, R_2 and the difference between R_2 and R_1 is the performance metric, Δ_R (R) to be compared to an application or specification requirement.

A few more words on the difference between conditioning and exposure may be appropriate. Durability cycling is a conditioning step if it is followed by an exposure which is intended to validate the integrity of the contact plating, the current example. If the following step is a microscope evaluation to check for plating wear through the durability cycling is an exposure because it has caused the degradation process being monitored. Similarly, heat aging (Temperature Life), exposure to elevated temperatures, is a conditioning step if it is intended to drive stress relaxation which reduces the contact force and makes the connector more susceptible to motion and, therefore to corrosion effects. Heat aging is an exposure if it is intended to accelerate the rate of corrosion, that is, to drive the degradation directly.

Now comes the most important question:

What is the relationship between testing and performance?

Some of our previous discussion has indirectly addressed that question. Now we must be more specific. There are two major topics to be addressed. Simulation and acceleration factors.

Simulation

Does the test exposure replicate the selected field degradation mechanism? In other words, after the connector has been subjected to the test environment, the exposure, does the connector experience the same degradation mechanism it experiences in the field.

Acceleration Factor

Is the relationship between test severity/duration and lifetime in the field known? In other words, after the connector has been subjected to the test environment for a specified duration does the connector experience the same degradation as it would in the field after a known application lifetime?

Some examples of exposure environments and duration will clarify these statements.

Durability Cycling

The exposure for durability cycling is mating and unmating the connector. This, of course, is a direct simulation of the field as long as the test samples are fixtured appropriately. The acceleration factor is unity with respect to individual mating cycles. But the mating cycles can be performed repetitively in the tests to simulate the expected number of cycles in the application in a much shorter time period than would be experienced in the field so the exposure has a known acceleration factor. If, for example, a product is specified to have a lifetime of 50,000 hours and is rated for 500 mating cycles, the 500 mating cycles can be performed in the laboratory in, say, 5 hours giving an acceleration factor of 10,000. For completeness, it should be noted that the measurement in this test would be an evaluation for wear through of the contact finish.

Temperature Life (Heat Aging)

As noted previously, temperature life can be a conditioning step or an exposure. With respect to simulation, the simulation is direct, 105 °C is the same whether in the test chamber or in the application environment. The significant issue is acceleration factor. In this regard it is important to note that heat aging can drive a number of different degradation mechanisms. The three most relevant to connector performance are stress relaxation, corrosion and InterMetallic Compound (IMC) growth. These degradation mechanisms are all diffusion controlled and, therefore, vary exponentially with temperature.

A "rule of thumb" for acceleration factors for diffusion-controlled reactions is that the reaction rate doubles for every 10 °C increment in temperature.

That is, the stress relaxation rate at 115 °C is double that at 105 °C. With respect to stress relaxation, data for many materials is available from suppliers to allow calculation of stress relaxation for given time/temperature relationships.

Corrosion

Both simulation and acceleration factors are more problematic in corrosion exposures. The issue with simulation is whether or not the exposure to the laboratory environment produces the same corrosion mechanisms and corrosion products that are experienced in the field. As noted in Chapter II/2.1.2.1 Noble Metal Finish Degradation Mechanisms, a number of copper alloy corrosion products, oxides, sulfides and chlorides, occur in the field. An appropriate corrosion test environment will produce the same mix of corrosion products as is experienced in the field. Acceleration factors for corrosion exposures are more complex because they include contributions due to chemical concentrations, humidity and temperature. The MFG environments developed over years of testing meet the simulation requirement and acceleration factors have been determined empirically for selected environments as will be discussed.

Mechanical Exposures: Shock and Vibration

For mechanical exposures, in particular shock and vibration, there are significant issues for both simulation and acceleration factor. The issues arise from the fact that mechanical disturbances can drive fretting motions which, in turn, can lead to the various fretting related degradation mechanisms discussed in Chapter II/2.1.2.3 Non-noble Finish Degradation Mechanisms. Simulation issues are application dependent. What sorts of excitations are active in the field; shock, short time high magnitude forces, or vibration, repetitive forces of lower and varying magnitude for example.

Acceleration factors are even more problematic given that the frequencies and magnitudes of these events are variable with the application environment. In essence shock and vibration exposures are intended to assess whether a given mechanical disturbance, derived from application conditions, does or does not cause motion of the contact interface.

Let us turn now from this discussion of testing fundamentals to documentation of testing results and requirements, in other words, specifications and standards.

Specifications/Standards

Specifications/standards are developed for different purposes by different organizations. For the purposes of this discussion, specifications are generally narrow in scope, being intended to define a specific product or product family. Standards are broader in scope to address market or industry needs. Some of the more important specification/ standard types are: product, customer, industry/consortium and national/international. These specifications/standards are developed by the product manufacturer, a product user, a group of product users in a certain industry and, by national or international standards organizations respectively. For example, connector manufacturers develop and test to their own product specifications to document the basic performance capabilities of their products and product families. Customers such as Dell, IBM and Intel define product specifications to address the specific needs of their applications. Industry organizations such as the Electronic Industry Association (EIA) develop standards to ensure compatibility among multiple sources to meet their market or industry needs. The next step in standards extends the concern for compatibility to national and international levels through Standards Organizations (SO) such as, nationally, Electronic Industries Alliance (EIA), Verband der Elektrotechnik (VDE), British Standards Institution (BSI), and, internationally, through International Electrotechnical Commission (IEC) and International Organization for Standardization (ISO). Naturally, there is a commonality in philosophy and practice across these specification/standards documents as indicated through the use of common specification sequences.

1.18.3 Generic Test Plan

A basic connector product specification, for example, will call out a number of test groups to assess connector performance with respect to important degradation mechanisms and identify appropriate measurement, conditioning, and exposure sequences. A generic testing program is illustrated below in Figure 1.191.

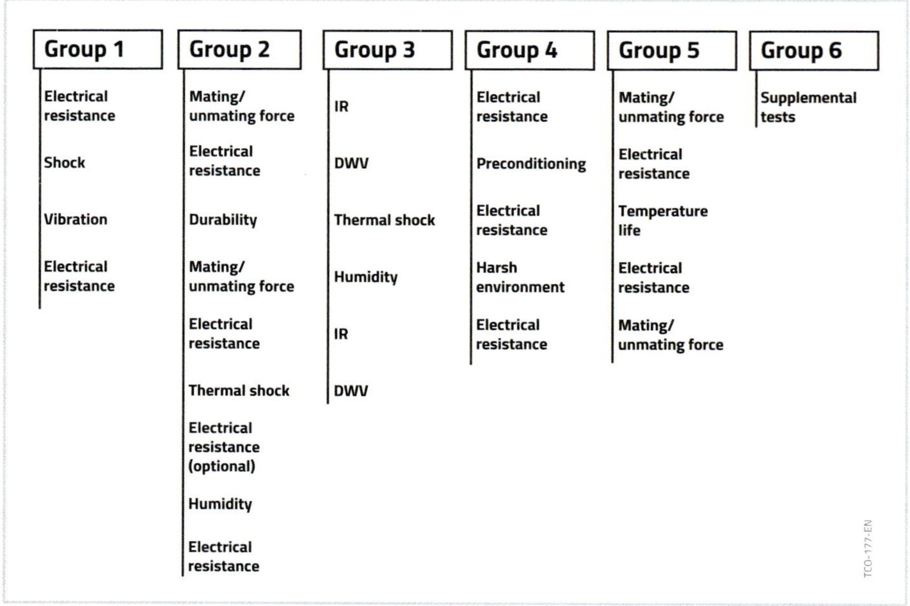

Fig. 1.191: Generic test plan following EIA 364

In this test program, **Group 1** is a mechanical group consisting of shock and vibration exposures. The shock exposure could be either mechanical or thermal and the vibration exposure could use a variety of test procedures, sine or random vibration applied to one or more product orientations. The measurement called out is electrical resistance which could also be of two types, Low Level Contact Resistance (LLCR) and Contact Resistance at Rated Current (CRRC). Details on these and other conditioning/ exposures and measurements will be provided later.

The **Group 2** test sequence includes mechanical (durability) and corrosion (humidity) exposures and mechanical (mating/unmating force) and electrical (contact resistance) measurements. This sequence will be discussed in some detail because it is simple in principle, but illustrates some more sophisticated interactions. The first mating/unmating force is a baseline measurement as is the first electrical resistance measurement. Durability cycling is an exposure because it is immediately followed by mating/unmating force and electrical resistance measurements. These measurements will indicate whether the selected number of mating/unmating cycles has affected contact spring mechanical performance. In this case, durability cycling is also a conditioning step in the sense that the second electrical resistance measurement provides the baseline for the next conditioning/exposure sequences. Thermal shock is actually a mechanical conditioning/exposure because rapid differential thermal expansion changes can result in mechanical disturbances of the contact interface. Note the third electrical measurement is optional. If it is made, the thermal shock can be considered an exposure; if it is not the thermal shock is a conditioning step. The samples are then subjected to a humidity (corrosion) exposure and the sequence ends with a resistance measurement. The difference between the second (or third) electrical resistance measurements and the final measurement provides the data for performance evaluation.

It is important to note at this point that test sequences identify appropriate measurements, but generally do not include requirements for evaluation of performance.

Group 3 testing is directed towards evaluation of the housing to validate that the molding processes used have not degraded the polymer electrical performance to an unacceptable level. It includes two measurements, Insulation Resistance (IR) and Dielectric Withstanding Voltage (DWV). The order of these measurements is important because the test voltage for DWV may degrade the polymer over and above any molding degradation and invalidate the original IR measurement causing "good" product to be rejected prior to testing. Thermal shock (mechanical) and humidity (corrosion) are both exposures because mechanical damage to, and chemical degradation of, the polymer can degrade both IR and DWV.

Group 4 addresses environmental degradation and includes electrical resistance measurements, preconditioning step(s) and harsh environment (corrosion). Preconditioning steps for corrosion exposures include durability cycling, to validate contact finish performance, and heat aging, to validate the stress relaxation resistance of the contact spring system. The harsh environment is selected to suit the intended applications of the connector system. For noble metal contact systems MFG exposures are appropriate while tin contact finishes are generally tested using humidity or temperature/humidity exposures. The second electrical resistance is both a screen and a baseline measurement. If the samples show excessive R at this point the test may be terminated, if not the measurements provides the baseline for the corrosion exposure.

Group 5 addresses thermal stability. It includes two measurements, mating/unmating force and electrical resistance. Again, the order is important because mating/unmating can degrade the connector performance so the electrical resistance measurement is a screen and the baseline for performance evaluation. The exposure is temperature life, or heat aging, with the temperature and test exposure times selected on the basis of the product design and application lifetime. Details concerning Temperature/time parameter selection will be discussed later. The final electrical resistance measurement data assesses the connector performance during the test.

Test **Group 6** provides an opportunity to specify additional testing or more severe testing of product for specific applications. For example, automotive under-hood applications may require more demanding shock/vibration and temperature life exposures. Marine applications may require additional exposures, for underwater applications or seawater corrosion mechanisms typically not tested for conventional connectors.

Traditionally product specifications use parallel test groups in which a set of samples is subjected to a single series of measurement, conditioning, exposure and measurement sequences as discussed. This, of course, does not simulate field conditions where several degradation mechanisms are active, or potentially active, over the application lifetime of the connector. For example, on/off (temperature) cycles are distributed over the product lifetime, unmating/mating (durability) cycles may occur periodically, temperature and corrosion processes may be continuous with variations in severity depending on environmental conditions. For these conditions the appropriate test sequence would be to subject a single set of samples to all the appropriate degradation mechanisms, a parallel test sequence. The complexity of product application lifecycle conditions precludes the definition of a "standard" parallel test sequence, which is one argument for the serial approach which is well defined and allows, at least, a comparative evaluation of connector performance. Parallel testing programs, however, are under consideration.

1.18.4 Test Explanations

Given that the stability of the overall resistance of a connector is arguably the most critical connector performance parameter, electrical resistance measurements are one of the most critical measurement methods.

1.18.4.1 Connector Resistance

For convenience a review of the components of connector resistance, first presented in 1.6.3 The Electrical Interface: Contact Resistance, will be provided. Figure 1.192 shows a cross section of a connector with the three major connector resistance components highlighted. The overall resistance of a connector (RO) will be given by the sum of the permanent connection (RPC), bulk (RB) and contact interface (RC) resistances.

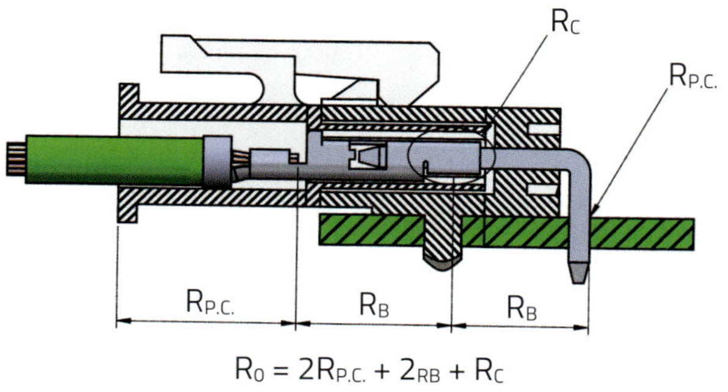

Fig. 1.192: Connector contact resistance

Permanent Connection Resistance

The dominant permanent connection resistances are those of crimped, insulation displacement, compliant pin and soldered connections. Figure 1.192 illustrates crimped and compliant pin permanent connections. The resistance of permanent connections is generally small, fractions of a mΩ, and stable. The low magnitude of the resistance is due to the large contact areas typical of permanent connections. The stability is provided, for mechanical connections, by the large deformations and high forces used during the connection process. For soldered connections the solder volumes also typically provide large and stable contact areas. A typical connector will have two permanent connection resistances, one on the receptacle and one on the plug.

Bulk Resistance

The bulk resistance of a connector is due to the material and geometry of the contact springs. The most important material parameter is the electrical resistivity of the contact spring material.

1.18 Connector Testing

A simple equation illustrates the importance of contact spring geometry. The resistance of a conductor is given by:

$$R = \rho \left(\frac{L}{A}\right) \qquad \text{(EQ 20)}$$

R: Overall resistance of the conductor [Ω]
ρ: Electrical resistivity of the conductor material
L: Length of the conductor
A: Cross sectional area of the conductor

A typical connector will have two bulk resistances, that of the receptacle and plug contacts. The bulk resistance contributions generally account for the majority of the connector overall resistance in typical connectors.

Contact Interface Resistance

The contact interface resistance, RC, is that due to the metallic contact areas created when the receptacle and plug contacts come together during mating. The important parameters determining RC are:

$$R_C = \rho \left(\frac{H}{F_n}\right)^{\frac{1}{2}} \qquad \text{(EQ 21)}$$

R_C: Contact interface resistance [Ω]
ρ: Electrical resistivity of the contact spring material
H: Hardness of the contact surface materials
F_N: Contact normal force [N].

While multiple beam contacts may have more than one contact interface, R_C is generally taken as a single value. The magnitude of R_C is between that of permanent and bulk resistances, being closer to permanent connections in signal connectors and closer to bulk resistances in power contacts.

The magnitude of connector overall resistances falls in the range of fractional milliohms (power applications) to a few tens of milliohms (signal applications).

Contact Resistance Measurement Methods

There are two commonly used electrical resistance measurements: Low Level Contact Resistance (LLCR) and Contact Resistance at Rated Current (CRRC). These methods differ primarily in the means and magnitudes of the voltages and currents applied during the measurement.

LLCR measurements are voltage controlled with an open circuit voltage of the order of tens of millivolts. Some specifications require 20 mV maximum, while others allow 50 mV. These measurement voltages are intended to be so low that they are insufficient to cause dielectric breakdown of any films which may be on the surface of the contacts at the contact interface. In other words, such resistance measurements indicate metal-to-metal contact interfaces. LLCR measurements are sometimes referred to as "dry circuit" measurements. This terminol-

ogy dates back to days when voltage spikes, "wetting voltages" were applied to break down surface films, the low millivolt open circuit voltage, therefore, was identified as "dry circuit" voltage. The test current is typically 100 milliamperes though lower currents are sometimes specified.

LLCR measurements are generally made using what is called a "4-wire" or Kelvin measurement circuit. A basic 4-wire circuit is shown in Figure 1.193. Note that two wires, one source and one sense wire feed into the top and the bottom of the sample.

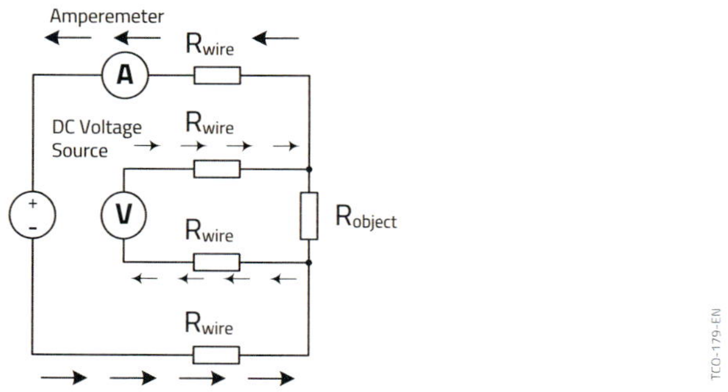

Fig. 1.193: Schematic of 4-wire (Kelvin) resistance measurement circuit

Figure 1.194 shows this circuit implemented in a Kelvin Clip to facilitate resistance measurements. The respective A and V tips are mechanically coupled to clip to the sample, but electrically insulated from one another.

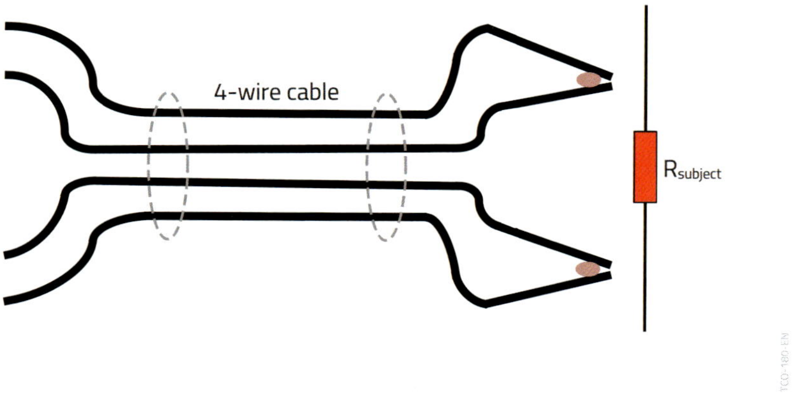

Fig. 1.194: Kelvin clip implementation of the Kelvin circuit

Kelvin clips are generally used to measure the resistance of individual contacts as shown in Figure 1.195. The same basic schematic, of course, can be expanded to enable the measurement of multiple contact positions in a housing. When board mounted connectors are being measured, attention must be paid to ensure that the requirements of the Kelvin schematic are realized on the PCB.

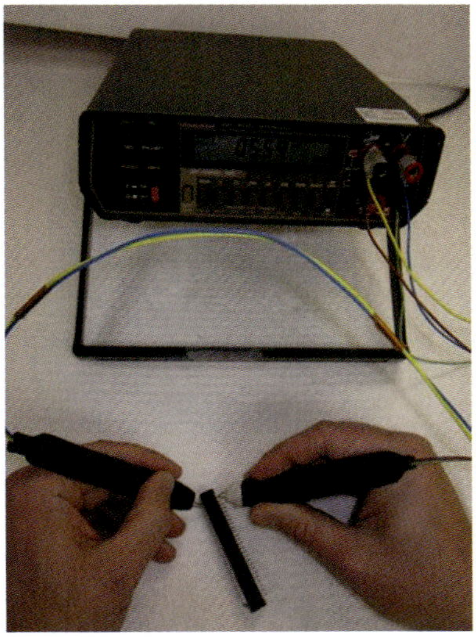

Fig. 1.195: Contact resistance measurement using Kelvin clips

CRRC resistance measurements use a similar (4-wire) circuit, but the current is controlled to the rated current of the connector, several to tens of amperes. The millivolt drop across the contact is then measured and the resistance calculated from Ohm's law. The driving voltage to realize the rated current will generally exceed the few tens of millivolts allowed for LLCR measurements. CRRC should not be measured until thermal stability has been achieved, for example until two successive contact temperature measurements after the specified time interval show the same temperature.

In cases where both resistance measurements may be required, such as power connectors, the following measurement sequence should be used:

CRRC, LLCR, Conditioning-Exposure, LLCR, CRRC. CRRC precedes LLCR initially because the LLCR measurement provides the base reference for evaluating resistance. After the required conditioning/exposure sequences have been run LLCR is done first because the higher applied voltages used in CRRC measurements could disrupt any surface films which have formed during the exposures negating the effect of the exposures.

While details of attaching test probes to the sample connectors are application dependent it is always good practice to place the probes as close to the sources of resistance as possible to minimize any bulk resistance contributions of the input leads to the measurement. Special test boards are often necessary to test printed wiring board connectors to accomplish this end.

1.18.4.2 Exposures

Attention now turns to exposures, with examples of thermal, mechanical and environmental exposures.

Temperature Life

Temperature life, or heat aging, exposures drive diffusion-controlled reactions which include corrosion and stress relaxation. Our discussion begins with stress relaxation with corrosion to be discussed in a later section. The significance of stress relaxation is that it results in a decrease in contact force. As noted in 1.10.2.4, repeated here for convenience,

$$F=\left(\frac{\sigma}{6}\right) w \left(\frac{t}{l^2}\right) \qquad \text{(EQ 22)}$$

F: Contact force in a cantilever beam [N]
σ: Stress in the beam [N/mm²]
w:width of the cantilever beam
l^2: thickness and length of the cantilever beam.

So the contact force decreases as stress relaxation reduces the stress in the beam. Recall that, according to Equation 2.10, the contact interface resistance depends inversely on the contact normal force, F, which creates the contact interface area. Consider now Figure 1.193 which shows the contact interface resistance as a function of contact force for both loading and unloading. The loading curve is different from that shown in the main text because it is for a clean metal surface while Figure 1.196 is for a film covered surface, copper alloy 725, a copper-nickel-tin alloy. The initial high resistance region is where film effects are present and the resistance then drops rapidly as the films are disrupted at increased contact force.

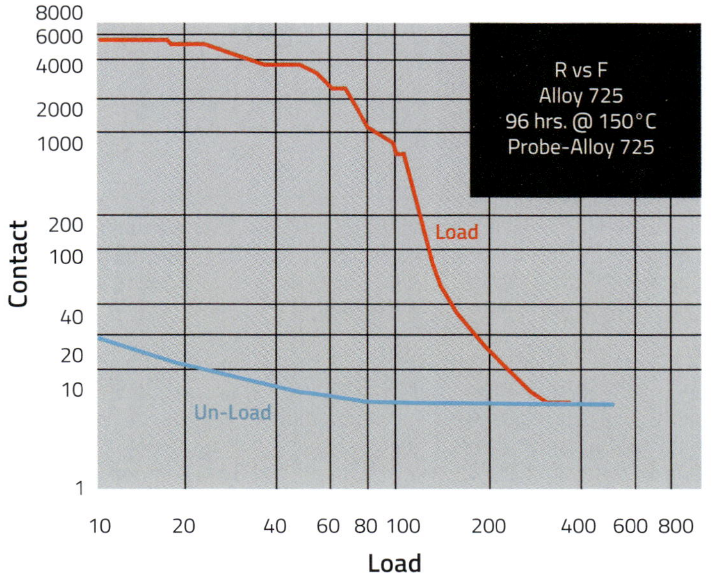

Fig. 1.196: Contact interface resistance as a function of contact force: loading and unloading

It is the unloading curve that is of interest in this discussion. Note that the resistance does not change during the initial stages of unloading. This is because the contact area was created at the maximum normal force and is maintained as the force decreases until elastic recovery forces come into play and the contact area does begin to decrease. Generally the contact force reductions due to stress relaxation are of the order of those in the flat section of the unloading curve. Therefore, stress relaxation does not impact the contact resistance directly. Instead, the effect of contact normal force reduction due to stress relaxation comes into play with respect to contact interface resistance in that the mechanical stability of the contact interface is reduced making the interface more susceptible to mechanical disturbance. If the contact interface is disturbed, that is, moves to a new location, the newly created contact interface will be determined by the stress relaxed contact force and, therefore, will be lower than the original contact interface. It is this two-step process, force reduction followed by disturbance that causes the contact resistance to increase as a result of stress relaxation. This is why temperature life is generally a conditioning rather than an exposure.

As noted in 1.18.3, however, there is an additional long-term effect of stress relaxation in that the loss in contact force is permanent because the contact beams takes a permanent set due to the kinetics of stress relaxation. This means that all subsequent mating of the connector will have a lower deflection and, therefore, a lower contact force than the original connector prior to temperature life conditioning. When temperature life is used to drive stress relaxation, the temperature and duration of the conditioning is based primarily on the application temperature and life requirements.

For example, say an application has an intended life of 5 years (60 months) at an application temperature of 55 °C. Using the doubling every 10 °C rule of thumb, that time/temperature profile can be realized in 2 months at 105 °C (32 x for ΔT of 50 °C and 2 months' time).

It is important to keep in mind, however, that increases in temperature may bring different degradation mechanisms into play when making such extrapolations. For a twin beam contact the cited acceleration factor may be appropriate, but for a single beam contact latched into a housing, polymer creep may be excessive at the elevated temperature.

1.18.4.3 Mechanical Conditioning

Mechanical conditioning are intended to assess the mechanical stability of the contact interface under field conditions. As noted previously in this chapter, disturbances of the contact interface can be driven by either thermal or mechanical forces. The thermal forces result indirectly from thermal expansion mismatch stresses, thermal cycling or thermal shock, while the mechanical forces are applied directly, mechanical shock or vibration.

The difference between thermal cycling and thermal shock is the rate of change of temperature between the upper and lower temperature limits. Thermal cycling is generally a slow change in temperature, often by cycling either the applied current or the oven temperature, while thermal shock is generally rapid by moving the samples between chambers at different temperatures. An oven suitable for thermal cycling is shown in Figure 1.197 and a dual chamber, thermal shock oven, in Figure 1.198.

Fig. 1.197: Temperature life, thermal cycling oven

Fig. 1.198: Dual chamber thermal shock oven

Thermal shock is further differentiated by the different thermal response times of the connector and system components. While the magnitude of the applied stresses may be the same for a given temperature change, the rate of application of the stresses in thermal shock is generally a more demanding environment. The thermal stresses realized are dependent on the size of the connector because the differential change in length is dependent on the length of the connector or system. This fact should be taken into consideration when selecting the samples to be tested. Table. 1.8 lists typical conditions of thermal shock.

Typical test conditions for thermal shock following EIA 364 D, TP32:	
a) Number of cycles	5 to 1000 cycles
b) Hot extreme	+85 °C to +150 °C +3 °C/0 °C
c) Cold extreme	−65 °C to −40 °C +0 °C/−3 °C
d) Time at temperature	30 minutes (contingent on mass)
e) Mating conditions	mated or unmated
f) Mounting conditions	mounted or unmounted
g) Transfer time	within 5 minutes

Tab. 1.8: Typical thermal shock parameters

Mechanical shock is generally intended to simulate transportation stresses, though there are some connector environments, e.g. automotive that experience mechanical shocks as a part of their application environment. The G values and number of shocks applied are application dependent.

1.18.4.4 Vibration

Vibration parameters are arguably the most problematic of the conditioning/exposure testing parameters. The most significant issue is that the vibration conditions in the field are unknown except for the recognition that they are application dependent and highly variable. Thus vibration testing is standardized to allow comparative rather than performance evaluations.

Vibration testing is performed in three axes as illustrated in Figure 1.199

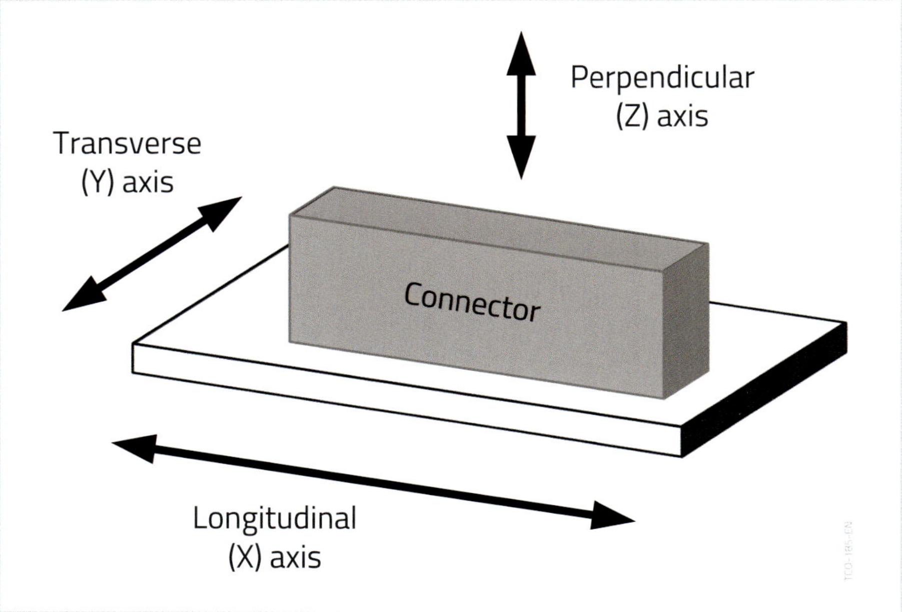

Fig. 1.199: Vibration testing axis definition

In the configuration shown in Figure 1.199, the Y and Z axes are the most significant. The Y axis is sensitive to rocking motions of the connector and, therefore, fretting motions and degradation, and the height of the connector can result in amplification of the rocking forces. The Z axis is the mating direction of the connector, and, therefore, is also susceptible to fretting degradation. Clearly how the connector is fixtured, and the orientation of the fixtured connection on the vibration table can have a significant influence on the connector performance in the test. A typical vibration table for fixturing of samples is shown in Figure 1.200.

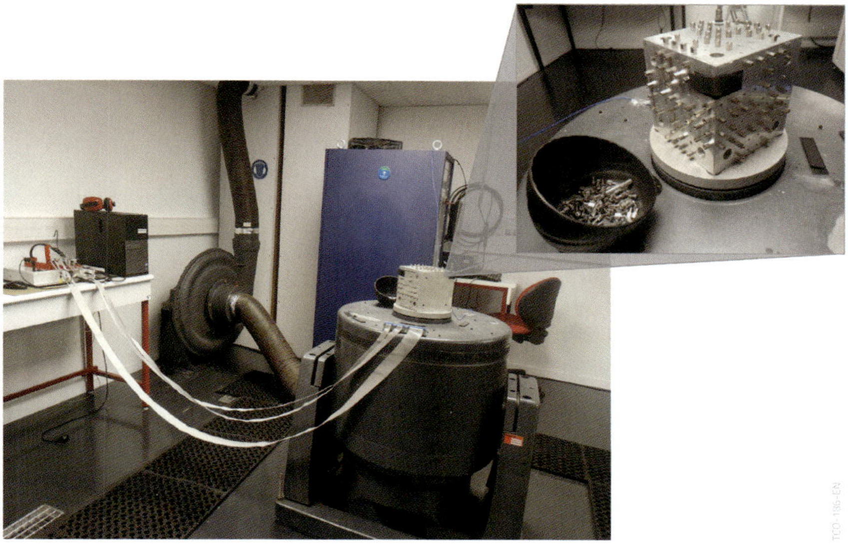

Fig. 1.200: Typical vibration table (Photo courtesy T. Peel, Contech Research)

The two major vibration methodologies are sine and random vibration. Sine vibration sweeps through the selected frequency levels at a selected amplitude and for a selected duration. Random vibration, as the name implies, varies the frequencies and amplitudes of vibration within the selected frequency spectrum. While sine vibration is the most commonly specified method, the random vibration is more representative of field conditions. Table. 1.9 shows typical vibration testing.

Typical test conditions for vibration as described in EIA 364, TP28:	
Sine Vibration:	
Frequency	10 to 2000 to 10 Hz
Amplitude	0.06" da or 15 G's
Duration	4.0 hrs/axis, 3 axis total
Test current	100 ma
Sweep time	20 minutes
Random Vibration:	
Power spectral density	0.02 g^2/Hz
G 'RMS'	5.35
Frequency	50 to 2000 Hz
Duration	45 minutes to 8 hours/axis
Sweep time	20 minutes

Tab. 1.9: Typical vibration parameters

Definition of a vibration environment involves consideration of at least the following factors: severity levels (amplitude and frequency), duration (time or cycles per axis), attributes to be monitored (resistance, discontinuities) and fixturing. The fixturing decisions must take a systems approach to look for factors that could amplify the effects of vibrational forces on the connector. For example, in an edge card connector the length of the board above the connector can have a significant effect on the vibrational impact on the Y axis.

LLCR is a commonly used measurement for vibration testing. It is used primarily for assessment of fretting degradation and can be done continuously or periodically depending on the data desired. Sophisticated data gathering techniques may be required for this purpose.

In addition to assessment of mechanical stability, vibration is also used to determine whether electrical discontinuities may occur during vibration. There are two levels of discontinuity monitoring. One microsecond discontinuity monitoring has been the most commonly used approach. Recently shorter discontinuities have become of interest and discontinuities in the range from 50 nanoseconds to 2 nanoseconds have been monitored with increasingly sophisticated measurement and fixturing requirements. Discussion of these methods are outside the scope of this chapter.

1.18.4.5 Corrosion

As noted previously, the environmental exposures for non-noble (tin) and noble (gold) plated contacts are different. The dominant degradation mechanism for tin connectors is fretting corrosion, a build-up of oxide debris at the contact interface due to small scale motions (tens of microns). The oxidation of tin is a given and the degradation process is dominated by the mechanical stability of the contact interface with respect to inhibiting fretting motions. Tin corrosion exposures are predominantly temperature-humidity related and there is no generally accepted correlation between such exposures and field performance.

The situation with noble metal finishes, in particular gold over nickel, is very different. There were a number of investigations into the environmental stability of gold contact finishes in the '70s and a significant amount of data relating test environments to field performance was developed. The details of these studies are outside the scope, but suffice it to say that there are two test environments that are generally accepted as indicative of field performance. As shown in Table. 1.10, taken from EIA 364 TP 65 they are:

Pollutant	Controlled	Uncontrolled
Cl_2	10 ± 3 ppb	20 ± 5 ppb
NO_2	200 ± 50 ppb	200 ± 50 ppb
H_2S	10 ± 5 ppb	100 ± 20 ppb
SO_2	100 ± 20 ppb	200 ± 50 ppb
Temp	30 ± 1 °C	30 ± 1 °C
RH	70 ± 2 %	70 ± 2, %
Corrosion rate	12–16 mgm/cm^2/day	36–46 mgm/cm^2/day

Tab. 1.10: Mixed flowing gas exposure parameters

While other test environments are specified, they have in common the important pollutants, Cl_2, NO_2, H_2S and SO_2, and the low level of pollutant concentrations, parts per billion and the temperature and humidity levels.

Some of the major findings about noble metal corrosion in mixed flowing gas environments are suggested in Figure 1.201 and Figure 1.202.

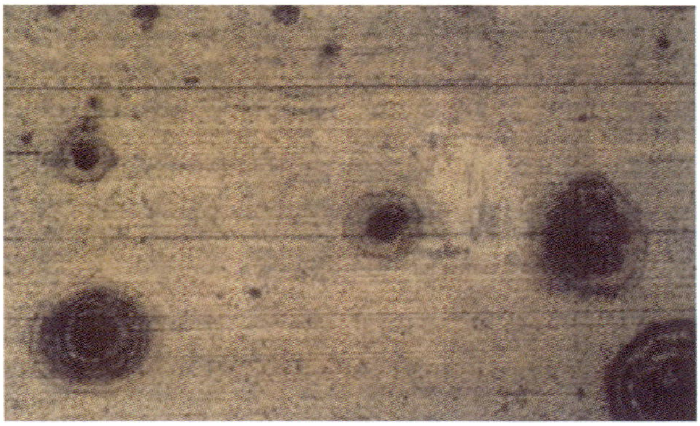

Fig. 1.201: Pore corrosion and corrosion migration from pore sites

First, Figure 1.201 illustrates pore corrosion and the migration of corrosion products from the pore sites over the contact surface, the rings around the pore sites. The corrosion source, in all cases, is the copper alloy of the contact spring. The corrosion products that are migrating are identified in Figure 1.202. The X-ray maps shown in Figure 2.121 show that the contact finish under test was a gold flash over nickel over copper alloy. The pore site in this case originated at the copper alloy as indicated by the intensity of the copper signal at the upper right in the figure. The corrosion products, the lower row of X-ray maps consisted of copper, oxygen, chlorine and sulphur. The corrosion synergy between chlorine and sulphur distinguishes mild from harsh environments. Note the difference in corrosion rates between the controlled and uncontrolled environments and the chlorine and sulphur concentrations in Table. 1.10. The corrosion rates are, in fact, used to validate the exposure by measuring the sample weight gain. There are acceleration factors for these two environments with a two day exposure to the test environment being equivalent to a year of exposure in the field environment.

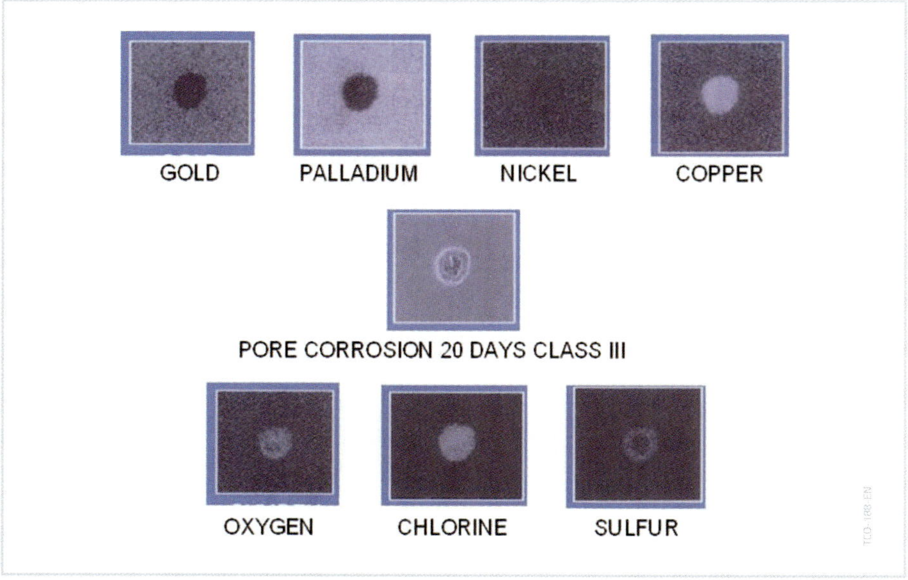

Fig. 1.202: X-ray maps identifying corrosion product migration from pore sites

1.18.5 Glow-Wire

In today's world, safety has become a major concern for everyone, from individuals to country governments, non-governmental organizations, and so on. A large number of regulations in a variety of fields aim to make everyone's environment as safe as possible.

Apart from electrocution, the main risk in the electronic industry is fire. Even nowadays, there are worldwide occurrences of fires reported to be caused by electrical appliances. The consequences are often serious, even fatal. The scope of the Glow-Wire is not to prevent causes of fire but rather to deal with them.

Fig. 1.203: Coffee machine on fire

1.18.5.1 Purpose

The Glow-Wire particularly applies to connectors used in household and similar electrical appliances. Basically, this refers to white goods such as refrigerators, microwave ovens, washing machines, coffee machines and so on, whereas audio, video and similar electronic devices are all excluded, as well as computers.

However, only the plastic material of the connector is taken into consideration. If it turns really hot, i.e. beyond 500 °C, the aim is that it does not burn or that it rapidly self-extinguishes in order to prevent flame propagation and thus the start of a fire. The main reason why the plastic may reach such a temperature is a significant increase in current carried by the metal material of the connector, which is what the Glow-Wire simulates. Such a situation may happen with defective components, in case of electrical overloads, short circuits or improper use.

There are exceptions regarding connectors themselves – they are discussed in section 1.18.5.3. For these, other tests are considered to assess the risk of fire but they are not addressed in details in this paper.

It is worth mentioning that risks of fire are not addressed by the Glow-Wire only, but also by other means such as the UL94 standard which is well known in the connector industry. But their scope are not the same: UL94 does not particularly focus on household appliances; it applies to raw materials only, not to end-products. It enables to assess the behaviour of a plastic against a flame while the Glow-Wire assesses it against a heat source. Nevertheless, they both target user safety.

1.18.5.2 Test principle

The Glow-Wire test is easy to understand and to perform; it only requires adequate equipment as defined in the test standards beginning at 1.18.5.3. The Glow-Wire consists of a solid metal wire brought to a specific temperature before the test starts by sending a lot of current through it (making the wire glow, hence the name). A specific type of paper may also be placed underneath. The sample holder then moves to put the specimen to be tested in contact with the Glow-Wire for a certain period of time. An illustration is provided in Figure 1.204.

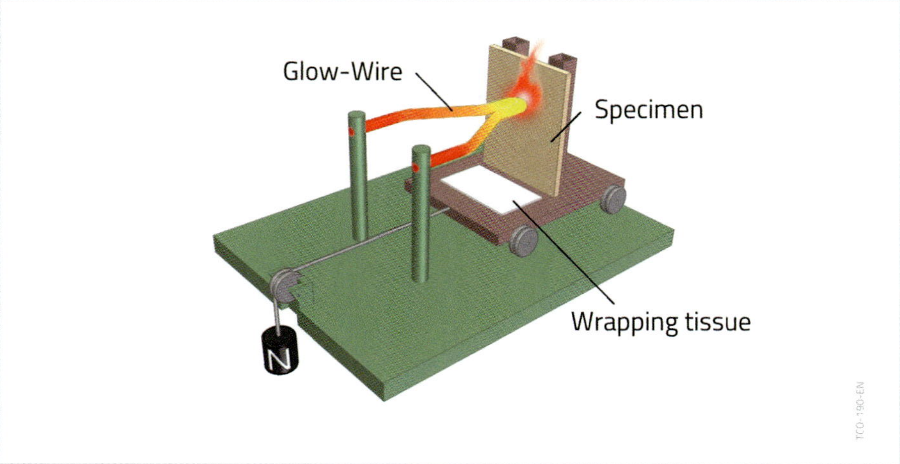

Fig. 1.204: Simplified apparatus for glow wire test

The test is passed or failed depending on the behaviour of the sample and of the paper underneath. Typically observed behaviours are glowing, burning, melting and dripping. The specimen may fall off the sample holder if it is too distorted by heat; it may even be totally consumed. On the contrary, the specimen may not be damaged at all.

1.18.5.3 Test standards

Four standards regulate Glow-Wire tests. One is dedicated to the test apparatus and common procedure and the three others each describe a different test. Glow-Wire compliance itself is defined in a fifth standard that sets minimum criteria for the three tests. They are all IEC (International Electrotechnical Commission) standards.

The four test standards presented in the current section are listed below:

- IEC 60695-2-10 Fire hazard testing – Part 2-10: Glowing/hot-wire based test methods – Glow-wire apparatus and common test procedure
- IEC 60695-2-11 Fire hazard testing – Part 2-11: Glowing/hot-wire based test methods – Glow-wire flammability test method for end-products (GWEPT)

- IEC 60695-2-12 Fire hazard testing – Part 2-12: Glowing/hot-wire based test methods – Glow-wire flammability index (GWFI) test method for materials
- IEC 60695-2-13 Fire hazard testing – Part 2-13: Glowing/hot-wire based test methods – Glow-wire ignition temperature (GWIT) test method for materials.

The fifth standard is:

- IEC 60335-1 Household and similar electrical appliances – Safety – Part 1: General requirements. This one is detailed in section 3.7.4.

This paper focuses on the above documents but there are many others:

- IEC 60335-2-xx Household and similar electrical appliances – Safety – Part 2: Particular requirements ...;

Each xx relating to one kind of appliance that may or may not contain instructions regarding Glow-Wire tests.

Test equipment and common procedure – IEC 60695-2-10

In order to avoid confusion and to focus on the test procedure itself, this article does not present all technical pcints of the standards. In particular, it does not address the apparatus compliance.

Equipment

The equipment consists in a power supply and its electric circuit, a Glow-Wire (held by a stud), a thermocouple and a thermometer, a wrapping tissue (when required) and a wooden board, a sample holder, a chamber and a timing device. The Glow-Wire must move toward the sample, or conversely.

The power supply must be a stable (±2 % rms maximum) and the measured rms current value must be ±1 % accurate. It must be noted that the temperature shall not be maintained by use of a feedback loop. This means that the operator can only adjust the current value to set the temperature; he cannot directly set the desired temperature. Also, care should be taken to make sure the electric circuit can handle the high currents required for the tests. Typical values range from 60 to 140 A.

The Glow-Wire must be made of more than 77 % of nickel and 20 ±1 % of chromium; before bending, its diameter must be 4.00 ±0.07 mm. After bending, it must look like in Figure 1.205 and it should always remain in the horizontal plane.

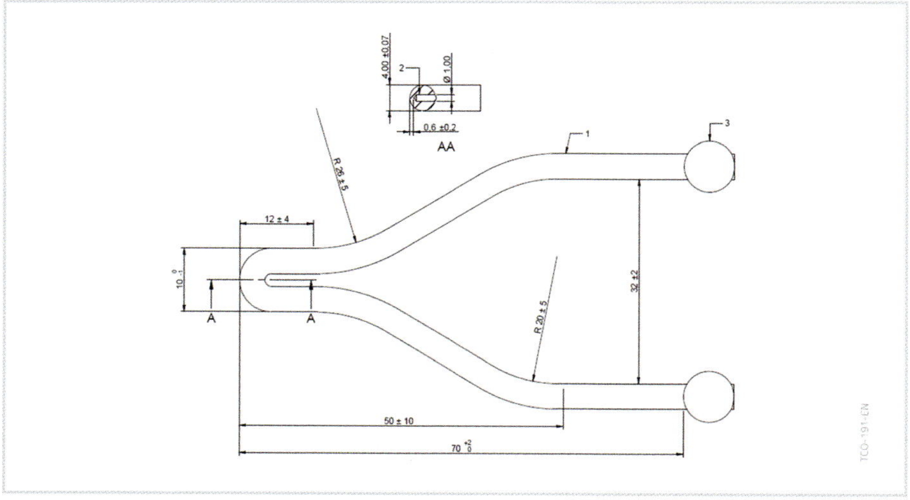

Fig. 1.205: Glow-Wire shape and dimensions

The thermocouple is inserted in the hole in the tip of the Glow-Wire. This thermocouple must be of class 1 (according to IEC 60584-2) and suitable for temperatures ranging from 500 to 960 ±5 °C. For protection purposes, it must be insulated with a mineral material sheathed by a metal that can handle at least 1050 °C. See Figure 3.50 for a better understanding. The nominal diameter is 1.0 mm.

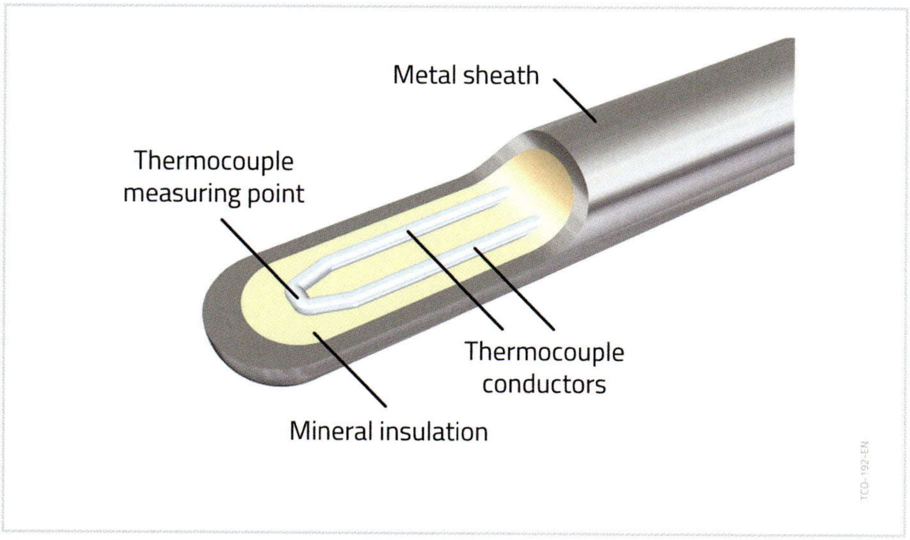

Fig. 1.206: Mineral-insulated metal-sheathed thermocouple

In the case described later in this chapter, the wrapping tissue (ISO 4046-4:2002 Clause 4.215) is not required. However, it is recommended to have one for ease of cleaning rather than just the wooden board where cooled down drops of plastic material are difficult to remove. This board must be at least 10 mm thick. The wrapping tissue must rest on it or be in close contact and placed under the tip of the glow-wire at a distance of 200 ±5 mm. Its mass per area unit should range from 12 to 30 g/m^2.

Neither shape nor dimensions are specified for the sample holder; the preferred fixing system is not specified either. The only requirements are insignificant heat losses and that the contact surface be vertical and centered on the tip of the Glow-Wire. See Figure 1.207 and Figure 1.208 for examples of sample holders.

Fig. 1.207: Sample holder for raw material specimen

Fig. 1.208: Sample holder for end-product specimen

The test chamber shall prevent any draught. Its volume must be large enough to make sure that specimens are 100 mm away from any of its surfaces and that there is enough oxygen in order not to disturb the test, considering that part of it can be consumed by flames. According to the standard, a volume of 0.5 m^3 should be sufficient. In order to enable the operator to clearly identify flaming or glowing, the light received by the specimen shall not exceed 20 lx (the light from the Glow-Wire is not to be taken into account). Such measurement is done with a lux meter placed in the sample holder, with its back facing the Glow-Wire. This is why a dark background colour is usually used for the chamber. Figure 1.209 shows an example of chamber.

TCO-195-EN

Fig. 1.209: Glow wire test chamber

As regards the timing device, the only requirement is the accuracy that shall not exceed 0.2 s.

Common procedure

First, the wrapping tissue and the wooden board must have been stored for at least 24 h in an environment with a relative humidity comprised between 45 and 75 % and a temperature ranging from +15 to +35 °C. The aim is to make sure that the test results will not be influenced by these parameters.

After making sure that the thermocouple is fully inserted and that the specimen is distant from at least 5 cm from the Glow-Wire, the latter is brought to the desired temperature by adjusting the current flowing through. Once the temperature is stable, i.e. it does not vary by more than ±5 °C over 60 s, the test can begin.

The specimen is brought toward the Glow-Wire (or conversely) at a speed ranging from 10 to 25 mm/s until the tip of the Glow-Wire contacts the specimen. However, the impact force must be inferior or equal to 1.05 N; the speed must therefore be reduced just before this moment. Moreover, the contact force must be 0.95 ±0.10 N but the penetration of the Glow-Wire tip into or through the specimen shall not exceed 7 ±0.5 mm; therefore, if the plastic around melts, there might be no more contact when the tip reaches this penetration limit. 30 ±1 s after the impact, the specimen or the Glow-Wire is withdrawn at the same speed mentioned above.

The standard appendix also describes the observation of flames. This is important because it may be really hard to tell whether the observed phenomenon is an ignition or not. The following statement applies: if there is a blue circular corona around the tip of or along the Glow-Wire, this is not an ignition as long as it does not lead to a flame directly on the specimen.

Test on end-products – IEC 60695-2-11 (GWEPT)

The first thing to clarify regarding GWEPT is the meaning of an end-product. As regards connectors, it is rather simple: it refers to the connector itself. But the metal parts, e.g. pins, solders tabs or screws shall be removed before testing. As mentioned in point 1.18.5.1, only the plastic material shall be taken into consideration. This relates to Glow-Wire compliance requirements described in point 1.18.5.4.

For other end-products, the material or components normally surrounding them are placed underneath the specimen. In the case of connectors, such elements are unknown since they vary from one application to another. Therefore, a wrapping tissue is used instead.

Small parts and insignificant mass

There are two restrictions to the testing of end-products. The first one refers to so called "small parts" and relates to the fact that in order to be tested, a specimen must be large enough to allow the penetration of the Glow-Wire, as well as thick enough in order not to be cut in two by the Glow-Wire.

The IEC also states that if the end-product including all its surfaces can fit completely in a 15 mm diameter circle as shown in Figure 1.210 (a), or if at least one of its surfaces cannot completely surround an 8-mm diameter circle while at least one of its dimensions is superior to 15 mm, as shown in Figure 1.210 (b), it should be considered as a small part. Thus, it is thus not suitable for GWEPT.

Figure 1.210 (c) is not a small part and can be tested.

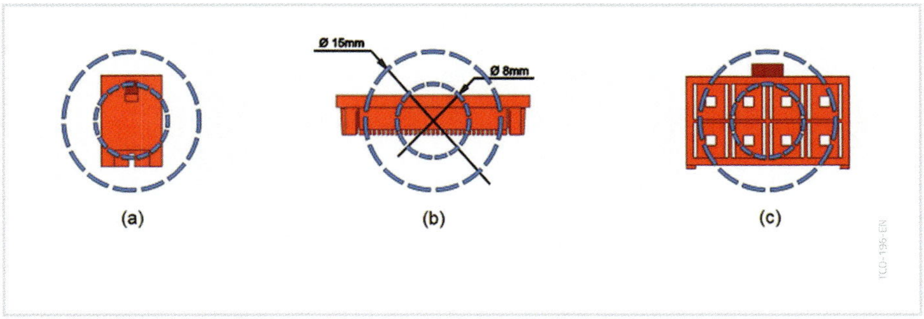

Fig. 1.210: Small parts definition

The second restriction is called insignificant mass. IEC 60335-1 considers that an end-product with a mass that does not exceed 0.5 g bears no risk of fire propagation. Consequently, GWEPT is not necessary for such an end-product.

Setting up

Specimens shall be stored in the same conditions as described in the common procedure for the wooden board and wrapping tissue and the testing must be completed within 30 minutes after they are removed from such conditions. The temperature of the laboratory itself must range from +15 to +35 °C and the relative humidity shall not exceed 75 %.

Moreover, the tip of the Glow-Wire shall be in contact with the part of the specimen that is most subject to thermal stress when in normal operating conditions. If this definition cannot be applied, then the tip shall be in contact with the thinnest part of the specimen but more than 15 mm away from its upper edge whenever possible.

Preferred test temperatures are +550, +600, +650, +700, +750, +800, +850, +900 and +960 °C. The duration of contact with the Glow-Wire is 30 ±1 s and there is an additional observation period of 30 s.

Results

The test is passed if the specimen does not ignite, or if it extinguishes within the observation time without igniting the wrapping tissue.

Tests on raw material – IEC 60695-2-12 (GWFI), -13 (GWIT)

While GWEPT focuses on the end-product, GWFI and GWIT are dedicated to raw materials. These two tests are very similar but their respective scopes are slightly different.

While GWFI tries to define the maximum temperature at which the specimen self-extinguishes, GWIT tries to define the minimum temperature at which the specimen ignites. Another difference is that GWIT does not require any wrapping tissue.

Specimen characteristics

A specimen can be molded by casting, injection, compression or transfer and if none of these methods is possible, it should be cut or sliced from a bigger piece of raw material. The height and width must be at least 60 mm and measured from inside the clamping area.

As regards thickness, preferred values are as follows: 0.10 ±0.02; 0.20 ±0.02; 0.40 ±0.05; 0.75 ±0.10; 1.50 ±0.15; 3.0 ±0.2; and 6.0 ±0.4 mm. The selected thickness depends on the dimensions of the end-product. For instance, it would not make sense to test a 6-mm thickness if the end-product biggest insulation thickness is 0.4 mm. Actually, IEC 60335-1 that is discussed in point 3.5.4 states that among the preferred values listed above, the value closest to but not greater than the end-product thickness must be used.

Setting up

The storage conditions of the specimen are more precise than for GWEPT. In this case, the specimen must remain at least 48 h at 23 ±2 °C with a relative humidity ranging from 40 to 60 %. Moreover, it must be used within 4 h after it has been removed from these conditions. The same applies for the wrapping tissue used for GWFI, except that it has to be used within 1 h. As for GWEPT, the temperature of the laboratory itself must range from +15 to +35 °C and the relative humidity shall not exceed 75 %.

The specimen must be perpendicular to the plane of the Glow-Wire and the tip of the latter shall contact the center of the specimen. The preferred temperatures range from +550 (+500 for GWIT) to +960 °C, with increments of +25 °C (and +30 °C beyond +900 °C). There again, the required accuracy is higher than for GWEPT.

The duration of contact with the Glow-Wire is 30 ±1 s and there is an additional observation period of 30 s for GWFI but of only 5 s for GWIT.

Results

There again, there are differences between GWEPT, GWFI and GWIT. While GWEPT only requires one specimen to be valid, GWFI and GWIT both require three specimens to pass the test. The other difference is that GWFI and GWIT require the specimen not to be totally consumed to pass the test, which is not required by GWEPT.

Otherwise, GWFI is passed under the same conditions than GWEPT and GWIT is passed if there is no ignition or if each flame event lasts for no more than 5 s.

Last point to mention: when passed, GWIT reported value is +25 °C higher than the actual test temperature.

Figure 1.211 and Figure 1.212 show examples of different raw material behaviours during the test.

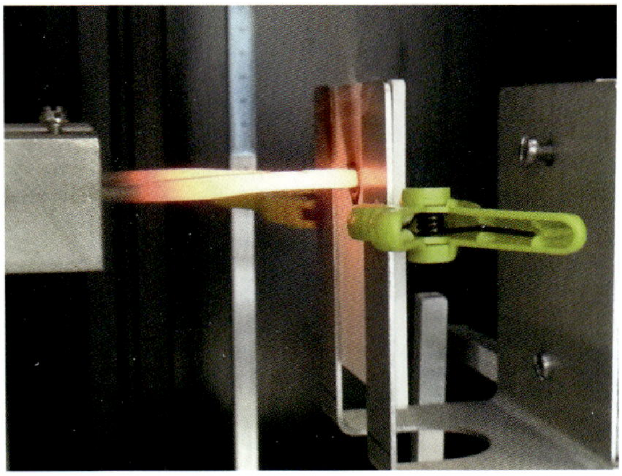

Fig. 1.211: Raw material preventing fire

Fig. 1.212: Raw material causing fire

1.18.5.4 Glow-Wire compliance – IEC 60335-1

Various tests have been addressed so far, but not Glow-Wire compliance requirements themselves. This actually refers to another standard, IEC 60335-1, which does not define how to perform the test or how to assess if it is passed or failed; instead, it indicates the lowest acceptable temperatures for GWEPT, GWFI and GWIT.

Indeed, IEC 60335-1 requires various tests or temperatures depending on whether appliances are left unattended or not and on the current intensity in the connection. For unattended appliances with current above 0.2 A, it even includes an additional time limitation for GWEPT when tested at +750 °C. Components which are not application-specific, like Würth Elektronik connectors, must be compliant with the worst case scenario, i.e. unattended appliances with connection carrying more than 0.2 A. This is summarized in Figure 1.213.

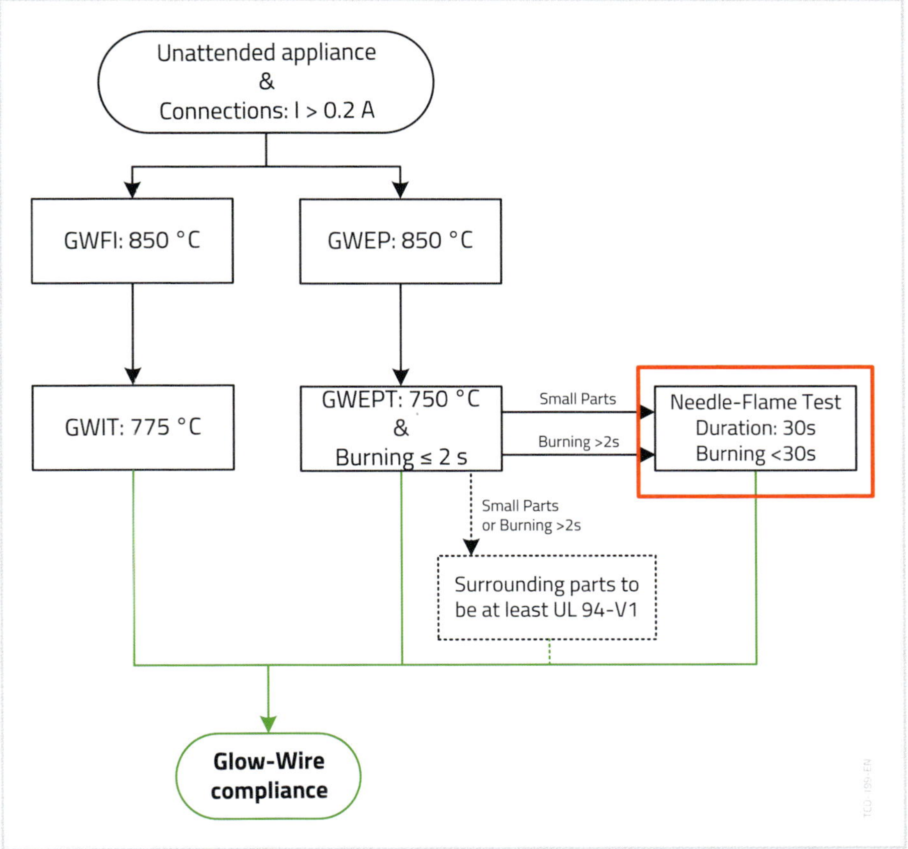

Fig. 1.213: Compliance process

The most important thing shown in this diagram is that end-product testing is not required to be Glow-Wire compliant as long as the raw material passed a +850 °C GWFI and a +775 °C GWIT. This is very important for connector manufacturers because testing every single product is obviously not realistic. Just consider that at the date of publication of this Trilogy, the Würth Elektronik catalogue of connectors already contains several thousands of part numbers.

It must be noted that there are other possibilities for approval if the parts surrounding the component in the appliance match some other criteria. But since this paper is only considering connectors that are not application-specific, it is not worth mentioning them.

Würth Elektronik is dedicated to providing customers with the best solutions and is thus already supplying Glow-Wire compliant products with the Wire-To-Board and Terminal Block ranges. For further information, please contact your sales representative.

1.18.5.5 References and disclaimer

This is a summary of the various IEC standards listed in point 1.18.5.3 and 1.18.5.4. Parts of these standards are not discussed here and some details are not mentioned. In no way Würth Elektronik can be liable for any problem that might arise from the reading of this article. It is the responsibility of the reader to check these standards.

The following versions of the standards are in force at the moment of the writing:

- IEC 60695-2-10 edition 2.0 from April 2013;
- IEC 60695-2-11 edition 2.0 from February 2014;
- IEC 60695-2-12 edition 2.1 from February 2014, consolidated with amendment 1;
- IEC 60695-2-13 edition 2.1 from February 2014, consolidated with amendment 1;
- IEC 60335-1 edition 5.0 from May 2010.

1.18.6 IP Classification

Mechanical housings and electrical enclosures protect on the one hand electrical and electronics systems against intrusion, dust, accidental contact and water and on the other hand people from touching accidentally voltage carrying or hot parts. IEC 60529 defines the so called IP Code or Ingress Protection Code, which classifies the degree of protection provided by an enclosure. This rating also applies to cables and connectors.

For example, a cellular phone rated at IP67 is "dust resistant" and can be "immersed in 1 meters of freshwater for up to 30 minutes". Similarly, an electrical socket rated IP22 is protected against insertion of fingers and will not become unsafe during a specified test in which it is exposed to vertically or nearly vertically dripping water. IP22 or IP2X are typical minimum requirements for the design of electrical accessories for indoor use.

The IP Code consists basically of two digits, where the first digit indicates the level of protection an enclosure provides against access to hazardous parts (e.g. electrical conductors, moving or hot parts) and the ingress of solid foreign objects. The second digit indicates the level of protection that the enclosure provides against harmful ingress of water (Fig. 1.176).

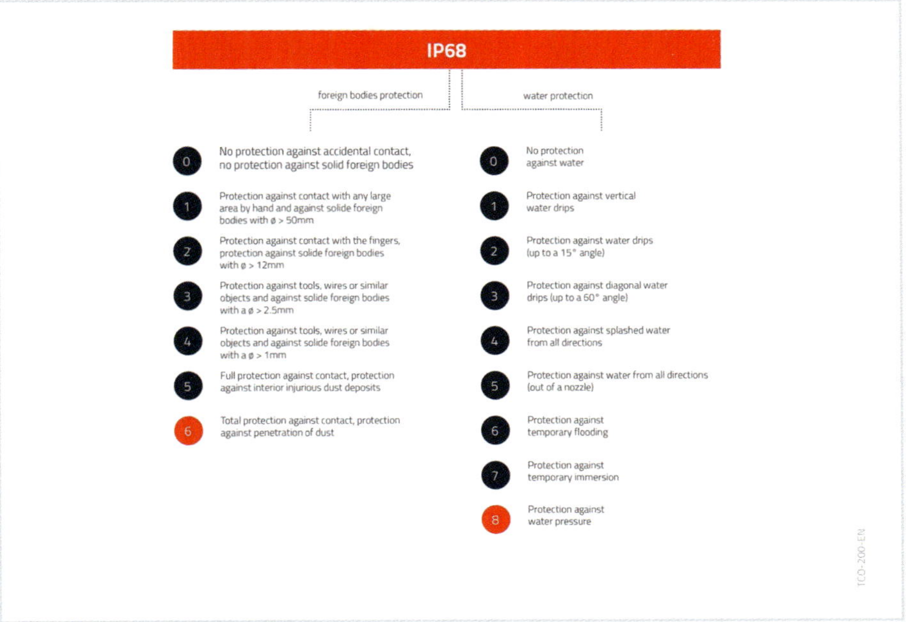

Fig. 1.214: IP Classification Code Scheme.

On top of the 2-digit-code there are further optional classifications by additional (A, B, C, D) und supplementary letters (F, H, M, S, W). The additional letters specify the protection of personnel against access to hazardous parts, whereas the supplementary letters refer to the protection of equipment (e.g. F = Oil resistant).

A special case is the IP69K rating, which specifies protection against high-temperature water jets. The letter K is defined by ISO 20653 not IEC 60529.

In the area of connectors, e.g. circular connectors, typical IP ratings are IP67 or IP68.

Trilogy of Connectors

Components

Part 2: Components

This chapter will provide an overview of design and material requirements for contact finishes, contact springs and connector housings as well as the major degradation mechanisms for these connector components. Material selection criteria for each will also be reviewed.

2.1 Levels of Interconnection

Levels of Interconnection addresses where the connector is used within an electronic system. The location of the connector in the system influences the environmental exposure and the assembly/mating durability requirements the connector must support. These factors, in turn influence the type of connector which will be used at a given system level. It should be noted that more than one type of connector may be used at some of the levels.

The Levels of Interconnection, LOI, system for characterizing interconnect devices, primarily connectors, in electronic systems was first developed around 1990. That is multiple generations ago in terms of the functionality of electronic devices and, therefore, the functional requirements on interconnect devices. Despite that fact, LOI has some benefits in terms of characterizing some of the most important connector performance requirements. This is because LOI focuses on the devices being interconnected and where they are in the electronic system, also this focus highlights some important connector performance requirements. An overview of an early five level version of LOI follows. Due to the proliferation of devices and connectors since 1990, only selected examples will be discussed.

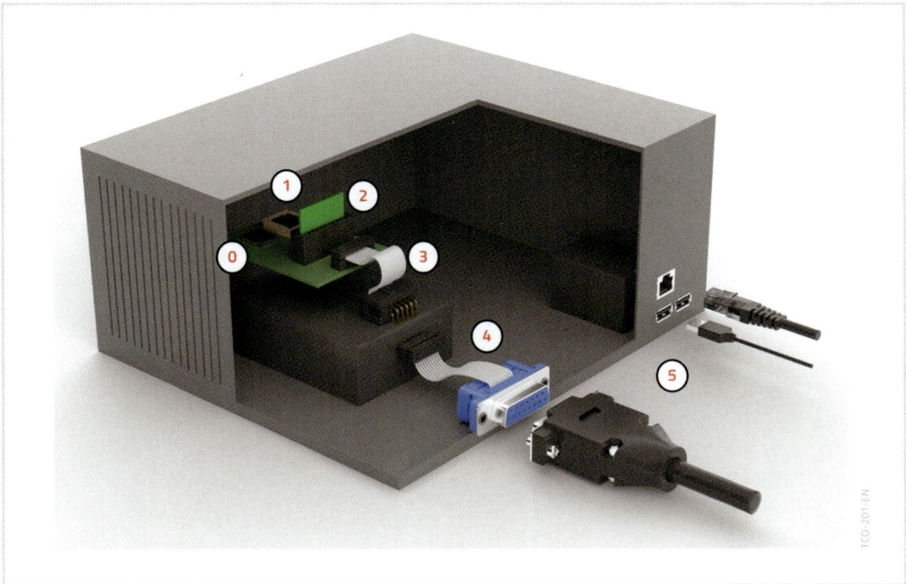

Fig. 1.215: Level Of Interconnection

Level 0

IC Chip or Chip-to Package. This is a device level interconnect from the devices on the chip, between the chip Input/Output leads and the lead frame or package. Level 0 has, obviously, seen the greatest amount of change since its establishment, the device count and functionality of IC chips having increased dramatically since that time. These interconnects are all permanent connections.

Level 1

Level 1 includes chip carrier and DIP socket interconnects to a Printed Wiring Board, PWB. Chip carrier styles, pin counts and pin densities have increased to keep pace with the increased functionality of the devices. These interconnects may be permanently soldered or mechanically separable. Separable connections would be used until device design and reliability was established so that soldered connections could be considered.

Level 2

Level 2 includes board-to-board interconnects within a functional unit, a sub-assembly. Daughterboard to motherboard interconnects are typical with both one-piece, card edge, and two-piece connectors, depending on density and reliability concerns. Two-piece connectors can offer higher density and provide higher reliability.

These interconnects generally provide both permanent and separable connections, permanent to the individual boards and separable between the boards.

Level 3

Level 3 consists of interconnects between sub-assemblies within the equipment, e.g. power supply to system board and memory to CPU. A variety of interconnect types may be used at Level 3 including board-to-board and wire/cable-to-board. A number of subsystems may be interconnected at this level using different cable and connector configurations.

Level 4

Level 4 interconnections are between sub-assemblies and system Input/Output(I/O). A variety of connector systems are used at Level 4 because of the variety of I/O support required, i.e., protocols for video, audio, telecom etc. with a low level of power to push this signal. Level 4 is the first level at which the connectors, in this case the I/O connectors, are externally exposed.

Level 5

Level 5 consists of system-to-external system interconnects, the second half of I/O. These interconnects generally require an external cable assembly, e.g. printer-to-system or data-to-system interconnects.

The LOI approach can be applied to any system, although all levels may not exist in every system.

Some connector families can be used at different level, that's the example of the D-sub connectors which can be used at level 4 as well as level 5 (see Figure 1.215)

Consider now how these Levels impact on connector requirements such as robustness, environmental stability, mating force and mating durability.

Robustness

Robustness refers to the ability of the connector system to withstand improper handling and assembly. Levels 1 through 4 are internal to the system and, in most cases, are handled primarily by personnel trained in the proper treatment of the connectors. This does not mean that robustness is not a concern, however, because in some cases access to the connector may be awkward making the mating process problematic.

Level 5 connectors are accessible to all users and subject to improper mating/unmating and unintended external loads. To address these issues Level 5 connectors often include polarizing and lead-in features to minimize the potential for misalignment. A strain relief might also be added to minimize the potential for damage at the termination point, mating surface or weak housing feature when the lead is under tension.

The D shape of the D-sub connectors is certainly one of the most famous examples of connectors polarization.

Environmental Stability

Corrosion, thermal and mechanical effects must be considered. With respect to corrosion, Levels 1 through 4 are shielded from the external environment by the enclosure.

Keep in mind, however, that cooling fans may bring the external environment into the enclosure. There are two thermal effects to be considered since the temperature internal to the enclosure may be elevated above ambient by heat generating devices within the system, i.e. Joule heating and/or transformers. **All connector degradation mechanisms, in particular corrosion and stress relaxation, increase with increasing temperature.** Temperature differences, for example due to on/off cycling of the system may make thermal expansion mismatch a concern, especially for tin finished connectors due to the potential for fretting corrosion degradation. **Mechanical stresses such as vibration can also create a driving force for contact interface motion leading to fretting corrosion in tin finished connectors and fretting wear, and loss of the gold surface, in noble metal finished contacts.**

Mating force can be an issue in two ways at all levels. If, as noted above, the mating process is awkward due to the location of the connector in the enclosure, the potential for mating damage will be increased if mating forces are high. If the pin counts of the connectors are high this tendency will be increased. This potential negative, however, must be balanced against the mechanical stability that might be lost if mating force is reduced leading to motion induced degradation mechanisms as discussed elsewhere.

Mating durability is generally not a concern at Levels 1 through 3. It may be an issue at Levels 4 and 5 where the number of mating cycles could be high in the case of multi-use or portable equipment such as laptop computers where all I/O will be subjected to multiple mating cycles.

2.2 Connector Types

As noted previously, the definition of connector type is determined by several factors including the contact spring geometry, the housing geometry and the method of permanent connection. In this discussion the LOI levels where the connector systems are used will also be identified for reference. The discussion of connector types will be general. Many variations on each type exist with specifics depending on the industry being served. This is especially the case in computer/telecom applications.

Würth Elektronik clusters their connector portfolio to 5 different sections:

- Core Connectors
- Power Connectors
- High Power Connectors
- Data Connectors
- RF Connectors

Which will be shortly introduced in the following.

2.2.1 Core Connectors

2.2.1.1 ConBHD

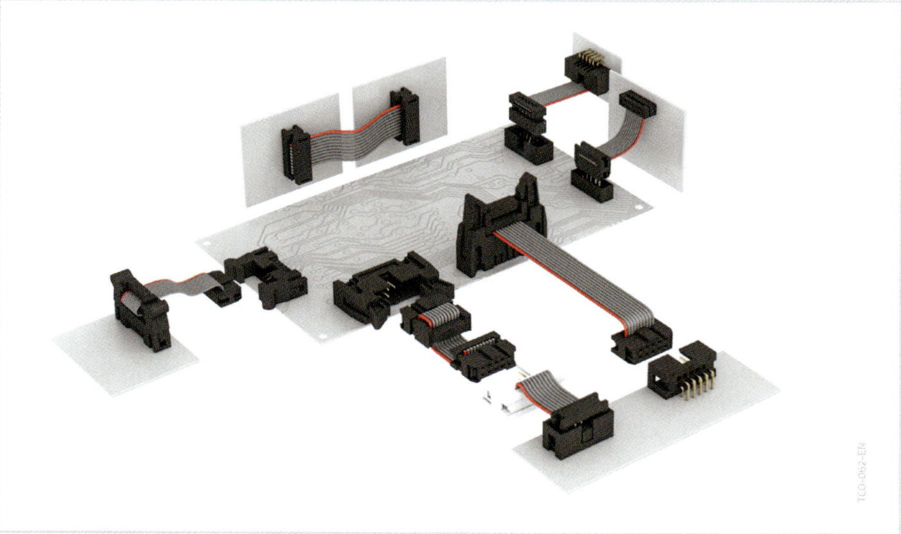

Fig. 1.216: WR-BHD overview

The WE series WR-BHD incorporates a large scale of BOX HEADERS and IDC CONNECTORS that can be found in almost every electronic device. This traditional and well known format of connectors suits well to interconnect typical stranded-ribbon cables to a printed-circuit-device (wire-to-board), pluggable or permanent.

The WR-BHD series can be found in a large scale of applications, for signal- as well as for moderate power-connections. It can be utilized for internal cabling, i.g. to attach sensor- or control-boards to a main-control-unit. But also to lead a connector to the outside of the appliance as an external port for various systems like I^2C, RS232/488, USB 2.0, etc, or proprietary I/O by using the MALE panel mount IDC Connector with flange EARS.

Quick facts

- Pitches 2,54 mm, 2.0 mm, 1,27 mm designed to use with flat ribbon cables of 1,27 mm, 1,00 mm and 0,635 mm pitch
- Polarities 6 to 64pins, gold plated
- Shrouded plastic frame for reinforce mating
- Orientation straight and right-angled
- THT- and SMT mounting (PCB)
- Cable-connectors with insulation-displacement-connector crimp-technology (IDC)

In addition to the standard THT-mounted connectors, a "High-Temperature"-Type is available, suitable for double-wave-soldering processing.

To achieve the best mechanical strength the IDC cable connectors are supplied with an additional strain relief if required. The PCB sided receptacle connectors are available with two different sizes of holding levers, to be used with- or without strain relief installed cables connectors.

The box-header-design offers a polarization to avoid wrong installation.

Another feature is offered by the "Transition"-connector which means a fixed cable attachment for one cable side – usually a pluggable type is then attached to other end of the flat-ribbon cable.

The latest miniature sized type comes with a small 1,27 mm pitch, offering high density of pins on a very small footprint. This type, often can be seen as a programming connector for micro-controller based systems.

Matching flat-ribbon-cables (WR-CAB) and press-in tools complete the portfolio.

2.2.1.2 ConPSHD

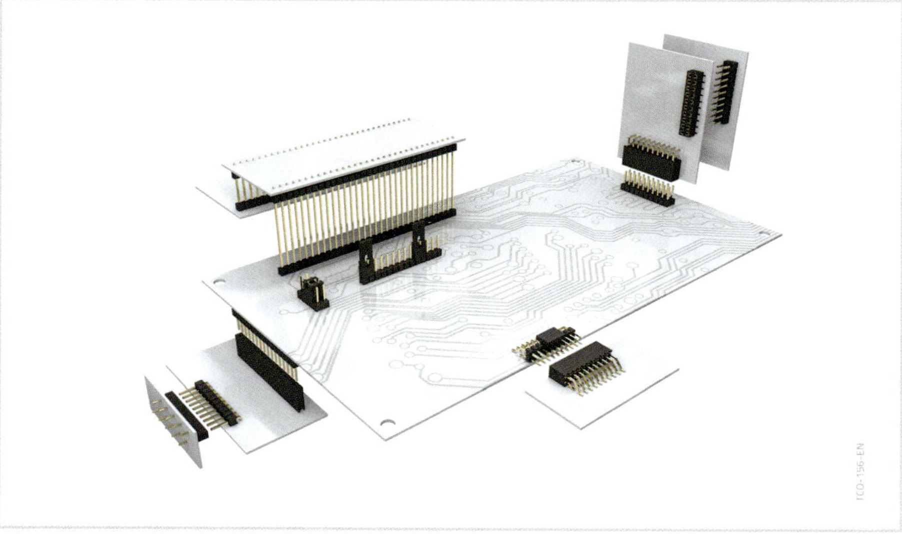

Fig. 1.217: WR-PHD overview

Pin- and Socket-Headers are obviousely the most common connectors and can be found on most of the assembled PCBs over decades

The most widely used method for electrically connecting printed circuit boards (PCBs) in a specific direction or across a certain distance is both easy and cost-effective. PCBs can be linked horizontally, at an angle, or stacked parallel to one another using Board-to-Board connections. In addition to their electrical functionality, pin and socket headers also provide solid mechanical support.

The standard 2.54 mm pitch size is highly favored due to its convenient conversion between imperial and metric measurements. However, the demand for space-saving designs has led to the adoption of smaller, yet still common, sizes like 2.00 mm and 1.27 mm.

WE offers a wide selection of different types of Pin-headers, plus Socket-Headers as matching counterparts. Several combinations are possible in order to achieve a horizontal-, angled-, or stacked connection of boards ; whether in single or double rows of pins.

THT Board Stackers enable the mounting of two PCB as a mezzanine using just a single connector. Various stacking heights are available to accommodate different design requirements.

Fig. 1.218: Bottom Entry Feature

To achieve a minimum distance PCB-stack, the WE Bottom-Entry-Types is the best choice. "Bottom-Entry" means that the pins connect from the bottom side – through the PCB – into the back side of the Socket header. B-t-B distance can be held as small as 2 mm (Figure 1.218)

Jumpers are a quick adjustable electrical bridge between two neighboured pins, often used for dedicated settings in a circuit.

Quick facts

- Pitches 2,54 mm, 2,00 mm and 1,27 mm
- Polarities 2 to 72 pins, in 1- or 2 rows
- Orientation straight and right-angled
- THT- and SMT mounting (PCB)
- Tape & Reel packaging available for automated assembly of SMT parts

2.2.1.3 ConMM

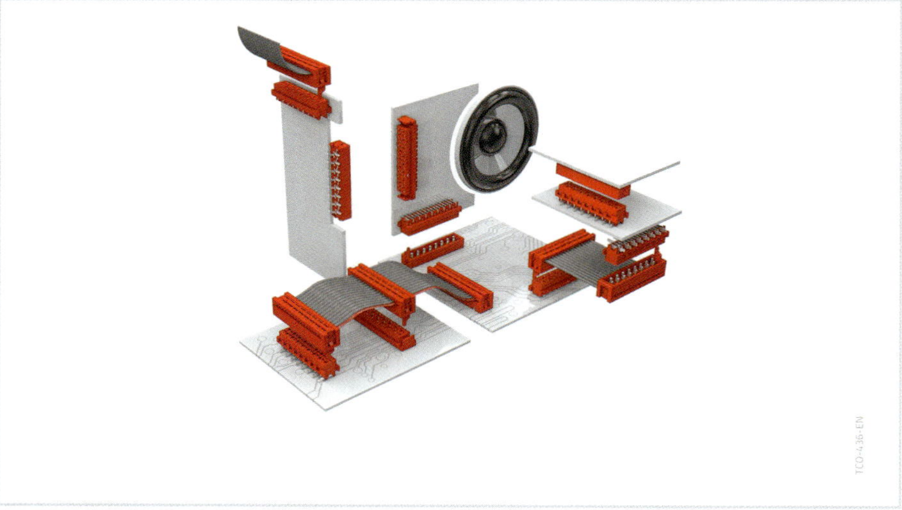

Fig. 1.219: WR-MM overview

The WE MiniModule family is another connector series, that, at a closer look, reveals some distinctive features that makes this family unique. The WR-MM series can be used to create board-to-board (BtB) connections of two PCBs as well as wire-to-board (WtB) connections utilizing standard flat-ribbon-cables.

The 2.54 mm "pseudo/staggered-pitch arrangement" of the contact pins, significantly reduces the real size of the footprint required on a PCB.

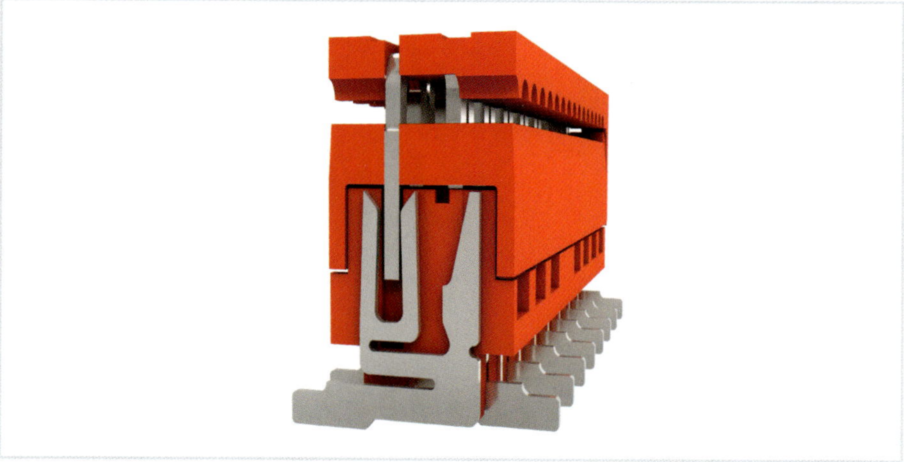

Fig. 1.220: WR-MM contact design

Arguably, the most beneficial feature of the MiniModule series is about its capabilities to cope with vibrations. Spring-loaded contacts in a tuning fork design (Figure 1.220) prevent frictional corrosion. Highly reliable contacts made of phosphor-bronze material guarantee fatigue-free elasticity over its full lifetime.

Board-to-Board connections can have vertical- or 90°-angled orientation of mounting; Wire-to-Board connections likewise, to the connection of the ribbon cable.

"Latching"-types of the WR-MM connectors offer a secure locking mechanism with snap tabs on the side of the interface

"Polarization" prevents plugging a connection in the wrong way by the use of a polarization tab

All features can be combined at a time, e.g. BtB-type with "Latching" and "Polarization"

This product also offers the "Transition"-connector which means a fixed cable attachment for one cable side – usually a pluggable type then is attached to other end of the flat-ribbon cable

Over all, the WR-MM series means a handy, reliable and fit-to-purpose selectable connector series for many different applications, such as for signal transfer or small power requirements up to 3 A.

Quick facts

- Pitches 2.54 mm pseudo pitch
- Polarities 4 to 26 pins, 2row arrangement
- Current 3 A max. for BtB-connections
  1 A max. for WtB-IDC connections, using 1.27 mm flat ribbon cable AWG28
- Voltage up to 250 V (AC)
- Pin-plating Tin over Nickel
- Orientation straight and right-angled
- THT- and SMT mounting (PCB)
- Reel-Type packaging for automated assembly of SMT parts
- UL/RoHS/REACH conformity

2.2.1.4 ConWTB

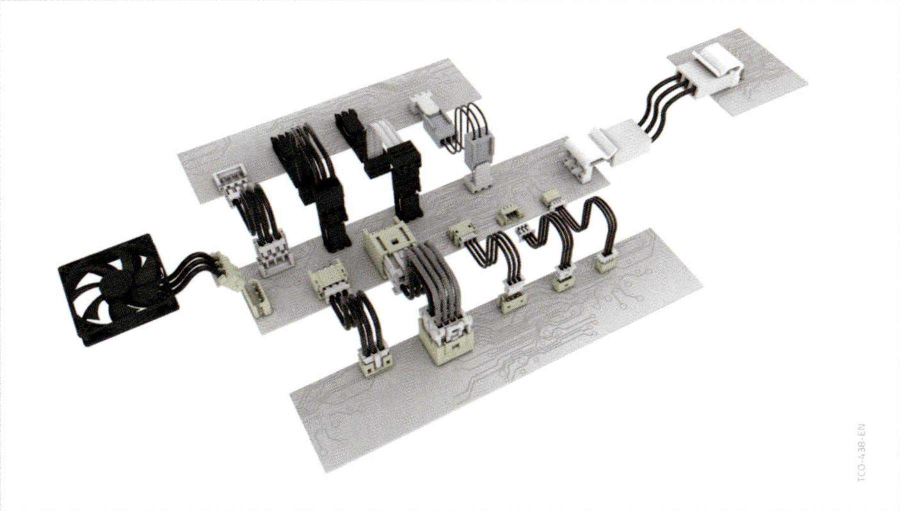

Fig. 1.221: WR-WTB Overview

The Würth Elektronik WR-WTB (Wire-To-Board) range comprises a versatile selection of wire-to-board connectors designed to provide reliable and secure connections between wires and printed circuit boards (PCBs). Offering a wide array of pitch options from 1.00 mm to 3.96 mm, these connectors cater to diverse electronic applications with varying space requirements and signal transmission needs.

These plastic headers utilise either Nylon (Polyamide) or LCP to ensure a good mechanical robustness and a high abrasion resistance. The electrical contacts are made using a copper alloy and tin plating to provide a high conductivity with excellent durability and corrosion resistance.

Using a smaller or larger pitch option within the Wire-To-Board range offers different benefits based on the specific requirements and constraints of the electronic application. Here are the advantages of each option:

Benefits of Smaller Pitch Option:

- Higher Density: Smaller pitch connectors allow for a higher number of pins or contacts within the same physical space on the PCB.
- Miniaturization: By using connectors with a smaller pitch, electronic devices and systems can be designed to be more compact and lightweight, crucial in portable devices or applications with space constraints.
- High-Speed Data Transmission: Smaller pitch connectors are often designed to support high-speed data transmission. They can reduce signal crosstalk and maintain signal integrity as signal and return pins are physically closer together which mitigates any inductive loops.
- Enhanced Performance: In some cases, connectors with smaller pitches may have improved electrical performance characteristics, such as lower inductance and an increased capacitance which can positively impact signal quality.

Benefits of Larger Pitch Option:

- Ease of Assembly: Connectors with a larger pitch provide more space between pins or contacts, making them easier to handle during the assembly process.
- Robustness and Durability: Larger pitch connectors generally have more substantial pins or contacts, which can make them more robust and resistant to mechanical stress and vibration.
- Better Thermal Performance: Connectors with larger pitches offer improved thermal performance, as there is more space for airflow and heat dissipation around each pin. Typically, this means a larger connector has a higher rated current.
- Tolerance to Misalignment: Connectors with larger pitches tend to be more forgiving of misalignment during the mating process, which can be beneficial in situations where accurate alignment is challenging.

Tolerance to Misalignment: Connectors with larger pitches tend to be more forgiving of misalignment during the mating process, which can be beneficial in situations where accurate alignment is challenging.

2.2.1.5 REDFIT IDC SKEDD

As a signal connector, the REDFIT IDC is the first connector from Würth Elektronik eiSos available with SKEDD technology (Figure 1.222). With a 1.27 mm pitch, the REDFIT IDC is no larger than comparable signal connectors such as box headers.

Fig. 1.222: REDFIT IDC SKEDD connector

The contact springs come with selective plating. A tin layer enables an optimal IDC connection. The SKEDD spring contacts are ENIG (Electroless Nickel Immersion Gold) plated to ensure optimal signal quality over long periods.

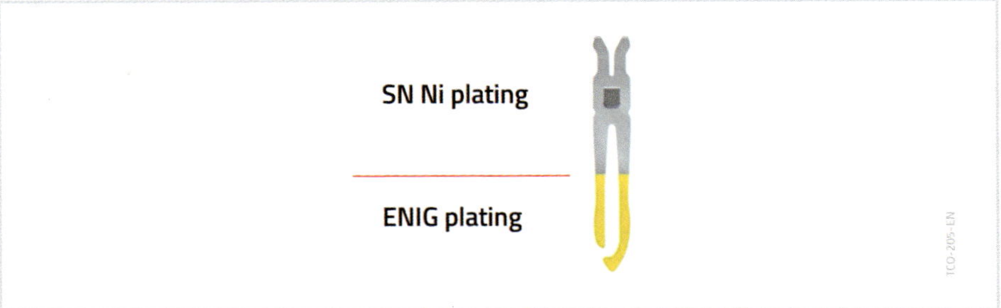

Fig. 1.223: Contact plating SKEDD

The housing of the REDFIT IDC comes with two additional plastic pins, which secure the connector to the board (Figure 1.224). As the diameter of the plastic pins is different, these pins provide reverse polarity protection.

Fig. 1.224: Protection against reverse polarity with plastic pins

Manufactured out of LCP plastic, the housing is flame-retardant. LCP plastic complies with the UL94 V0 rating.

Benefits of REDFIT SKEDD

- Pluggable connector to board, mating connectors no longer necessary
- Solderless connection technology
- Connector can be retrofitted without a soldering process
- Simple maintenance
- No thermal processes
- Can be hand-mounted
- PCB mountable on both sides
- Ideal debugging connector

2.2.1.6 ConDSUB – D-Sub Connectors

Though D-Sub is an I/O connector it is necessary to show it separately as the trend on the D-Sub market is not on the consumer market anymore therefore the trend is slightly different. Most new development show enhanced D-Sub used for industrial applications, including mixed signal and power contacts or waterproof D-Sub ...

Under the same name you find totally different products ranging from cost effective D-Sub for basic industrial applications; stamped pins, flash gold plated, through hole ...) up to advanced ones with turned pins 0.76 µm gold plating, SMT ...

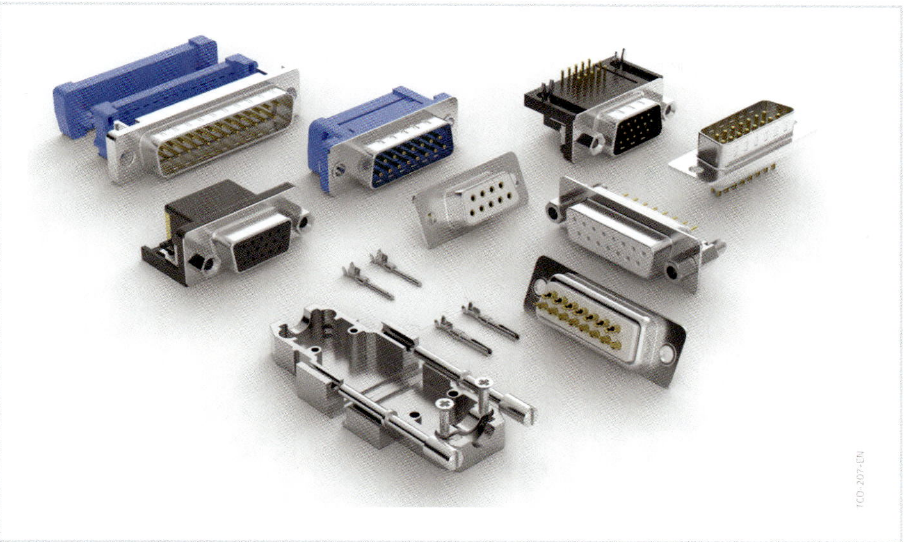

Fig. 1.225: Standard D-Sub connectors

Circular, rectangular and other typical military and aerospace dedicated connectors are not the main topic therefore they will not be detailed here.

2.2.1.7 ConCIRC

Applications and properties of circular connectors

Originally developed for military applications, circular connectors today are widely used in modern factories with a high degree of automation. However circular connectors are also increasingly used in telecommunications, renewable energy generation, security and surveillance technology, automotive and transportation or LED lighting. Their advantages are obvious: they are compact and thus take up little space, but offer a high contact density thanks to their round shape.

At the same time, circular connectors are often waterproof and dustproof, so they can be used in harsh environments. Typically, they comply with a high IP standard of IP65 to 68 (see chapter 1.18.6 on IP protection standards).

Circular connectors are designed to provide reliable connections, to thread into position and stay fixed. Thus, connections hold in a reliable way even under vibration, shock, moving cables, etc. A coding system ensures proper mating thanks to the error-proofing principle in the design standard.

Circular connectors typically carry three different types of electrical signals alone or in different combinations: Analog signals, digital signals or data, and electrical power as power supply. There are currently numerous metric sizes of circular connectors on the market, essentially ranging from M5 to M58 (diameter in mm). The basic structure of the different sizes is similar, pins are strategically spaced within the circular face It differs mainly by possible maximum number of contacts and the diameter of each contact, influencing electrical and mechanical performances.

Probably the largest market share for circular connectors is the M12 size, which is examined in more detail below.

M12 Circular Connectors – mating faces and coding

Within the family of M12 circular connectors there are numerous different versions, which differ in the number of contacts, their arrangement and coding. The number of contacts with defined mating faces ranges from 2 pins to 17 pins, with the most common versions having 3, 4, 5 and 8 pins.

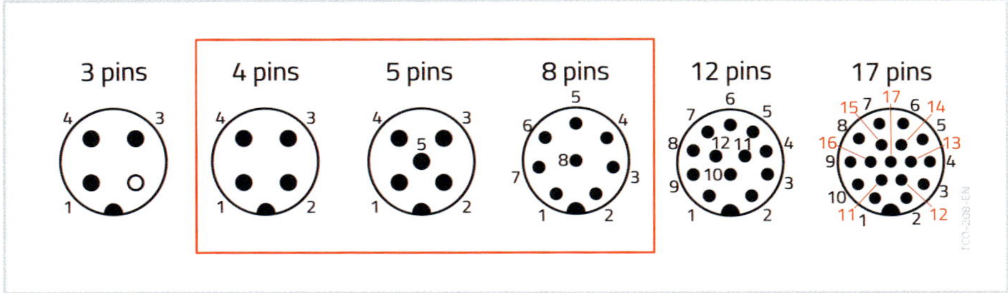

Fig. 1.226: M12 connectors include mating faces with 2 to 17 pins

To reduce the risk of incorrect mating in industrial installations and to fit to specific applications, various mechanical coding schemes are used, each assigned to one or two applications: "A-Coding" is reserved for sensors/actuators and DC power, while "B-Coding" covers Profibus and Interbus industrial fieldbus communication in automation. "D-Coding" is assigned to Ethernet and Profinet with data rates of 100 Mbit/s, and "X-Coding" to high-speed Ethernet with data rates of 10 Gbit/s. "L, S- and T-Coding" is reserved forhigher power supply.

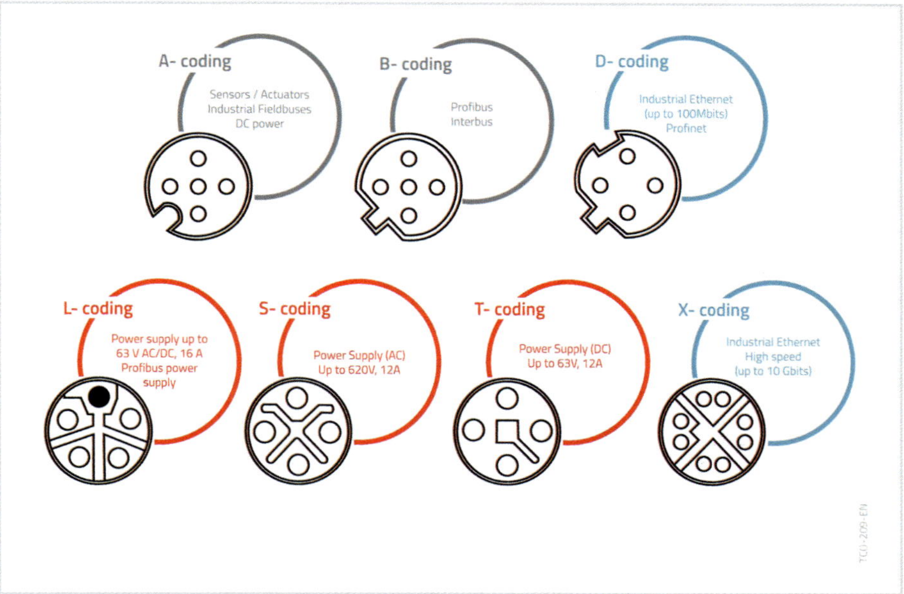

Fig. 1.227: There are various codings for M12 connectors which mechanically prevent incorrect mating of various functions.

M12 circular connectors – Coding A

The M12 circular connectors from Würth Elektronik comply with the IEC 61076-2-101 standard and are initially available in coding A for analog signals and power supply with 4, 5 and 8 pins. The product range includes sockets for panel mounting with sub-types for PCB mounting, for wires soldering and with assembled connection wires (0.5 m), as well as their counterparts – cable assembly and field attachable connectors – the latter optionally solderable or screwable.

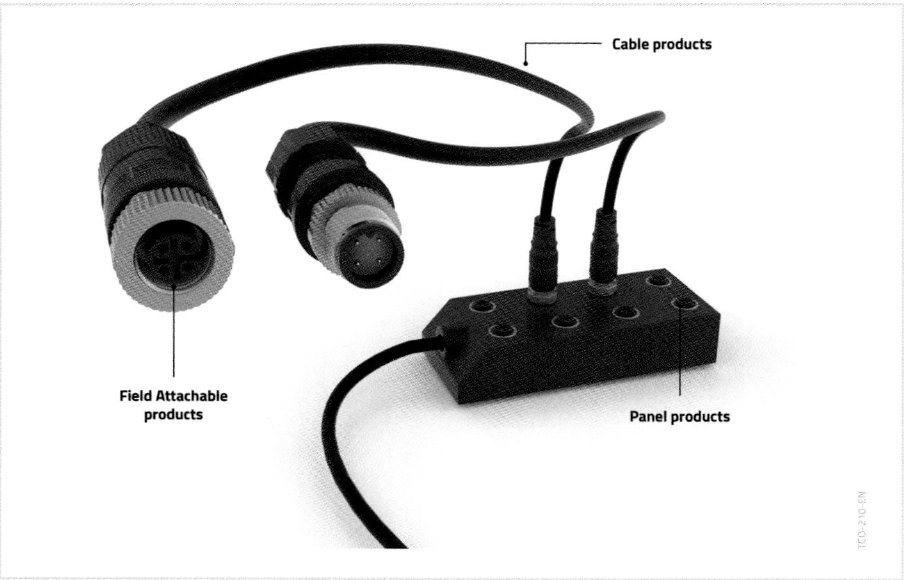

Fig. 1.228: M12 connectors are available in versions for mounting in a housing (panel), for cable assembly (cable) and for mounting in the field (field attachable).

All connector types are available in male and female versions, which allows to choose the appropriate gender for the application. The female version is naturally protected against contact from fingers and thus corresponds to IP20. Typically, the female connector is being used at the side of the device under voltage. Also, in some fieldbus systems for example masters are coded female and devices male.

Sockets for panel mounting

Within the sockets for panel mounting there are three sub-types:

1. The version for PCB mounting is available in metal or plastic versions, straight and angled, and with various thread sizes for panel mounting.
2. The solder version is also available in metal or plastic, with mounting nut on the front or rear side and with different thread sizes for panel mounting.
3. The version with prefabricated wires (0.5 m) is also available in metal or plastic version as well as with mounting nut on the front or back side.

1. Insert – PA66
2. Pins – Brass/Phosphor Bronze
3. Panel Nut – PA66/Zinc Alloy
4. Panel O-Ring – Silicone
5. Shell – PA66/Zinc Alloy
6. Heat Shrink Tube – PE
7. Wires – PVC – 0.5 m long

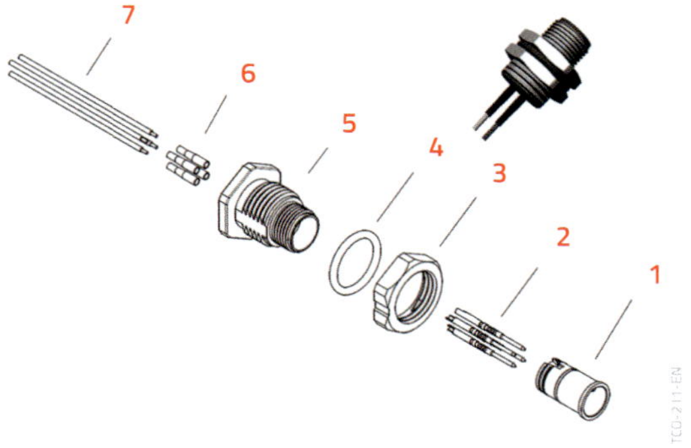

Fig. 1.229: M12 sockets for panel mounting: here the version with pre-assembled wires.

Sockets housings are always interfacing in a panel of an enclosure, LOI 5. There are several elements here that should ensure a good seal between the plug and the housing. The housing panel should always be located between the enclosure panel nut and the O-ring. The silicone O-ring guarantees that no dust or water can enter the enclosure.

Female connectors have a mating O-ring that is mandatory to ensure a seal between the two mating parts when they are mated together. With each screw/thread connection, there is also a sealing element to ensure the tightness of the connection. Epoxy resin fills into the housing and outer shell to increase water tightness and structural fixation.

The O-ring of the outer shell ensures the quality of the connection and the seal between the outer shell and the enclosure when they are assembled. A pin holder ensures good positioning of the pins on the PCB layout. The shielding pin is connected to the metal enclosure, helping to avoid signal interference.

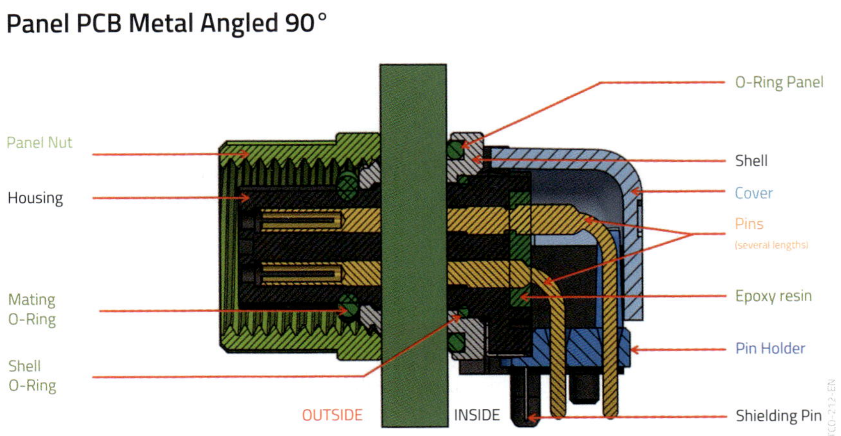

Fig. 1.230: Section through a socket for mounting in a housing: here the angled version for PCB mounting.

The electrical specifications of the sockets for panel mounting are identical to those of the plugs for cable assembly (Table. 1.11). Only the temperature range is wider at −40 °C to + 90 °C and the IP protection is higher at IP68.

Connectors for cable assembly

For cable assembly, M12 connectors are available with metal or plastic coupling nuts, straight and angled, and with two different shielding variants. Table 2.X.1 summarizes the specifications of the connectors for cable assembly.

Polarity	Current rating	Voltage rating	Contact Resist-ance (initial)	Contact Resist-ance (final)	Insulation resist-ance	Dielectric With-standing Voltage	Dura-bility	Temp. range	IP Rating (mated)
4 pins	5 A	250 V (AC)	≤ 10 mΩ	Rise in relation to initial values ≤ 15 mΩ	≥ 100 MΩ at 500 V (DC)	1400 V (AC)/min	100 mating cycles	−20 °C – +80 °C	IP67
5 pins		60 V (AC)				1000 V (AC)/min			
8 pins	2 A	30 V (AC)				650 V (AC)/min			

Tab. 1.11: Specifications for M12 connectors for cable assembly.

Field attachable connectors

For field assembly connectors, the solder versions are available in metal or plastic versions. The electrical specifications are identical to those of the cable assembly and panel connectors (Table. 1.11), except for the temperature range and IP protection. The solder versions operate from −40 °C to +90 °C and offer IP68 protection.

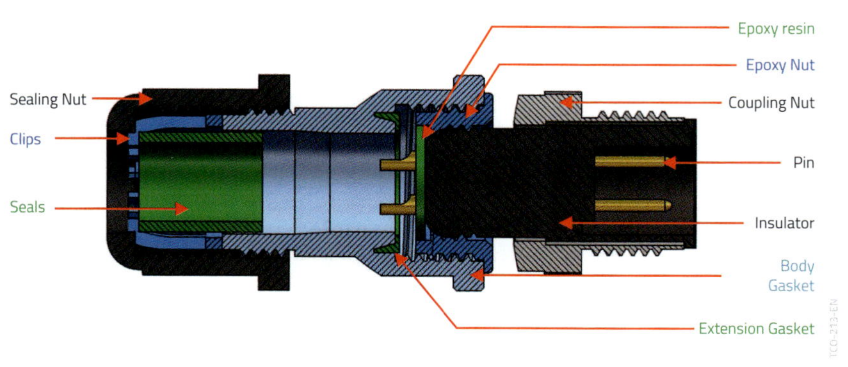

Fig. 1.231: Section through a M12 field attachable solder connector

After removing the cable insulation and stripping the wires to the appropriate length, the cable is passed through the sealing nut, compression clip, seal and unassembled body. The wires are fixed in the small cages with set screws to electrify the conductors to the interface pins. The stack of elements allows the conductors in the plug to be enclosed and, in particular, sealed from solids and fluids by compressing flexible materials to close gaps which would allow these environments to enter and effect the circuit. Thanks to the conical shape of the sealing nut and the clip, the diameter of the gasket seal decreases as the sealing nut is tightened. The more this diameter decreases, the tighter the fit with the cable. The body O-ring provides a sufficient seal between the body and the body nut, which are tightened together. The coupling nut, on the opposite side, provides the locking with the mating part. The seal between the two mating parts is created by the mating O-ring on the female connector.

2.2.2 Power Connectors

2.2.2.1 Autocom/Intercom

Fig. 1.232: Autocom/Intercom Connectors

There are two main advantages that the Wurth Elektronik 2.54 mm Intercom/Autocom part range offers when compared to similar wire-to-board part ranges. The first of which is the improved quality of materials used in the construction to allow the following technical benefits:

- High contact quality: The contact resistance of the contacts is typically less than 20 mΩ, which ensures good electrical conductivity and reduced losses.
- Secure crimping: The connectors are designed for secure crimping, even with heavy gauge wires, and the crimping force required for the connectors is low, making them easy to assemble.
- Wide temperature range: The connectors are rated for a wide temperature range, from −40 up to +125 °C, making them suitable for a variety of applications.
- Durable construction: The connectors are made from durable materials, such as brass and polyamide, that can withstand harsh environments.
- RoHS compliant: The connectors are RoHS compliant, which means they do not contain harmful substances such as lead and cadmium, making them environmentally friendly.

The second advantage is that there are a variety of output options available for the female mating half, including the conventional crimped connection, as well as multiple output configurations which use a combination of both IDC and discreet wires. This configuration proves very advantageous during the design's prototyping phase, offering an improved level of flexibility in cable options, in addition to offering more options for testing and debugging requirements in the field.

2.2.2.2 ConLECO

The **WR-LECO** (LED Connectors) family, offered by Wuerth Elektronik, is a special range of products dedicated to lighting applications.

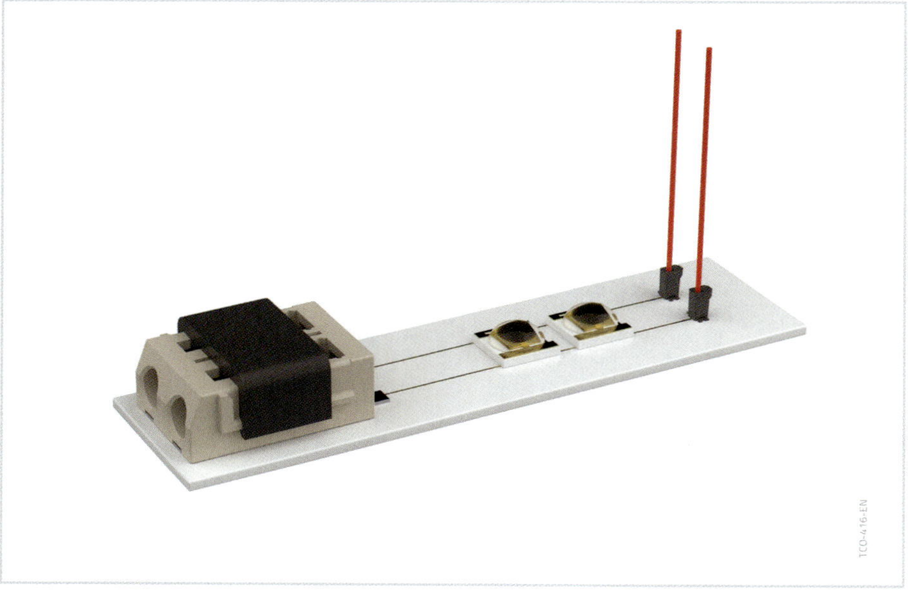

Fig. 1.233: WR-LECO connectors

The main features of this family are:

- Space saving low profile design
- Reflective white color preventing shading
- Environmentally compatible with Lead free and Rohs rating
- SMT mounting for quick automation

WR-LECO Wire-to-Board

About the Wire-to-Board we offer two solutions:

Single contact Poke-In Connector

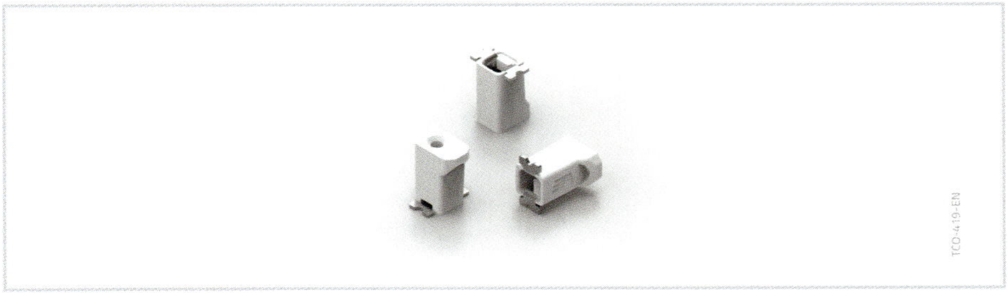

Fig. 1.234: 695 401 000 222

- Current Rating 7 A
- Solid wire only
- Vertical entry
- Rapid termination with back hooking barb to bite into the solid conductor

WR-LECO 300/310 series is a 4.00 mm pitch SMT Horizontal connector with slider actuator made for power supply connections, which offers the benefit of high current connection combined with an easier cable insertion thanks to the locking/unlocking technology of the slider. The main features of this family are:

Fig. 1.235: 695 40X 3X0 222

- Pitch: 4 mm
- Polarities :1 & 2 poles
- Current Rating : 9 A for 300 series & 6 A for 310 series
- Working Current: 600 V (AC) for 300 series & 250 V (AC) for 310 series

- Slider actuator
- Solid wire only
- Cable size: 20 to 18 AWG for 300 series and 24 to 22AWG (310 series)

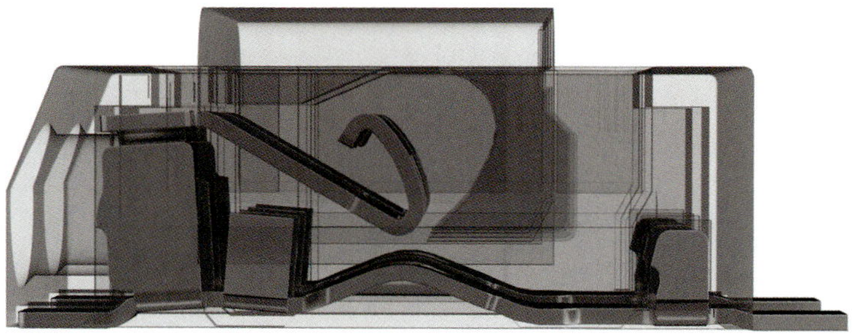

Fig. 1.236: Mechanical detail of locking slider

2.2.2.3 ConMPC

MPC families consist of Wire to Board (WtB) and Wire to Wire (WtW) power connectors: MPC3, or Micro Power Connectors, with a pitch of 3.00 mm, and MPC4, or Mini Power Connectors, with a pitch of 4.20 mm. Both types have positive locking mechanisms to secure the connection actively of the housings and also feature a coding system to prevent mismatching.

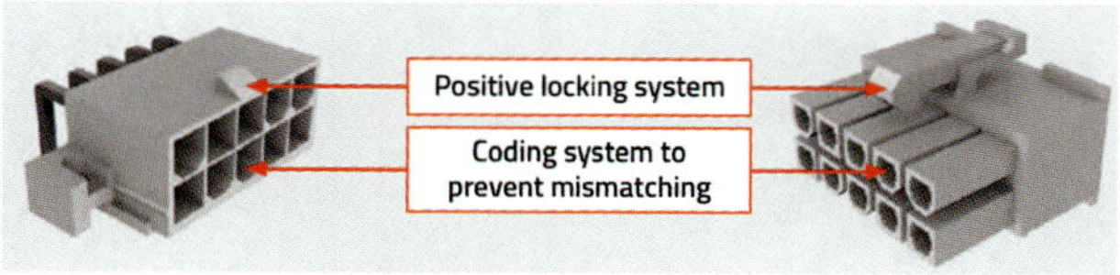

Fig. 1.237: WR-MPCx Series locking feature

For Wire-to-Board (WtB) connections, the matching is achieved by using a male PCB connector and a female housing with female crimping terminals. On the other hand, Wire-to-Wire (WtW) connections are made by using the same female housing but with a male housing featuring male crimping terminals to ensure compatibility between connectors.

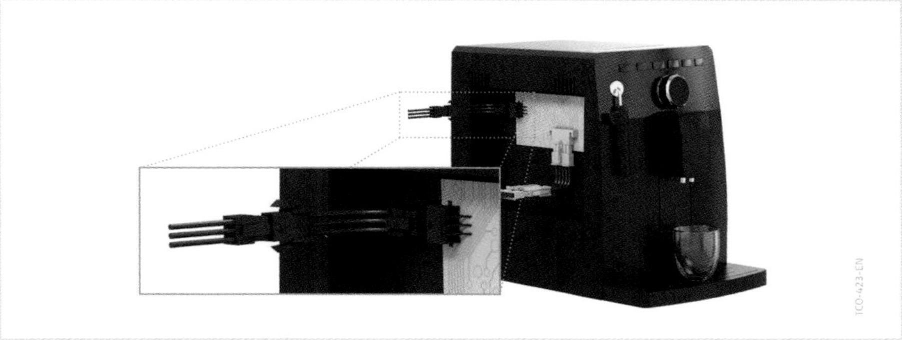

Fig. 1.238: Example of WtB and WtW connections with MPC3

MPC3

The MPC3 family is rated up to 5 A at 250 V and has a withstanding voltage of 1500 V. The temperature range for these connectors is −40 °C up to 105 °C. All the PCB versions are made of LCP plastic and are GWT compatible. Both crimp terminal housings, male and female, are made of PA66. The base material of the pins is copper alloy with tin or gold plating.

In the PCB male connectors, there are THT and SMT versions with different PCB retention systems, such as plastic pegs, retention clips, or SMT solder tabs.

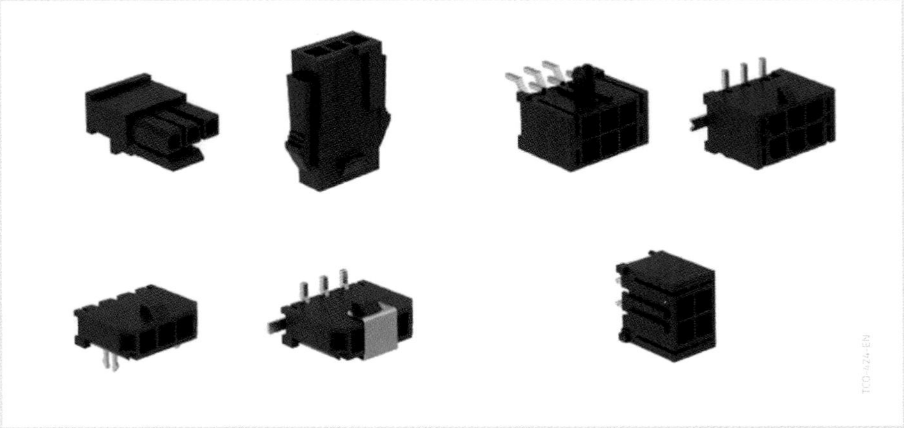

Fig. 1.239: Different versions of MPC3

Male and female crimping terminals are available for cable sections AWG30-26 and AWG24-20. Additionally, there is a female Low Force version that requires less insertion and extraction forces for the connectors, thanks to the design of the terminal. These low-force versions are particularly useful for a high number of poles. The packaging for these crimping terminals includes reels for automated processes and loose contacts for manual crimping to field wire or prototyping.

MPC4

The MPC4 connectors are rated up to 600 V, but the current capacity depends on the number of poles, as mentioned in Table. 1.12. The withstanding voltage for MPC4 connectors is 1500 V, and their temperature range is –40 °C up to 105 °C.

No. Circuits	2–3	4–6	7–10	12–24
Ampere	9	8	7	6

Tab. 1.12: MPC4 current rating based on pin numbers

PCB versions for MPC4 connectors are only available in THT with different PCB retention systems, such as plastic pegs or flanges. Male and female housings, as well as male PCB connectors, are made of PA66 plastic. The base material of the pins is copper alloy with tin plating.

Fig. 1.240: Different versions of MPC4

The single row MPC3 and MPC4 female housings show a special design with slightly larger center pin for power ground to enable a "first mate – last break" setup.

Male and female crimping terminals are available for cable sections AWG28-22, AWG24-18, and AWG16. The packaging for these crimping terminals includes reels for automated processes and loose contacts for manual crimping.

2.2.2.4 ConDCPJ

Our DC Powerjack series offers a wide range of options in terms of current and voltage ratings, making it suitable for various applications. The specifications we provide are as follows:

- Rated current: 2–5.5 A
- Working voltage: 20–30 V (DC)
- Mounting orientation: Horizontal and vertical
- PCB Mounting: Through-Hole Technology (THT) and Surface Mount Technology (SMT)

In addition to these specifications, there are other important considerations to consider when selecting the right DC power jack for your application. One crucial factor is to check the inner diameter of the DC plug (counterpart of the DC power jack) that will be used with it. To make the right choice, ensure that the DC power jack's center pin diameter matches the inner diameter of your chosen DC plug. For example, if the DC plug has an inner diameter of 2.05 mm, you need to select a DC power jack with a center pin diameter of 2 mm for a proper fit. See below explanation.

Ø A (mm)	2.0	2.5
XX	06	08
PLUG INNER Ø	2.05 $^{+0.10}_{-0.00}$	2.05 $^{+0.10}_{-0.00}$
PLUG OUTER Ø	5.5 ±0.05	5.5 ±0.05

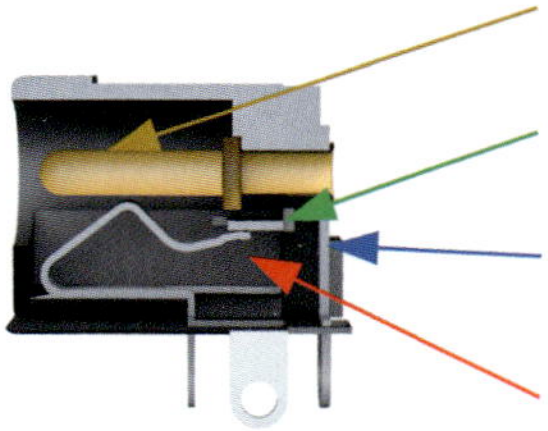

Fig. 1.241: DC Power Jack key elements

All DC power jacks come with on plug detection, an internally integrated normally closed switch. This feature allows the advantage of being able to detect when a plug is inserted. The state of the switch changes from closed to open when the plug is inserted, enabling the application to switch to the external power supply automatically. This functionality can be particularly useful in scenarios when switching between an external power supply and a battery pack seamlessly.

Considering these factors will help you make an informed decision and ensure that the chosen DC power jack perfectly suits the application's requirements.

2.2.3 High Power Connectors

2.2.3.1 ConTBL

We are offering an extensive range of terminal blocks, providing a wide array of options for current and voltage ratings. Our versatile product line includes various methods to mount cables to printed circuit boards. Before we go deeper into these mounting options, we first mention some specifications.

Specifications:

- Rated current: up to 57 A
- Working voltage: up to 750 V (AC)

- Mounting orientation: Horizontal and vertical, at 45 degrees and 55 degrees
- PCB Mounting: Through-Hole Technology (THT), Surface Mount Technology (SMT), and Through Hole Reflow (THR)

In addition to this essential parameters mentioned above, it's essential to consider the cable mounting on the printed circuit board. Below, we will explore the different cable mounting options we offer.

Screw Types:

Our screw-type connectors come in three options: wire protector, rising cage and excenter clamp. These options slightly differ in the way the cable is mounted. The wire protector connector indirectly clamps the cable using a screw.

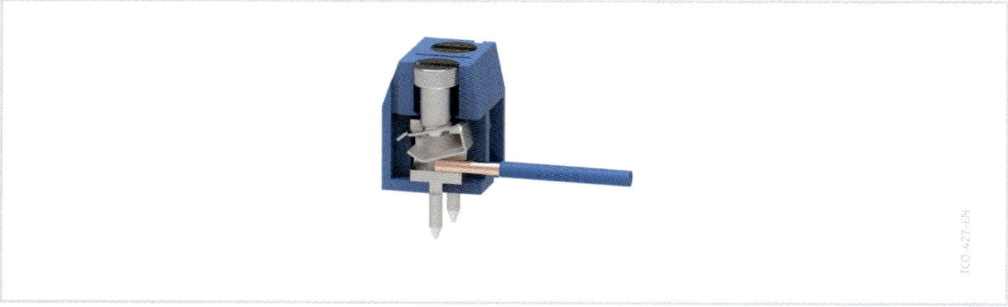

Fig. 1.242: Wire Protector

On the other hand, the rising cage mounting option operates differently. When tightening the screw, a cage moves up, clamping the cable between two metal parts.

Fig. 1.243: Rising Cage

The excenter clamp is a special design of fixing the cable with a front screwing feature.

Fig. 1.244: Excenter Clamp

For applications using these screw-type connectors, we recommend setting up a service plan. Over time, contact resistance between the cable and the connector may change due to copper wire relaxation or vibrations, resulting in lower pressure and higher contact resistance, potentially leading to increased power losses and temperature rise.

Screwless Mounting Options:

Our product portfolio includes two screwless options: spring clamp and push-in spring. With the spring clamp option, you need an actuator or tool/screwdriver to open the spring when mounting or releasing the cable.

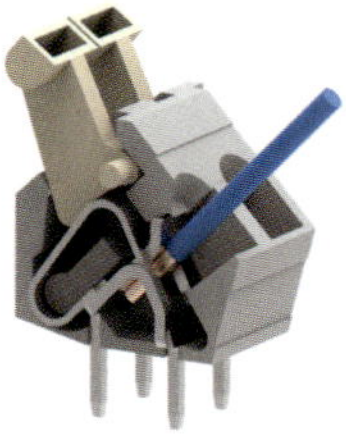

Fig. 1.245: Spring Clamp

Compared to the spring clamp connector, you don't need to open the spring when inserting a solid cable. However, for stranded wire cables, the spring needs to be opened during insertion. When combining a push-in connector with a solid wire cable, installation time is reduced, leading to lower labor costs – a cost-effective solution, especially for extensive cable installations.

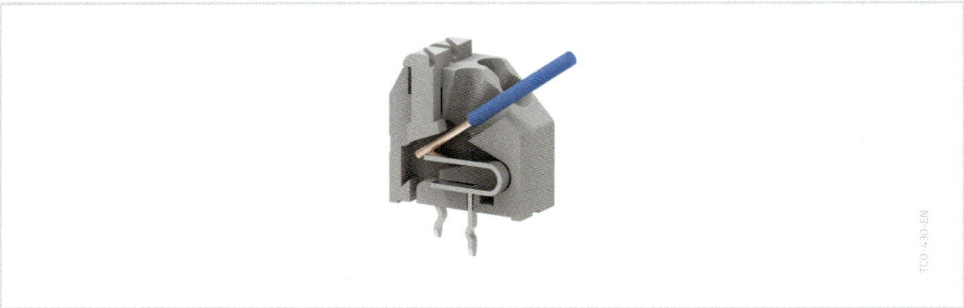

Fig. 1.246: Push-In Spring

Both screwless options can be used in environments with vibrations. Unlike screw-type connectors, the contact pressure remains constant when using the appropriate cable cross-section or American Wire Gauge (AWG). The spring automatically compensates for any relaxation of the copper wire, maintaining a constant contact resistance and temperature, eliminating the need for a service plan to check temperature rise.

Pluggable Solutions:

Our pluggable solutions are ideal for applications where quick replacement of the printed circuit board or cable harness is necessary, ensuring high-speed serviceability. We offer low-profile versions, combining pin headers and low-profile counterparts for easy mounting on pins.

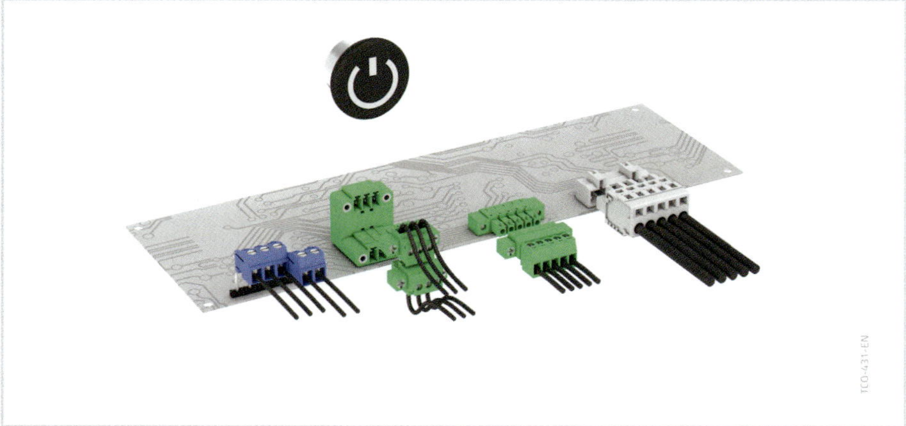

Fig. 1.247: Pluggable Solutions

We also provide double-level options for space-constrained environments with less height restrictions. Furthermore, our pluggable solutions for high-vibration environments come in various designs, including screw flanges, plastic latches, and friction locking between the PCB header and plug.

Additionally, our pluggable terminal block portfolio includes a unique solution where the printed circuit board supplies voltage/current to a cable. In this case, the PCB headers are female connectors, providing enhanced safety by reducing the risk of touching live parts. These connectors open up more possibilities, such as wire-to-wire and board-to-board solutions.

For all pluggable solutions, we offer both screw-type and screwless mounting options, with all the advantages mentioned earlier. Moreover, in the 3.81 mm and 5.08 mm pitch plug, we provide an IDC (Insulation Displacement Connection) solution. This eliminates the need to strip solid cables; instead, the cable can be directly inserted, and an actuator with a specially designed fork cuts through the insulation to create a nearly gas-tight connection with minimal temperature rise and excellent stability. However, please note that this connector can be used only once, because of the deformation of the fork in the contact area of the cable and contact.

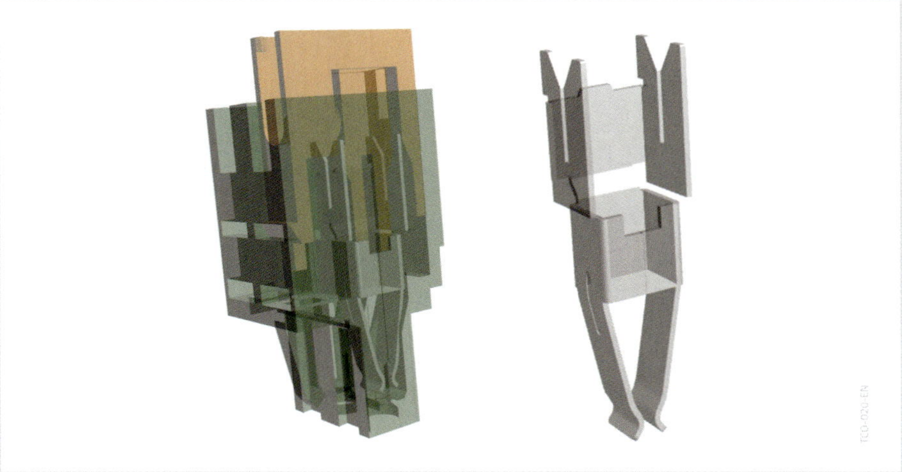

Fig. 1.248: IDC Plug

Terminal blocks can be used in numerous ways, particularly in applications requiring high serviceability. They are perfect for environments where using crimping tools to mount crimp contacts to a cable is challenging or where consistent contact pressure is necessary. The big diversity in dimensions, voltage, current ratings, as well as the various mounting options on printed circuit boards and cables make terminal blocks suitable for a wide range of applications. In fact, terminal blocks find use in almost every application that requires reliable electrical connections.

2.2.3.2 ConFAST

The Fast connectors are used when a simple, robust and easy connectable connector is needed for power supply applications up to 16 A. It is mainly used as a wire to board connector but there is also an option to use it as a board to board connector together with a female PCB receptacle.

The connector is designed with a flat metal tab which can be mated with a crimped cable lug, a standard flat push connector on the wire side. The cable lug can easily slide over the tab and create a reliable electrical connection.

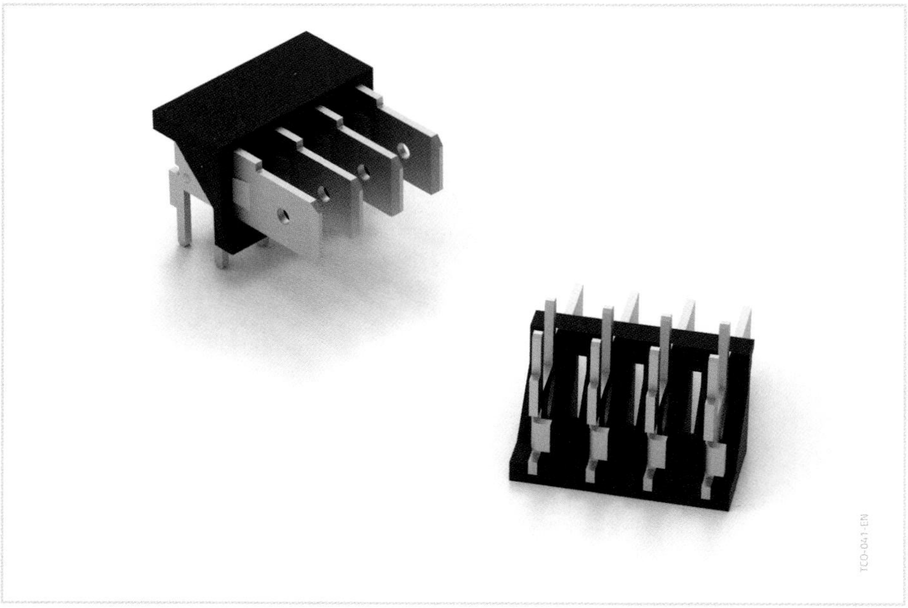

Fig. 1.249: WR-FAST Angled 6.3

Fig. 1.250: WR-FAST Vertical 2863

The Fast connectors are available in either 5.08 or 7.62 mm pitch and comes with two different blade designs. One is 6.3 and fits the cable lug with the mating blade width 6.3 mm. The other blade is splitted and therefore one or two cable lugs can be used at the same time. If two cable lugs are used, each of the cable lug have the mating blade width 2.8 mm.

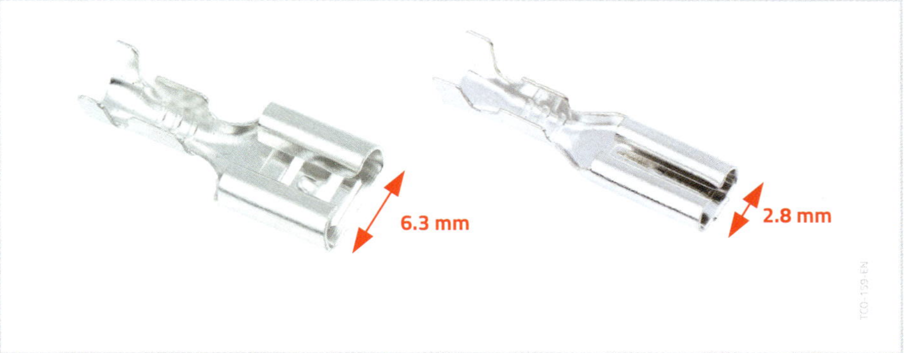

Fig. 1.251: FAST blade dimensions

The Fast connectors are common in Industrial machines and White goods applications. The plastic is rated UL94-V0 and meets glow wire IEC 60335-1.

Fig. 1.252: Application example of WR-FAST

They are available in 2–8 poles as standard. However there is also an option, if using only one single wire or if there is no need for the plastic house, to use loose mounting tabs, WA-MTAB.

2.2.3.3 ConFSH

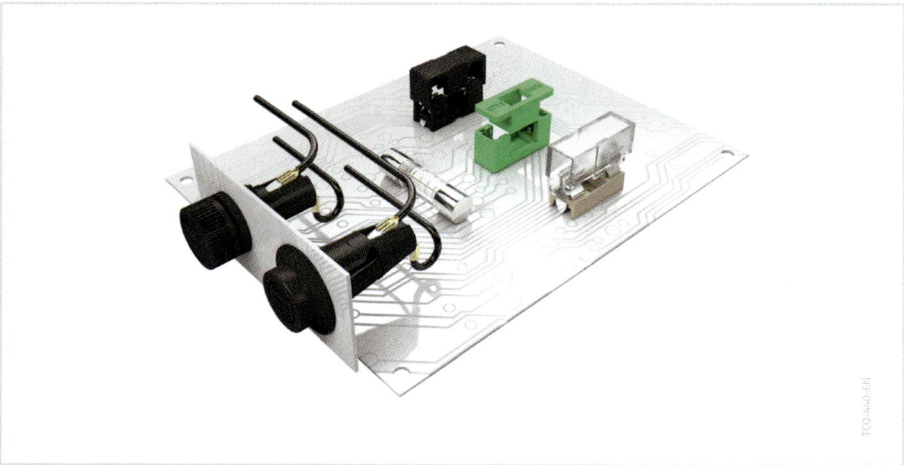

Fig. 1.253: WR-FSH

Fuse holders are connectors where miniature protection fuses are attached. The size of these holders is defined by the standard IEC 60127-1. For the ConFSH family, 5 x 20 mm and 6,3 x 32 mm fuses are compatible.

Fuse holders have different levels of protection against electrical shock depending on the type, as per the IEC 60127-6 Standard:

- PC1: Fuse without integral protection against electric shock (non IP2x).
- PC2: Fuse with integral protection against electric shock. IP2x with the fuse when putting or removing a fuse in/out of the fuse holder.
- PC3: Fuse with enhanced protection against electric shock. IP4x with the fuse when putting or removing a fuse in/out of the fuse holder.

Another important parameter for fuse holders is the Protection Class defined by the IEC 61140 standard, which explains the Protection Class of electrical products:

- Class 0: No protection, no link to earth.
- Class 1: Chassis must be linked to earth. The fuse must cut the circuit fast enough in case a phase touches the chassis for human protection.
- Class 2: Products that have double protection against electric shock for human protection.
- Class 3: Products that have a power supply with extra-low voltage that has no impact on human safety.

Two types of mounting can be found in the ConFSH family: PCB and Panel. PCB versions are also divided into two types: Clip Cover and Turn and Lock Cover. PCB clip cover fuse holders are PC1 and Class I, but when the cover is added, they become Class II due to the plastic casing. PCB turn and lock versions are PC2 and Class II. Meanwhile, all the panel versions are always PC2 and Class II.

2.2.4 Data Connectors

Coming from the computer market, all these connectors can now be found everywhere, especially in our living room where any LCD flat screen is now equipped with USB, HDMI and even sometimes with modular jacks.

They can also be found in their miniaturized version, in digital camera (mini USB) or mobile phone (micro USB).

These connectors are used to allow communication between different external modules. They exist in different soldering technologies (TH or SMT) depending on the final applications.

Here below (Figure 1.254) are some example of the most popular Data connectors.

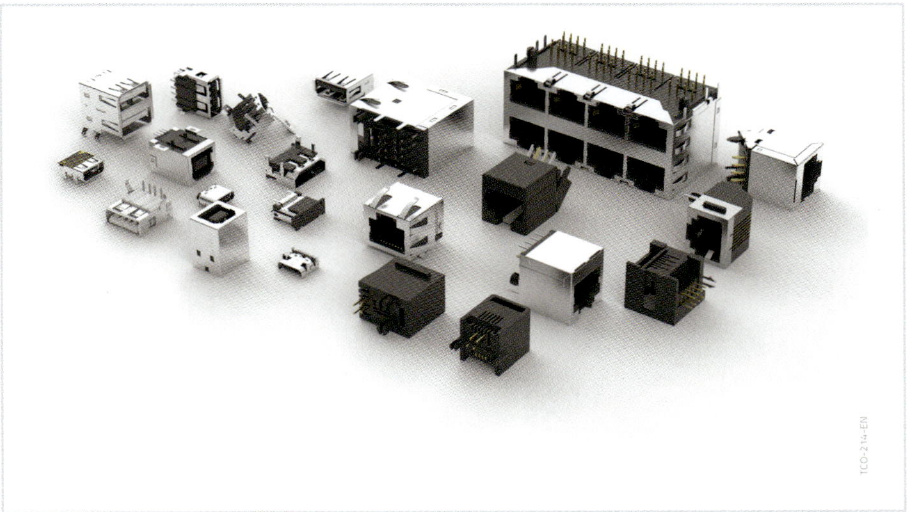

Fig. 1.254: Different Data connectors

2.2.4.1 ConBTB

This connector series facilitates the interconnection of printed circuit boards (PCBs) by stacking them in parallel. It offers a diverse range of features, including two different pitches, various pin counts, and multiple stacking heights.

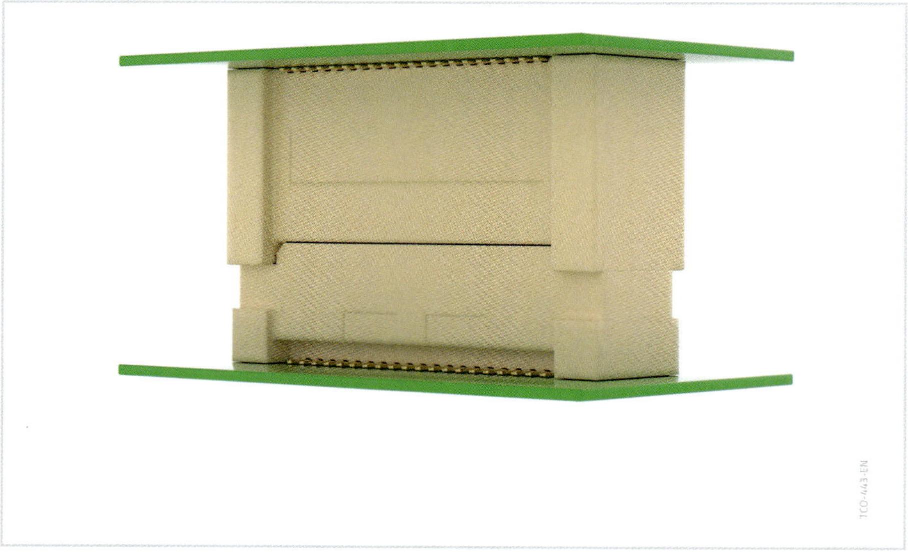

Fig. 1.255: BTB connection

The specifications we can offer are:

- Pitch: 0.8 mm and 1,0 mm
- Polarities up to 100 pins
- Polarized housings to avoid miss connecting .
- Rated current up to 1 A
- Working voltage up to 100 Vac
- PCB mounting: Surface Mount Technology (SMT)
- IEEE1386 compatibility for the 1 mm pitch BTB connector

One of the key design aspects is the floating design, which ensures that small misalignments during connection do not harm the connector contacts. This feature prevents damage to connectors and contacts, especially if soldered not at the exact place on the PCB, allowing for a small tolerance in the assembly process.

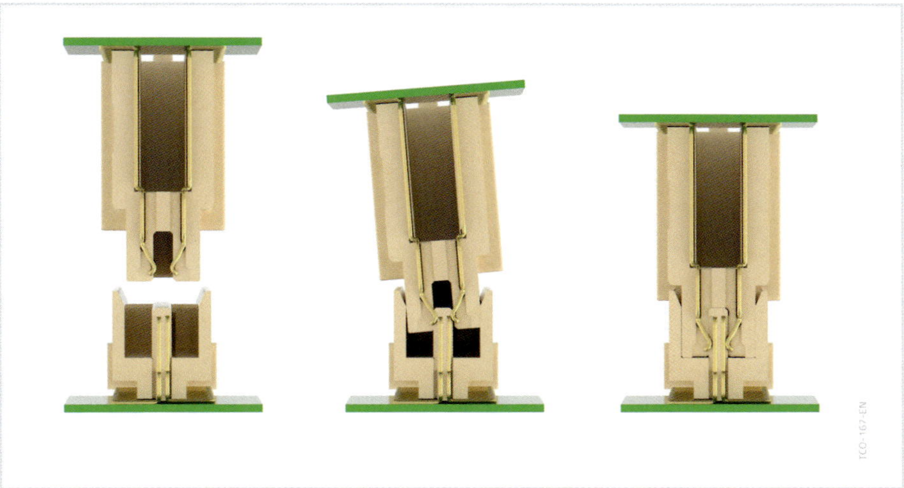

Fig. 1.256: Floating design BTB series

In today's electronic market, miniaturization is a growing demand, and this connector series can meet this need. However, with miniaturization comes the challenge of transmitting a large number of signals within limited space. The increasing signal speeds and data rates, such as USB, PCI Express, and SATA, require careful consideration of impedance characteristics.

Different protocols have varying requirements for line impedance, and some use differential pair signaling for high-speed data transmission. Therefore, it's essential to take these factors into account when using this connector series. Mechanical considerations are also critical in determining whether a specific signal or protocol can be efficiently transmitted.

2.2.4.2 Ethernet & Modular Connectors

Introduction

Today, modular jack connectors and Ethernet protocols can be found in most consumer and industrial applications, among others. There are so many options that it becomes difficult to choose the right one.

This section aims to give a brief overview of the Modular Jack connectors technology and Ethernet protocols in order to get a better understanding of potential applications, and to help design engineers to choose the right connector for their application.

From There to Now

Modular Jacks were initially a category of standardized connectors intended for phone communication. As the Ethernet adopted these connectors for its physical layer, new standards arose for high-speed data transmission and power transmission, beyond the needs for simple voice-level frequencies and power, fitting the new uses.

A History of Modular Connectors

Before Modular Connectors, telecommunication equipment worldwide was wired with all kinds of connectors, highly dependent on the equipment and the country of origin. No standard existed and cross-compatibility was non-existant.

From a Monopoly...

Historically, modular connectors were patented by Bell Telephone Laboratory and supplied by Western Electric Company (a subsidiary of AT&T, originally Bell Telephone Company) in the mid-60s). They started to replace hard wiring in the early seventies in the USA. The Bell System hold a total monopoly over phone network and equipment in the USA, and modular connectors were convenient to add some measure of interchangeability between cords, cables and equipment sold by AT&T PhoneCenter stores. Otherwise, Bell would not allow any third-party equipment on its lines. Those connectors would be roughly the same shape, with an even number of contacts (2 to 8) on the bottom and a latch on top. There was three different shapes, labeled as 4p, 6p, 8p, related to the maximum number of contacts those connectors could host.

... to a Universal Standard

In 1974, the US Justice Department began to break the monopoly of the Bell System over the USA. In 1976, the modular connectors were made a standard for the customer-side phone equipment in the USA, under the Registration Interface Program. This program named the different Registered Jacks, or RJ's, under the national law. In 1984, the Bell System would be definitely broken up.

Widespread adoption

In 1986, the IEEE 802.3e makes the unregistered modular connectors "unkeyed 8p8c" the standard for the nascent Ethernet over twisted pair protocol. The "RJ45" (the 8p8c – eight positions and 8 contacts modular connector) and other modular connectors can take over the world.

The Ethernet Revolution

Ethernet comes from the word Ether, the hypothetical medium supposed to propagate electromagnetic waves according to scientific concepts of the 18th and 19th centuries. This protocol was initially inspired by the ALOHAnet, in operation since 1971 and based on the radio protocol ALOHA (Hello).

Early History

At that time, manufacturers had all developed their own communication protocols (SNA for IBM, DECnet for DEC, Ethernet for Xerox, DSA for Bull, etc.). It rapidly became clear that it would be impossible to interconnect these systems without creating an international standard.

In 1977, the International Standards Organization (ISO) created a complete set of recommendations regarding compatibility requirements; the aim was to enable the connection between open systems, i.e. systems using the same protocols, in order to exchange data.

IEEE 802.3

The initial results obtained with Ethernet in 1980 drove Xerox, Intel and DEC to join their forces to develop this network and to put an end to all private protocols. The same year, the IEEE started project 802 to standardize local area networks. Ethernet was a candidate for LAN standardization.

In 1984, ISO published an update of their recommendations under the name: Reference OSI (Open Systems Interconnection). Detailed information about the OSI Model you will find in our design guide Trilogy of Magnetics. This update has become an international standard and serves as a guide. The same year, Ethernet over Twisted Pair was demonstrated as a valid technology by AT&T, using modular connectors.

In 1985, the IEEE finalized the standard 802.3 regarding Ethernet. In 1987, the IEEE released the 802.3e standard on Ethernet over Twisted Pair protocol, using modular connectors.

Protocols

Communication in an Ethernet network is achieved with the help of a protocol. This protocol will describe all the characteristics for transmission, following those guidelines will allow also a specified data rate to be achieved.

IEEE has provided over the years different standards called 802.3xx protocols. You can find some of the most common ones for modular connectors in Table. 1.13

Year	Standard	Protocol	Data rate	Applications
1990	802.3i	10BASE-T	10 Mbps	Phone, LAN
1995	802.3u	100BASE-T	100 Mbps	LAN
1999	802.3ab	1000BASE-T	1 Gbps	LAN, Gigabit Ethernet
2016	802.3bz	2.5GBASE-T	2,5 Gbps	
2016	802.3bz	5GBASE-T	5 Gbps	
2006	802.3an	10GBASE-T	10 Gbps	
2016	802.3bq	25GBASE-T	25 Gbps	Data centre, Video
2016	802.3bq	40GBase-T	40 Gbps	Data centre, Video

Tab. 1.13: Ethernet Protocols for modular connectors

Nowadays, the most popular ones are Fast Ethernet (100BASE-T4) and Gigabit Ethernet (1000BASE-T). Gigabit Ethernet allows even uncompressed video signal to be transmitted through Ethernet, making remote screens and cameras possible up to 100m from their bases.

Ethernet over Twisted Pair

Modular plugs are used with copper wire. It is a good compromise between cost, easiness of use and performance. However, single-wire copper circuits are limited in electromagnetic properties, and are vulnerable to interference. To accommodate the ever-higher performances demanded by phone, then network applications, cabling had to follow suit.

The Twisted Pair

In 1881, Alexander Graham Bell invented the twisted-pair cable, a solution to the disruption caused by electrification and power lines on the telephone and telegraph network. It was an evolution of the wire transposition as shown on From wire transposition to twisted pair. From the very beginning, the twisted pair cabling remains a very popular one in most applications, even if optical fiber and satellite communication have their own application fields.

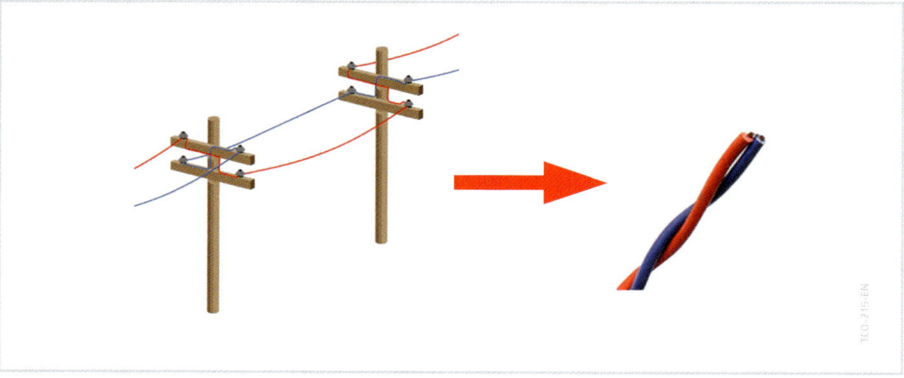

Fig. 1.257: From wire transposition to twisted pair

A twisted pair is made up of 2 copper conductors; each conductor is insulated by a dielectric (non-conductive material) such as PVC (Poly Vinyl Chloride), Polyethylene, Teflon, fluoropolymer, or any other low smoke, fire retardant substance. This insulation layer separates the conductors from each other in such a way that the electrical circuit is not shorted and protects conductors against physical damage.

As shown on Twisted pairs, insulated conductors are twisted in a helix with a constant pitch (or distance) in order to make a 360° twist. Each wire in the pair is exposed to the same interference sources, and this interference can be eliminated through common-mode rejection. Two cables of the twisted pair give same signal but in opposite voltage. Addition of electromagnetic waves given by the two wires is close to 0. It easily avoid interference between twisted pairs. Therefore, a twisted-pair design is both more resilient to outside interference, but also less prone to cause interferences in other devices.

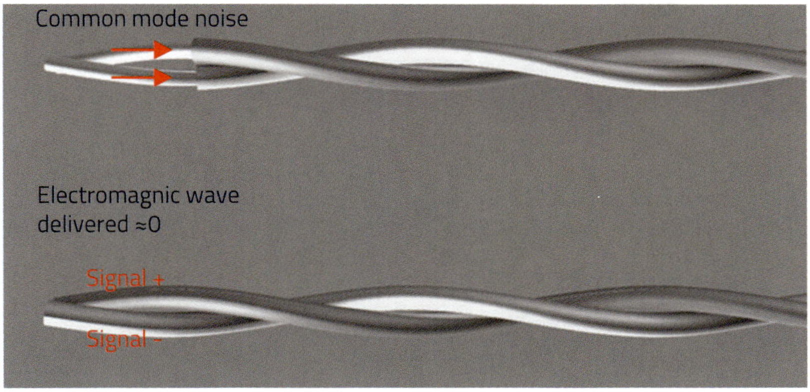

Fig. 1.258: Twisted pairs wires

A cable can have multiple pairs. All pairs will have a different twist rate to avoid interferences between them. As seen on Figure 1.259, the pars will have different numbers of twists per 10 cm:

- 1st pair: 5,5 twists
- 2nd pair: 6,5 twists
- 3rd pair: 7,5 twists
- 4th pair: 8,5 twists

The higher the twist number, the higher the cable quality.

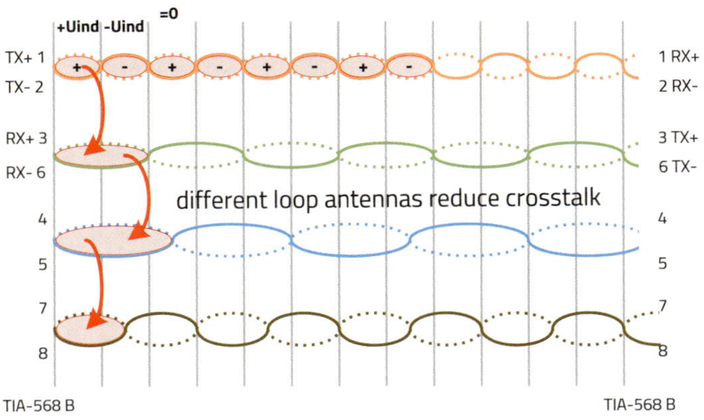

Fig. 1.259: 4-pair twist length comparison

Ethernet protocols can use 1, 2 or 4 pairs. At high speeds, the quality of the twisted pair becomes paramount as inner and outer interferences can destroy the signal integrity. Thus, we can define a Cable Rating, which is the maximum frequency a cable can be used for.

Shielding

Depending on the required signal integrity, noise-level of the environment (high noise or not) and on the signal frequency, one can decide to use shielded cables or not. Thus, different twisted-pair cable classes exist.

Different naming conventions for twisted pair cables exist. In the description below, we decided to use the old convention in plain and the new one in parentheses. Here are the most popular designation according to ISO/IEC 11801, and their schematics in Twisted-Pair Cabling Categories:

- UTP (U/UTP): Unshielded Twisted Pair (unshielded cable/unshielded twisted pairs)
- FTP (F/UTP): Foiled Twisted Pair (foiled cable/unshielded twisted pair)
- STP (U/FTP): Shielded Twisted Pair (unshielded cable/foiled twisted pairs)
- SFTP (SF/UTP): Shielded Foiled Twisted Pair (shielded&foiled cable/unshielded twisted pair)
- SSTP (S/FTP): Shielded Foiled Twisted Pair (shielded cable/foiled twisted pair)
- FFTP (F/FTP): Foiled Twisted Pairs (foiled cable/foiled twisted pair)

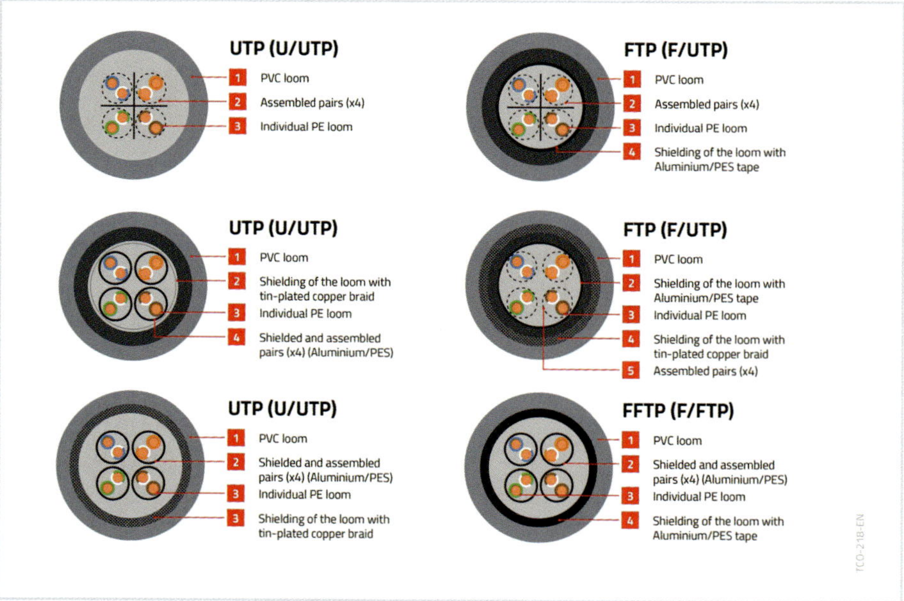

Fig. 1.260: Twisted-Pair Cabling Categories

Deciding whether you need to shield the wires or not is not easy. It's not only about the higher cost that it implies, but also about the cable rigidity and its installation which becomes drastically more complicated with the shielding-induced rigidity. Shielding must also be carefully protected during installation in order not to get damaged. Therefore, choosing a modular connector cabling system with or without shielding is an important decision and not a no-brainer.

TIA-568 Cable Categories

The EIA (Electronic Industries Alliance) standardized the use of twisted pair cables and connectors in the TIA-568 standard by defining several categories depending on cable types, system architectures, distances, connectors, and performance characteristics. The scope of this chapter does not enable to discuss everything; therefore, we will summarize the cable types and performances associated with modular connectors:

Protocol	Data rate	Protocol Bandwidth	Cable Category	Cable Rating	Max. distance (m)	Pairs
10BASE-T	10 Mbps	10 MHz	Cat.3	16 MHz	100	2 pairs
100BASE-T	100 Mbps	31.25 MHz	Cat.5	100 MHz	100	2 pairs
1000BASE-T	1 Gbps	62.5 MHz	Cat.5	100 MHz	100	4 pairs
2.5GBASE-T	2,5 Gbps	100 MHz	Cat.5e	100 MHz	100	4 pairs
5GBASE-T	5 Gbps	200 MHz	Cat.6	250 MHz	100	4 pairs
10GBASE-T	10 Gbps	400 MHz	Cat.6A	500 MHz	100	4 pairs
25GBASE-T	25 Gbps	1000 MHz	Cat.8	2000 MHz	30	4 pairs
40GBase-T	40 Gbps	1600 MHz	Cat.8	2000 MHz	30	4 pairs

Tab. 1.14: Ethernet Protocol VS cable categories

TIA-568 generally does not enforce the use of shielding, except for the Category 8 cables, where shielding is mandatory. Therefore, choosing a shielded cable, connector and jack is mostly a matter of expected external noise.

To follow protocols like Ethernet each pin needs to be wired with specific colors. In this case there are two standard layouts as follow:

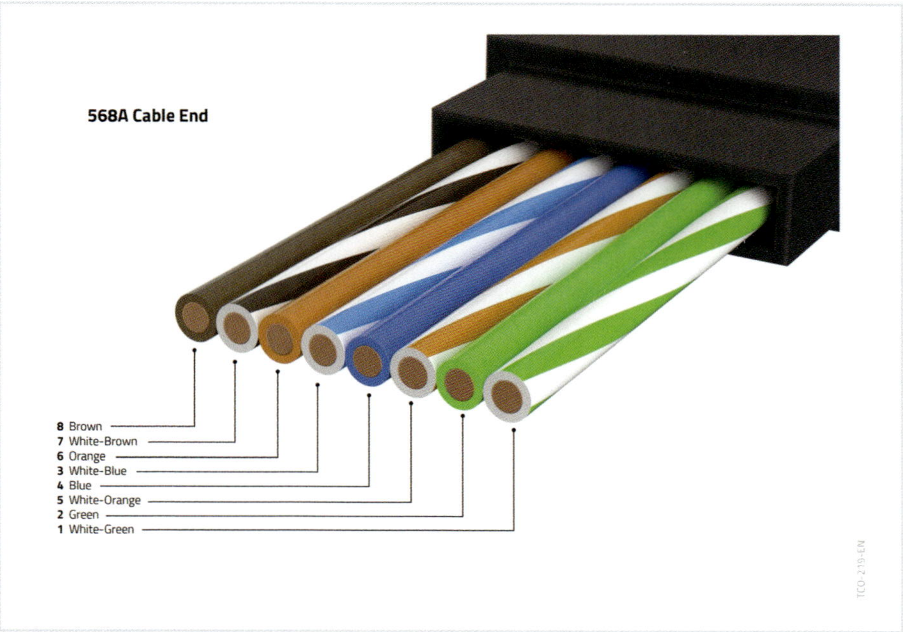

Fig. 1.261: TIE/EIA 568A cabling

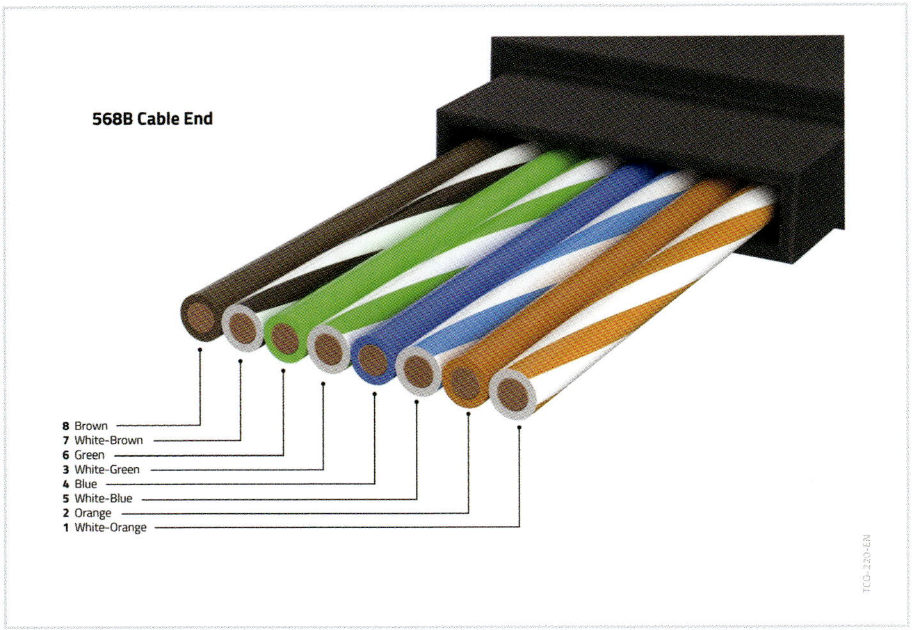

Fig. 1.262: TIE/EIA 568B CABLING

If you compare both wiring, you can easily see that the only difference is that orange and green pairs are swapped, nothing else. Using a crossover cable means that Tx and Rx pairs are reversed. The reason is that before auto MDI/MDIX capabilities became widespread, a cross over cable was necessary to interconnect similar network equipment (hub to hub ...). These cables are still used sometimes to connect two computers or to link industrial equipment which are not able to swap Tx and Rx when the signal arrives on Tx. However, most of today's communication equipment can automatically re-configure themselves (switching Tx and Rx).

The Modular Connector

Modular connectors are available in a male version named Plugs and in a female version named Jacks. The plug is used for the cable termination while the jack is used on fixed locations such as Electronic boards (PCB).

Modular Connector Features

The main features of the modular connector technology are as follows:

Cable Termination

One of the main reasons for the success of modular jacks/plugs is the easy way to perform a cable assembly. Wires are presented into the plug as shown below and then pressed with a crimping tool. The sharp edge of the contact will perform a so-called IDC termination (insulation displacement).

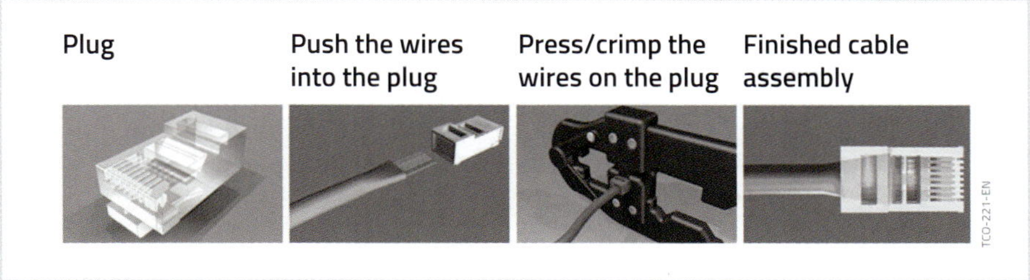

Fig. 1.263: Example of modular plug wiring

This can either be performed with a cheap hand tool or with an efficient automatic tool for higher volumes.

The Latch

Modular connectors are designed to latch securely plug and jacks together thanks to a "tab" which snaps into the jack when connected, so that the plug cannot be pulled out. To be able to unplug, the tab must be pushed down/up while pulling on the plug.

Electrical Properties

Modular jacks share the same electrical properties:

- Rated Current: 1.5 A
- Working Voltage: 120 V_{AC}
- Withstanding Voltage (pin-to-pin): 1000 V_{AC} for 1 minute
- Withstanding Voltage (pin-to-shielding): 1500 V_{AC} for 1 minute
- Contact Resistance: 20 mΩ maximum per pin
- Insulation Resistance: 500 MΩ

Size and Contacts

Modular connectors have standardized designs. Their dimensions make them compatible across manufacturers and is one of the major reasons of their success.

Modular jacks can be described in 2 ways, the first one being a xPxC nomenclature, as follows:

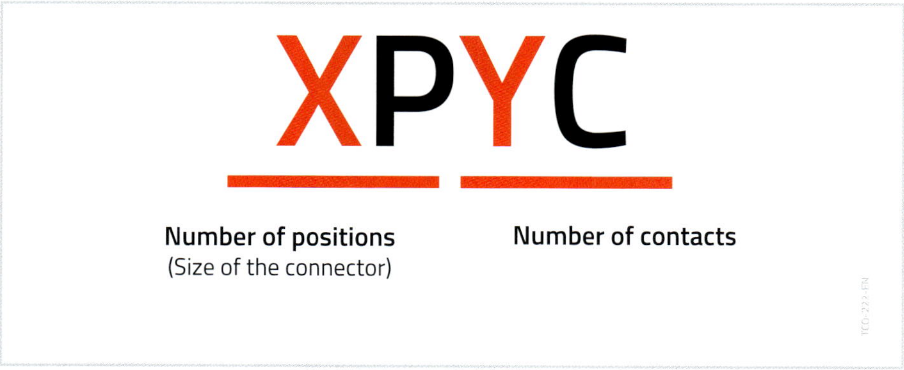

Fig. 1.264: XPXC nomenclature explanation

XP directly correlates with the external dimensions of the modular plug, as follows:

Connector	Length (mm)	Width (mm)	Height (mm)
4P	–	7.7	–
6P	12.4	9.7	6.60
8P	21.5	11.7	8.30

Tab. 1.15: Modular plug dimensions

YC is the number of contacts installed inside of the connector. Y can never be above X and contacts are generally placed from the centermost positions to the external positions.

The Ethernet Jack 8P8C stands for 8 positions, filled with 8 contacts (no empty pole).

The other option to describe these connectors is to use a Registered Jack (RJ) number. It is a standard to name network interface wiring and jacks. A few examples are RJ11, RJ45, RJ21, RJ28, through some of them are unofficial names.

Below you can find a 6P2C modular jack, with 6 visible positions and 2 visible contacts. This is a Registered Jack under the name of RJ11:

Fig. 1.265: A RJ11, a 6p2c modular jack

The 10P10C connector is commonly referred to as the RJ50 connector although it's not a standard registered jack. It has 10 positions fitted with 10 contacts. It is mainly used in proprietary data transfer systems; therefore, we do not find it quite often, except in special applications.

4P Connectors

The 4P4C connector is popularly but most often incorrectly called RJ9. This is not correct because registered connectors only classify the connection between different phones, or a phone and a network, and not between 2 parts of a phone as it is often the case here. Indeed, this connector is also called the handset connector since its most common use is to plug the telephone handset to the receiver:

Fig. 1.266: 4P4C (RJ9) telephone cord

6P Connectors

6P2C (RJ11) is certainly the most familiar registered jack since it is used for single line telephone jacks in most homes across the world. 6P4C (RJ14) is similar but for 2 lines and 6P6C is for 3.

The most common configuration of Jacks is 6P4C but often only connected to a 6P2C cable (RJ11 cable):

Fig. 1.267: (Left) French T-Plug (Phone wall socket adapter) with 6P4C Line (Right) 6P2C RJ11 cable

8P Connectors

8P8C connectors are generally used for Ethernet applications. Their pinning is then set as follows:

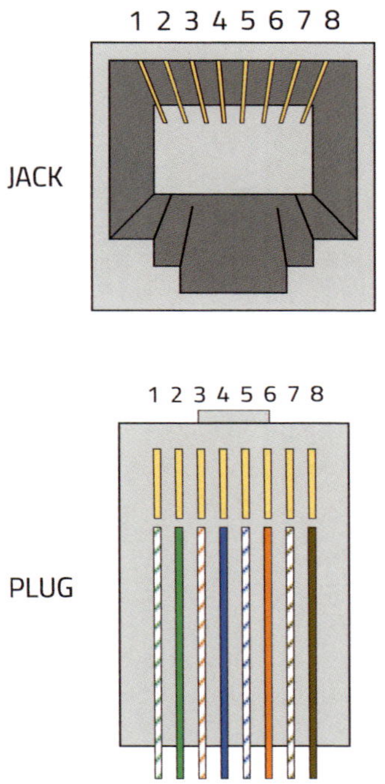

Fig. 1.268: 8P8C (RJ45) pinning

Even if they are popularly called RJ45 in the context of Ethernet and cat5 cables, it's not correct to identify 8P8C connectors as RJ45. In telephone systems, a standard RJ45S socket has a key which excludes insertion of a standard 8P8C plug. However, the unkeyed 8P8C version has spread so widely thanks to the computer market that the name RJ45 often designates the 8P8C without keying (which is not its original/ official definition).

8P8C jacks support a physical connection with RJ11/RJ14 & RJ25 plugs but only RJ11 and RJ14 have full electrical compatibility. Ethernet compatible pinout split the third pair of RJ25 into two separate cable pairs making it unusable by similar phones (it was necessary to preserve the signal integrity for Ethernet, which operates at much higher frequencies than telephone lines).

Design Specificities

In the Würth Elektronik catalog, a wide range of products is offered to designers to help them select the right Modular Jack for their applications.

Tab Up/Down

The most common way to install a jack, is to do it with the tab down; this makes it easier to operate (with thumb on top & index on tab). It's also better because dust will tend to fall down instead of stacking on the contact area (tab down means contact up!). However, since modular jacks can be mounted on the bottom of a PCB and that every product is different, modular jacks exist in up, down, vertical or even angled versions.

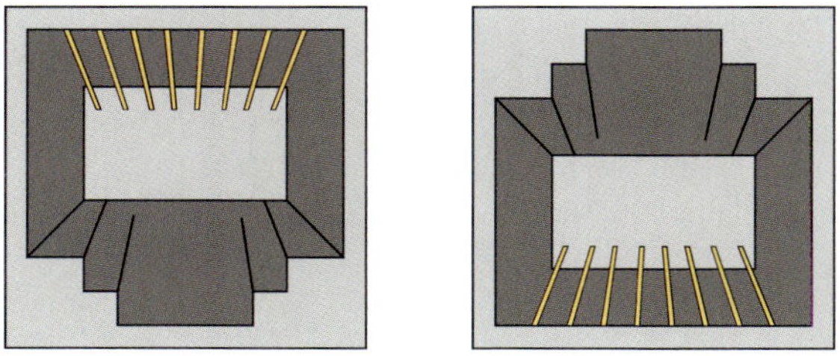

Fig. 1.269: Tab up/down conversion

Ethernet Categories and Signal Integrity

We have seen in our first part that the trend for Ethernet standards is to increase in speed, and frequency of signals, even if a better coding allows more bits of data to be transmitted in regards to frequency.

Protocol	Signal Data Rate	Bits/ frequency ratio	Protocol Bandwidth	Connector Category
10BASE-T	10 Mb/s	1 b/Hz	10 MHz	Cat.3
100BASE-T	100 Mb/s	3.2 b/Hz	31.25 MHz	Cat.5
1000BASE-T	1000 Mb/s	4 b/Hz	62,5 MHz	Cat.5
2.5GBASE-T	2500 Mb/s	6.25 b/Hz	100 MHz	Cat.5e
5GBASE-T	5000 Mb/s	6.25 b/Hz	200 MHz	Cat.6
10GBASE-T	10000 Mb/s	6.25 b/Hz	400 MHz	Cat.6A
25GBASE-T	25000 Mb/s	6.25 b/Hz	1000 MHz	Cat.8
40GBase-T	40000 Mb/s	6.25 b/Hz	1600 MHz	Cat.8

Tab. 1.16: Ethernet Protocols and associated 8P8C connectors categories

As transmission rates increase toward the gigabit per second, several well-known phenomena become problematic. One is the generation of electromagnetic interference at physical interfaces. We observe a coupling phenomenon that is the result of an energy transfer from one channel to another causing interference. Regarding signal transmission, we take into account two kinds of coupling: the Inductive coupling and the Capacitive coupling.

At high frequency, the capacitive and inductive coupling will have an equivalent behavior. This is discussed in detail in the Basics chapter as Impedance. TIA-568 Standard asks us to measure this crosstalk effect as part of the Scattering Parameters, a set of measurement in the frequency domain. The S-parameters are defined by the matrix linking the incident power waves and the exiting power waves, for any component. In our case, as we study a 4-port device, there should be 16 S-Parameters

$$\begin{pmatrix} b_1 \\ b_2 \\ b_3 \\ b_4 \end{pmatrix} = \begin{pmatrix} S_{11} & S_{12} & S_{13} & S_{14} \\ S_{21} & S_{22} & S_{23} & S_{24} \\ S_{31} & S_{32} & S_{33} & S_{34} \\ S_{41} & S_{42} & S_{43} & S_{44} \end{pmatrix} \begin{pmatrix} a_1 \\ a_2 \\ a_3 \\ a_4 \end{pmatrix} \qquad \text{(EQ 23)}$$

Ideally, only the differential mode parameters need to be examined, and these are consolidated into four distinct range values:

- Insertion Loss
- Return Loss
- Near-end Crosstalk (NEXT)
- Far-end Crosstalk (FEXT)

Insertion and return loss are the characteristics of the transmission line, they will mostly depend on the Impedance of the line. FEXT and NEXT will depend on the parasitic coupling between pins and the actual interference from one pair to the other pairs. The impedance of the line is, by standard, 100 Ohm.

If looking at the whole chain between the two chips transmitting Ethernet information from one device to the other, the connectors and cables have to adhere altogether to a certain loss budget from crosstalk and return loss. Since the cables are 100 m long, most of the signal losses will happen in the cables, leaving tight S-parameters requirements for the jacks (Figure 1.270). The plug losses are measured with the cables. EIA does not define return loss and FEXT for cat 3

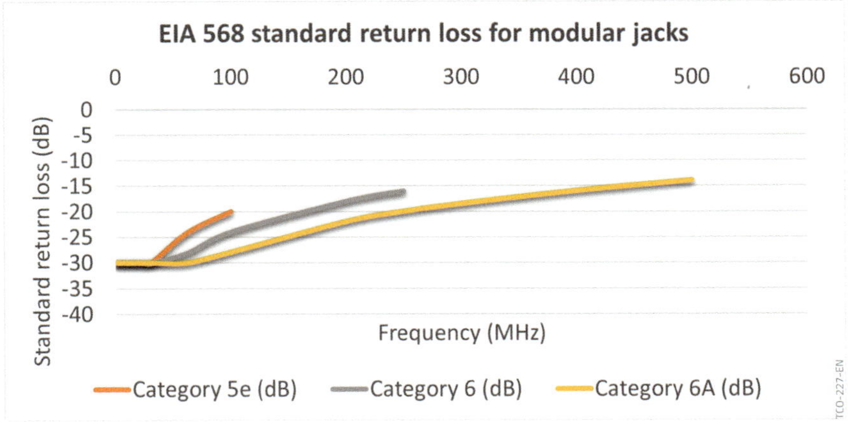

Fig. 1.270: EIA standard return loss for modular jacks

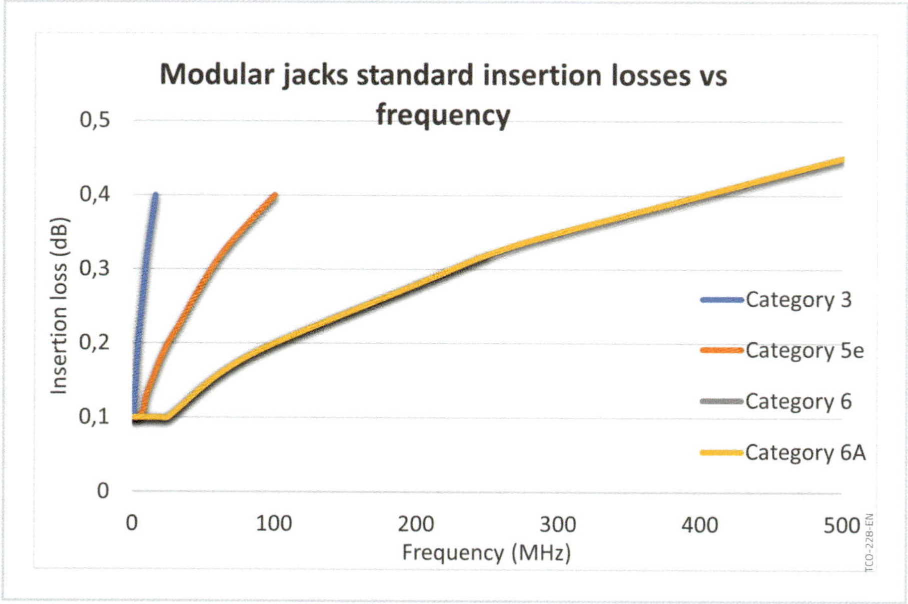

Fig. 1.271: Modular Jacks standard insertion losses vs frequency

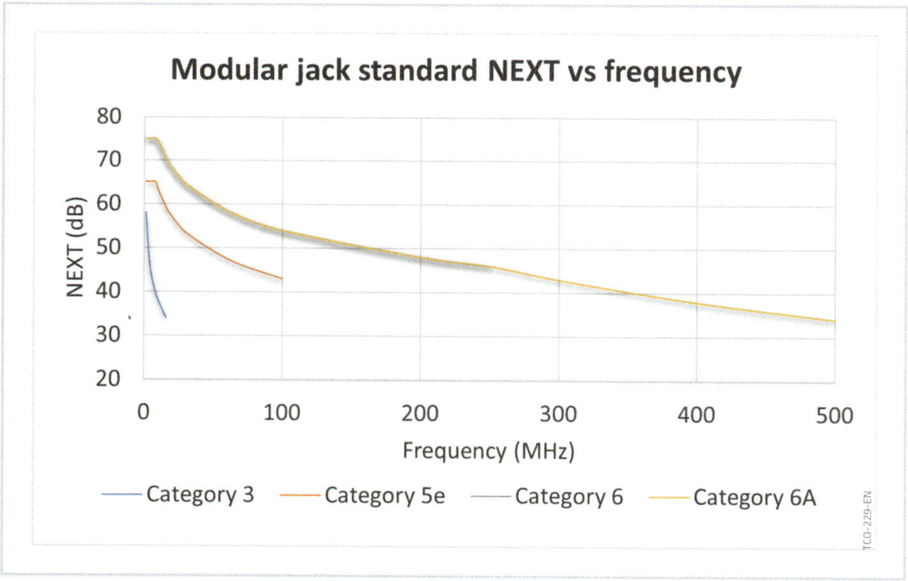

Fig. 1.272: Comparison of NEXT vs. frequency

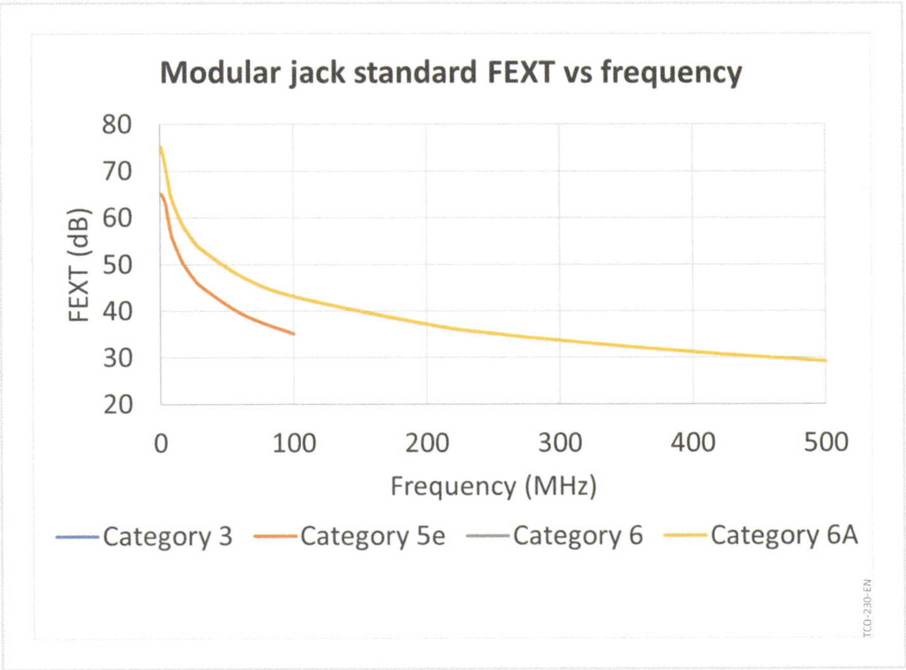

Fig. 1.273: Comparison of FEXT vs. frequency

As a way to keep the S-Parameters in line with the TIA-568 standards, modular jack design has been adapted. Impedance adaptation is necessary above Cat.5, and two technologies compete with each other: pin bending and internal PCB.

Fig. 1.274: Impedance adaptation in 8p8c modular jacks

In addition, to help the PCB designers with the delay skew between the signal pairs and especially for the interpair-skew, connector manufacturers created special cat5 connectors with a different pinning. As seen below, pins 1/2 and 4/5 have been shifted to compensate potential delays and to make it easier to remain below the maximum 45ns delay skew allowed by EIA.

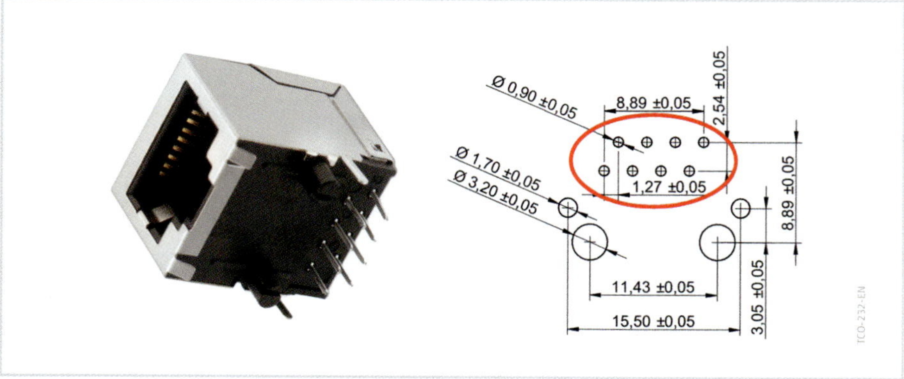

Fig. 1.275: 615008145521 8P8C Modular Jack with optimised delay-skew PCB

LEDs and Shielding

Manufacturers of PHY (according to the Ethernet standard) enable visual control via LEDs on the modular jacks.

Shielding of the jack allows continuity with shielded modular plugs, which in turn allows continuity with shielded or foiled cables. Thus signal integrity is protected. However, since cable shielding is not a feature related to the Ethernet category (up to Cat.8) but the expected EM interferences in the system, the choice of having a shielding or not is purely up to the application designer.

The Physical Layer

The 8P8C modular Jack (improperly called RJ45) is the first element of the Circuit Board Physical layer often named MDI (Medium Dependent Interface).

The PHY

An active device called the PHY controls this physical layer. Further descriptions about the physical layer you can find in the design guide Trilogy of Magnetics.

Below is the basic diagram of a physical layer for the 100BASE-T:

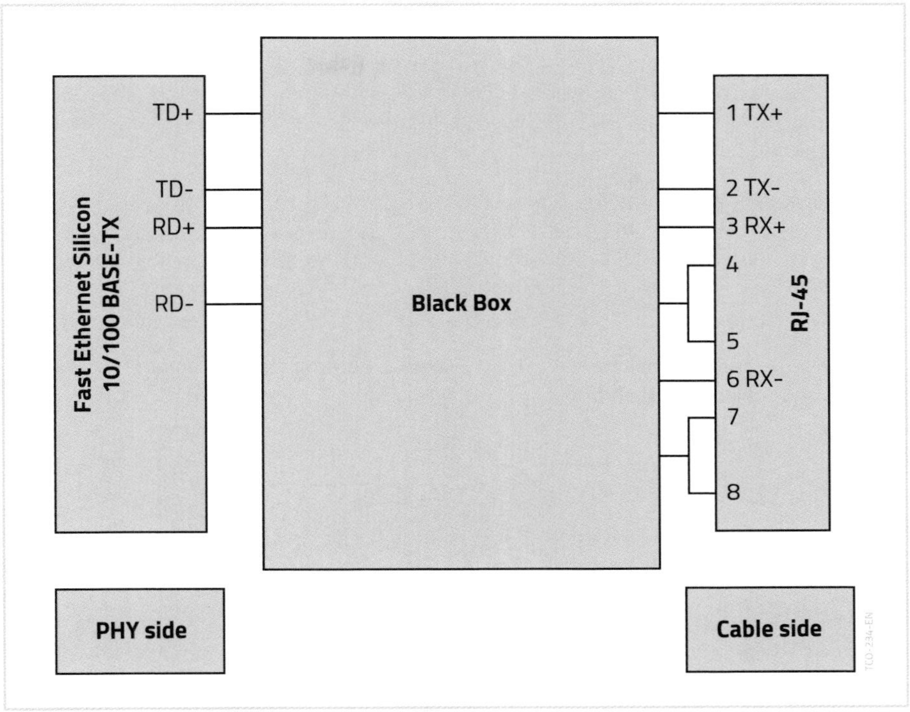

Fig. 1.276: 100BASE-T Basic diagram of the physical layer

As we can see on Figure 1.276, the pin assignment is pair 1-2 for TX and 3-6 for RX. Pairs 4-5 and 7-8 are not used for 100BASE-T (but they are used for Gigabit).

The black box will be explained in detail later on.

Visual Signaling

Within an 'office' type application it may be useful to quickly check the online/offline status; it is even more useful with on-board applications. Therefore, it is not surprising to find modular jacks with several LED options.

LEDs are controlled directly from the PHY.

The LAN Transformer

As explained before, the modular jack is the first element of a physical layer, just between the cable and the electronic board. According to the specifications of the Ethernet standard (802.3), it is necessary to withstand 1500 VRMS for 1min between port and chassis ground. To endure voltage spikes, a transformer is used.

High-speed signal transmission is really sensitive to noise. Noise can easily be generated by the cable and create EMI (Electromagnetic Interferences). On an Ethernet port, noise is often coming in common mode. The best solution to attenuate it without damaging the signal is to use a common mode choke.

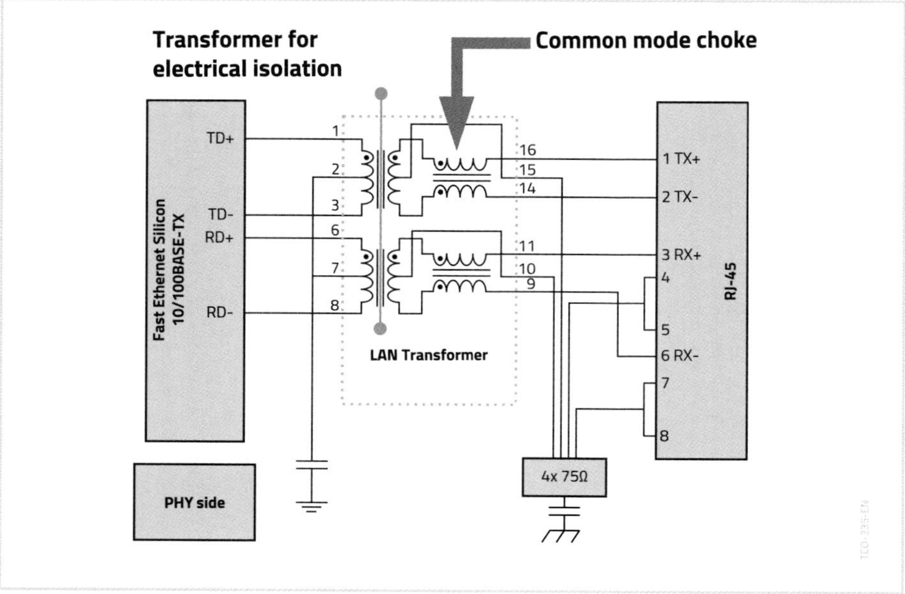

Fig. 1.277: 100BASE-T Physical Layer

This is why we always find a transformer and a common mode choke together (the black box in Figure 1.277). This set of components is called a LAN transformer. Würth Elektronik has a range of LAN transformers available (WE-LAN).

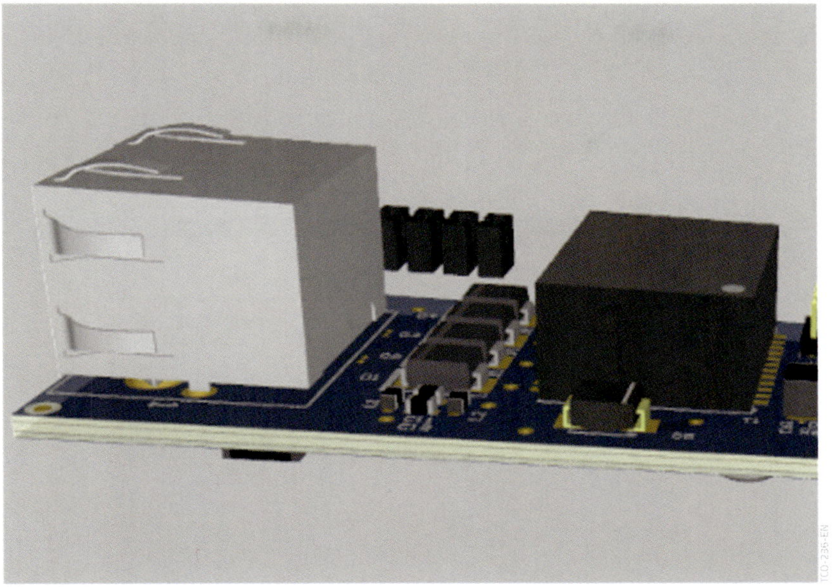

Fig. 1.278: 8P8C Modular Jack (left) and LAN Transformer (right) on PCB

Note: The four 75 Ohms resistors together with the capacitor on the bottom right side form a so-called "Bob Smith Termination" used to reduce EMI.

Component Integration

To achieve building the MDI, we need a certain number of components. But there is two ways of doing it: placing all the elements separately, or choosing an all-in-one solution.

Discrete Magnetics

Discrete magnetics means placing the modular jack, the LAN transformer, the Bob Smith Termination and possibly the Surge Protection Diode separately.

Advantages of this option is the price, but also the easiness to repair. However, it has a larger footprint, notably due to the required isolation distance as shown in Figure 1.279

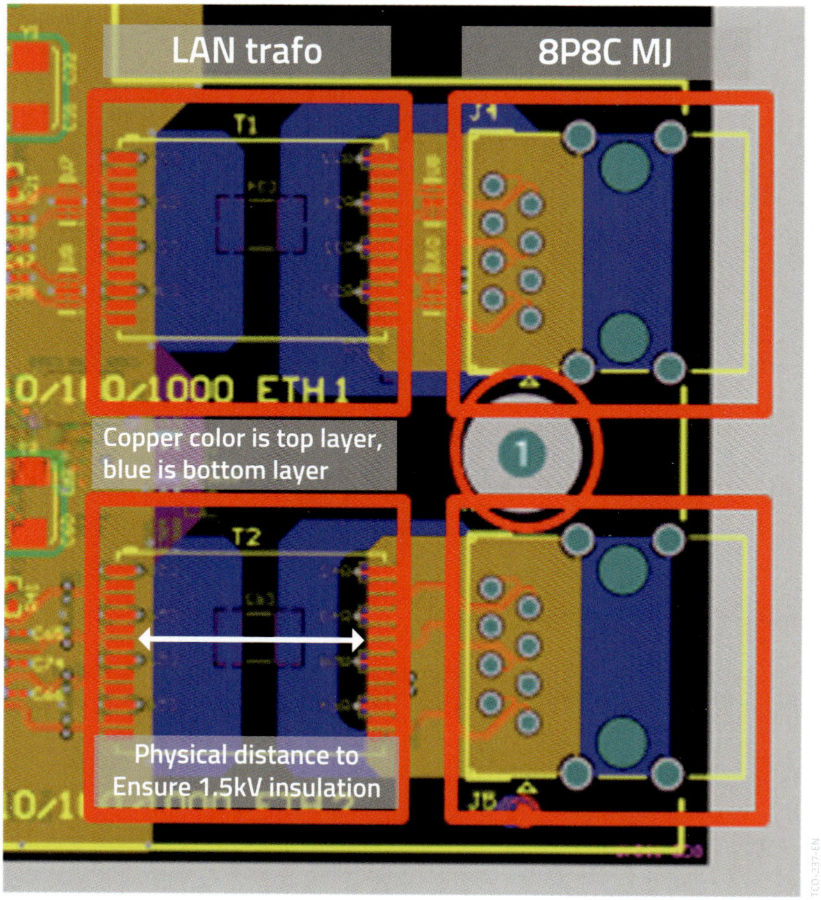

Fig. 1.279: Isolation distance in discrete LAN Transformer – 8P8C modular jack

Würth Elektronik has a range of surge protection diodes available (WE-TVS).

Integrated Magnetics

Integrating magnetics means having the modular jack, the LAN transformer and the Bob Smith Termination in a single component. The surge protection and the physical layer stay out of the component.

Advantages of the integrated option are the lower number of components (lower assembly cost) and the smaller foot print, but the solution is more expensive than the discrete elements. Its smaller size could also increase coupling and therefore reduce the common mode effect.

Both solutions have advantages and disadvantages. It is not a trivial choice and each application will need to be considered on an individual basis. Würth Elektronik has a range of integrated LAN transformers available (WE-RJ45LAN).

Integrated Physical Layer

Advantages and disadvantages for an integrated physical layer are both even more important than between discrete and integrated magnetics. Peak electrical performances are below what can be achieved with discrete components and integrated LAN transformers.

Power-Over-Ethernet

Power-Over-Ethernet (POE) technology allows devices to receive power as well as data over existing LAN cabling without modifying the existing system. Establishment of an international standard in 2003 (IEEE 802.3af) allowed an explosion of POE installation all over the world, which was not possible while old proprietary standards were existing together.

PoE Basics

Today, a lot of electronic applications require both data connectivity and power supply. The most familiar example is telephone, which is powered via the same twisted pair that carries the voice since its invention. The aim of PoE standard is to standardize how to do the same with Ethernet devices without modifying the communication protocol.

On a technical point of view, PoE builds on the voltage ranges used in phone communication. PoE can reach up to 100 W delivered from the Power Supply Equipment (PSE) and 71 W delivered at the Powered Device (PD). There is 7 classes of PoE, but they are grouped in 4 types based on material requirements:

	Type 1	Type 2	Type 3	Type 4
Commercial Name	PoE	PoE+	4PPoE/PoE++	4PPoE/PoE++
IEEE Standard	802.3af	802.3at	802.3bt	802.3bt
Maximum Power Delivered	15.40 W	30.0 W	60 W	100 W
Minimum Power Received	12.95 W	25.50 W	51 W	71 W
Voltage range (at PSE)	44.0–57.0 V	50.0–57.0 V	50.0–57.0 V	52.0–57.0 V
Voltage range (at PD)	37.0–57.0 V	42.5–57.0 V	42.5–57.0 V	41.1–57.0 V
Maximum current Imax	350 mA	600 mA	600 mA per pairset	960 mA per pairset
Max. cable resistance/ pairset	20 Ω	12.5 Ω	12.5 Ω	12.5 Ω
Supported cabling	Cat.3 and Cat.5	Cat.5	Cat.5	Cat.5
Supported modes	Mode A , Mode B	Mode A, Mode B	Mode A, Mode B, 4-pair Mode	4-pair Mode

Tab. 1.17: PoE types

As you can see, PoE can use 2 or 4 pairs to transmit power at 57 V max. 2 pairs is a pair set: since the PD is not normally grounded, the first pair is the power supply, the second is the ground return. Therefore, the provided supply is equal to the number of pair sets multiplied by IMAX multiplied by the voltage.

The PoE-ready devices run a complete recognition protocol to detect whether connected systems could receive power (without damaging them), classify their power ratings, and define the way of transmitting voltage and current. This is controlled by the PHY.

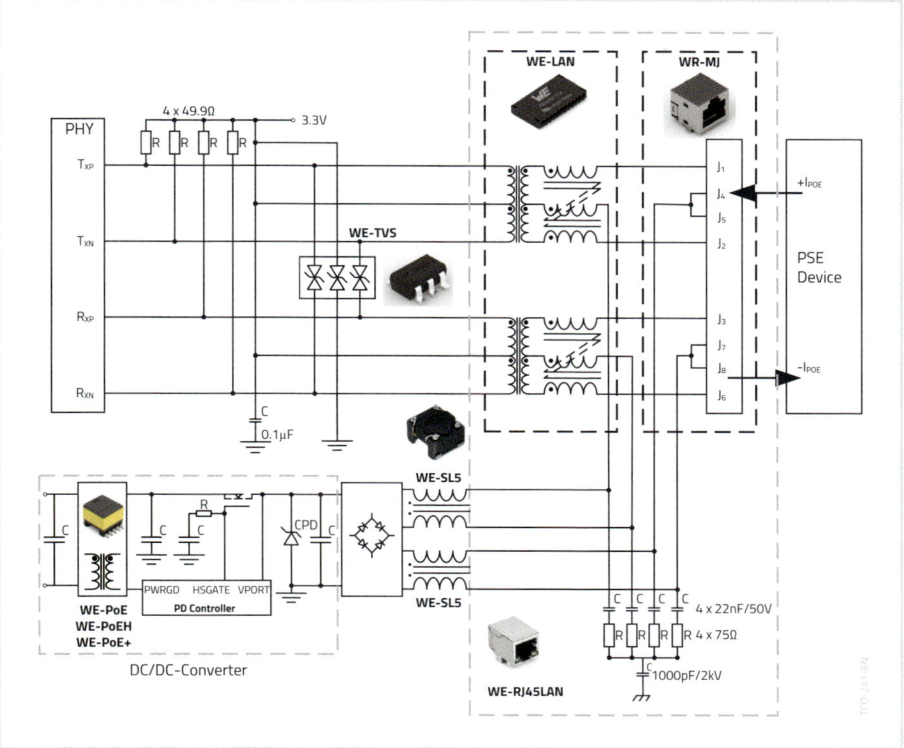

Fig. 1.280: Schematic example of PoE

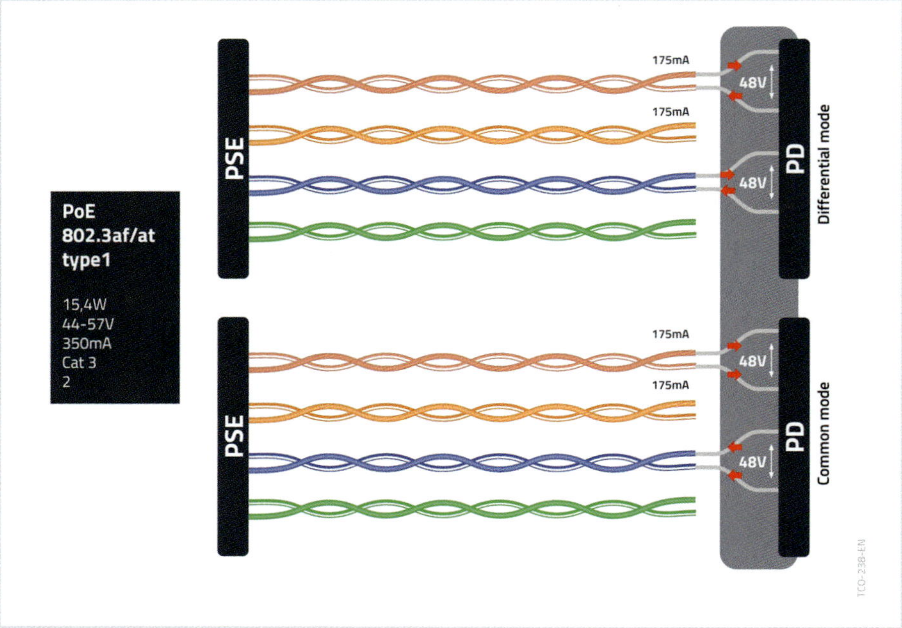

Fig. 1.281: PoE 802.3af/at wiring

PoE Integration

This sets of requirements are within characteristic values and tolerance for modular connectors and twisted-pair cables. However, it is beyond their original use as phone connectors. Thus, cables must be kept apart to avoid too much heating and bundles of powered cables should be avoided.

For modular jacks, a stricter current rating test has been enacted to make sure the connectors do not generate and concentrate too much heat when used for PoE at maximum level. Some impedance-adjusted designs may be prone to heating:

Fig. 1.282: 8P8C Cat5e Modular Jack PCB-type impedance adjustment

In a same fashion, additional care must be taken during soldering of the LEDs, as a too short distance between the LEDs and the signal pins can damage them during an unplugging under load.

In a more practical point of view, the following table shows what will be going through each pin of a modular jack used for PoE, depending on the Mode (A, B or 4-pair). When the data transfer protocol uses all 4 pairs (4-pair mode or NBASE-T4), both DC current and data will pass through the same pins:

Pin	Mode A		Mode A (NBASE-T4)		Mode B		Mode B (NBASE-T4)		4-pair Mode	
	Power	Data	Power	Data	Power	Data	Power	Data	Power	Data
1	DC+	Rx+	DC+	TRxA+		Rx+		TRxA+	DC-	TRxA+
2	DC+	Rx-	DC+	TRxA-		Rx-		TRxA-	DC-	TRxA-
3	DC-	Tx+	DC-	TRxB+		Tx+		TRxB+	DC+	TRxB+
4	Unused		TRxC+	DC+		DC+	TRxC+	DC+	TRxC+	
5	Unused		TRxC-	DC+		DC+	TRxC-	DC+	TRxC-	
6	DC-	Tx-	DC-	TRxB-		Tx-		TRxB-	DC+	TRxB-
7	Unused		TRxD+	DC-		DC-	TRxD+	DC-	TRxD+	
8	Unused		TRxD-	DC-		DC-	TRxD-	DC-	TRxD-	

Tab. 1.18: PoE pinning configurations

To transmit DC power and data at the same time in the same wires, it is necessary to use galvanic insulation. The galvanic insulation will transmit the high-frequency Ethernet signal but keep the DC current on the incoming side. In the same way, a galvanic insulation on the PSE side will be used to add the DC current to the Ethernet signal.

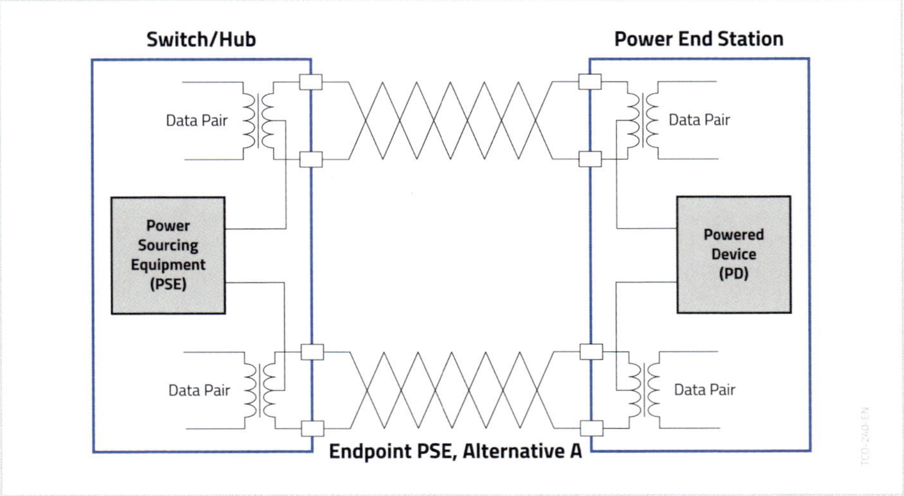

Fig. 1.283: Galvanic Insulation in PoE

As galvanic insulation is also used in LAN transformers as surge protection, they can also be used in PoE in discrete of integrated components:

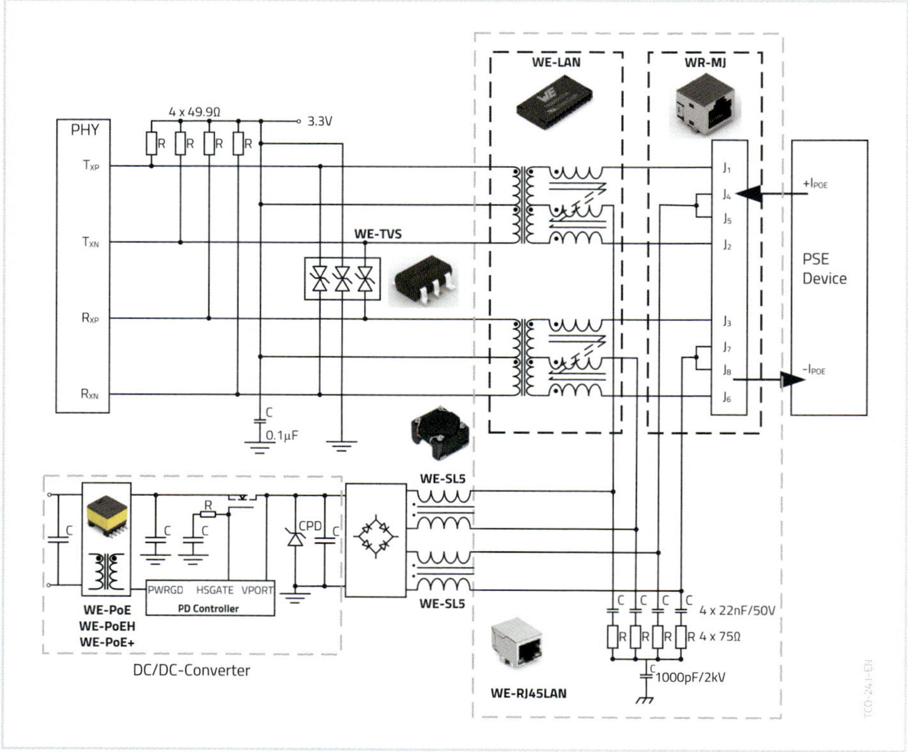

Fig. 1.284: Würth Elektronik LAN transformer PoE integration

PoE Applications

Typical PoE applications range from simple interrupt devices (like RFID readers or smart homes buttons) requiring very little power supply or data transfer to IP Cameras or screens. In the latter case, the power required can be very high in regards to the capacity of the PoE. Uncompressed video can also be transmitted through Ethernet, removing the need from a decoder in the camera or the screen. It requires gigabit Ethernet.

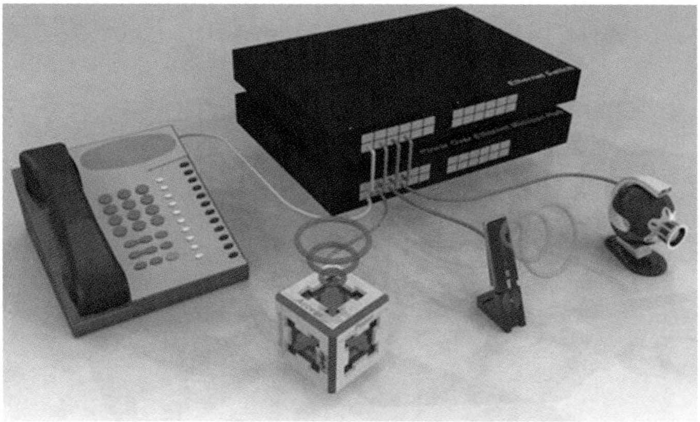

Fig. 1.285: Example of a PoE Switch and devices

Conclusion

Modular connectors are still the top solution for the Ethernet physical layer, thanks to many improvements in their use-cases. However, it is still important for an application designer to weight their uses. Even with new single-pair solutions, Modular Jacks are the perfect middle ground between price and performance, data and power delivery. The RJ45 is a staple of our homes and offices, and thanks to Power-over-Ethernet it is even finding a way into industrial applications. Nearly 60 years after their invention, there is no end in sight for them.

2.2.4.3 SPE – Single Pair Ethernet – From four pairs of wires to one (Draft Gerhard)

The Ethernet standard provides for regular copper cables with 2 pairs for Fast Ethernet and 4 pairs for Gigabit Ethernet. Single Pair Ethernet (SPE) simplifies wiring with a single pair of wires. With the standard borrowed from the automotive industry, many new use cases are possible that operate at low data rates.

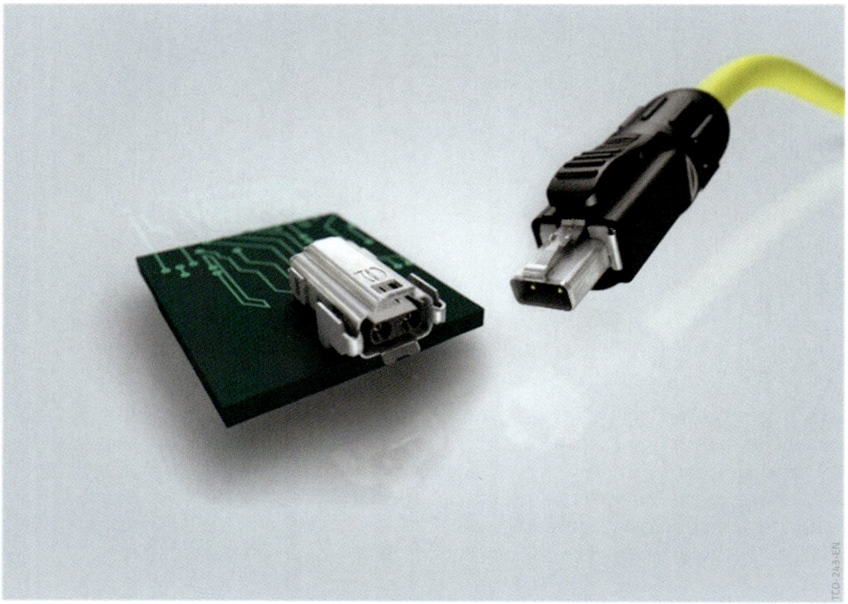

Fig. 1.286: IP20 version of the SPE connector according to IEC 63171-6.

Besides cable length, (cable) weight, mechanical stability of the connectors and reduction of PCB size were the drivers to create a new standard besides RJ45-based multi-pair Ethernet. Single Pair Ethernet (SPE) is a suitable answer to these market requirements and enables seamless IP-based communication from the cloud to any sensor or actuator.

SPE components

The new "physical layer" for SPE requires new components such as cables, connectors, inductive components, semiconductors and devices. The international standardization organizations and the driving companies have invested a lot of time and money in the last years to make all these parts available. The most important standards are already publicly available or are in the final stages. The first components are also ready for deployment, enabling the design of new devices with SPE connectivity.

The electrical requirements for single pair Ethernet are defined in the following IEEE standards:

- IEEE802.3cg (10BASE-T1) with bandwidth from 0.1 to 20 MHz and range up to 1000 m.
- IEEE802.3bw (100BASE-T1) with bandwidth of 0.3 with 66 MHz and range up to 40 m.
- IEEE802.3bp, (1000BASE-T1) with bandwidth from 1 to 600 MHz and range up to 40 m.

Cable

Based on the required transmission speed and link length, two basic types of SPE cables are available and standardized. For 10 Mbit/s networks up to 1000 m cable length, these standards specify the cable design:

- IEC 61156-13 – SPE data cable up to 20 MHz bandwidth for fixed installation,
- IEC 61156-14 – SPE data cable up to 20 MHz bandwidth for more flexible installation.
- For 1 Gbit/s networks up to 40 m cable length these standards are available:
- IEC 61156-11 – SPE data cable up to 600 MHz bandwidth for fixed installation,
- IEC 61156-12 – SPE data cable up to 600 MHz bandwidth for more flexible installation.

Compared to conventional Category 5e Industrial Ethernet cables with four pairs for 1 Gbit/s transmission, the size, space and weight of the cable are significantly reduced.

Dimensions	Industrial Ethernet Cat 5e (4x2x24 AWG)	SPE (1x2x22 AWG)	savings
Outside diameter	7.8 mm	5.8 mm	26 %
weight	79 kg/km	42 kg/km	47 %

Tab. 1.19: Comparison of dimensions and weight of the SPE cables.

All of these cables are shielded to provide the required crosstalk resistance for the 40 m 1GBASE-T1 and the 1,000 m 10BASE-T1L channel.

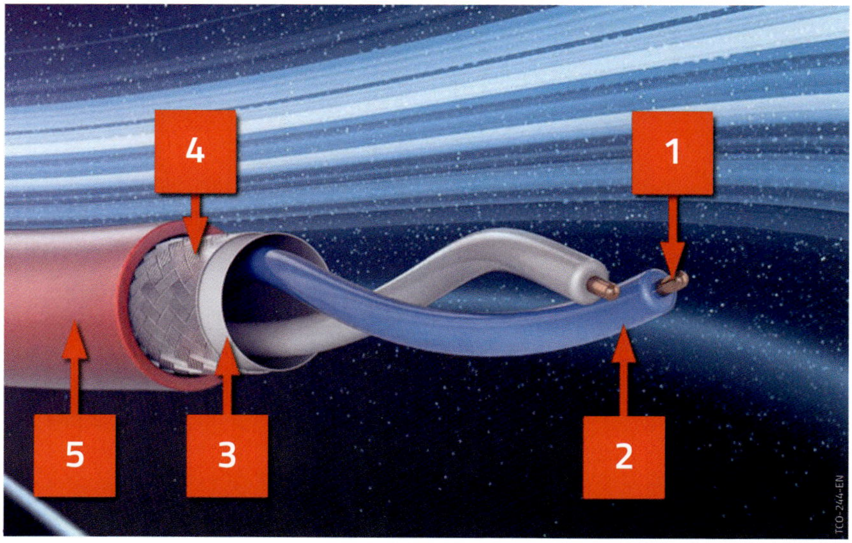

Fig. 1.287: Design for a typical SPE cable (1-copper wire, 2-wire insulation, 3-shield foil, 4-braided shield, 5-cable sheath).

Depending on the application, different cable sheath materials are possible. The copper cross-section of the cable must be selected according to the required link length and the Power-over-Data-Line (PoDL) requirement. For link lengths up to approx. 20 m, typical AWG-26 and up to approx. 150 m AWG-22 copper cross-sections are a suitable choice. For longer link lengths up to 1,000 m, AWG-16 or AWG-18 cables must be used.

To realize the transmission rate of 1 Gbit/s over a single pair, the standards impose high requirements on the electrical properties of an SPE cable.

These include the S-parameters insertion loss (IL), return loss (RL) and alien crosstalk (ANEXT) over a frequency range up to 600 MHz.

Insertion loss describes the logarithmic relationship between the power injected into the cable and the power transmitted through the line. The high IL requirements are necessary to realize the long transmission distances of SPE.

The return loss and impedance of the cable are important to the reflection characteristics of the overall system. Reflections are interferences on the cable. These interferences can interfere with the transmitter and receiver. To minimize reflections, the entire SPE system should have the same characteristic impedance of 100 Ω and low RL values.

SPE has the advantage that there can be no crosstalk from other pairs, but SPE must be able to handle alien crosstalk. ANEXT is crosstalk from other cables in the immediate vicinity. To protect the transmission from interfering ANEXT industrial SPE, cables should be designed with a combined foil and braid shield.

The aluminum foil shield develops a high shielding effect for high-frequency electromagnetic fields. The braided shield provides mechanical stabilization and shielding of low-frequency electromagnetic fields. The effect of a braid depends on the thickness of the individual wires and the degree of coverage. SPE cables for industrial use should have a coverage of at least 85 %.

The shielding effect of a cable works in both directions, i.e. the shielding attenuation reduces both the radiation of interference from the cable signal and interference from other devices acting on the cable from outside.

Important note: Cable shields must be connected on both sides to the housing/ground of the device before the transformers.

Connectors

Completely new types of connectors are required for SPE. These connectors are much smaller compared to the typical RJ45 connectors and offer the same robustness as the often used industrially manufactured M12 D- and X-coded connectors. This new SPE interface is defined in the IEC 63171-6 standard and includes various M8/M12 versions for very harsh industrial applications and an IP20 interface for use in control cabinets (Fig. 2.2.4.Y). All these connector types are based on the same basic bodies and use a robust pin and socket contact system. This modular design concept, with identical data inserts in all versions, allows IP20 connectors to be mated into IP65/67 receptacles for testing or setup.

This SPE connector series is specified for 60 VDC at 4 A @ 60 °C, making the connectors ready for all PoDL classes. For harsh industrial environments with strong EMC interference, the connector has a 360° shield housing to provide a good shield connection from the cable shield to the PCB with four shield pins. These THR solder pins also provide a robust connection between the sockets and the PCB.

Fig. 1.288: Different SPE connector types according to IEC 63171-6.

The connector mating face design is absolutely symmetrical and the contacts are arranged in parallel with the same contact length. The RF compliant connector technology allows signal transmissions up to 1000BASE-T1.

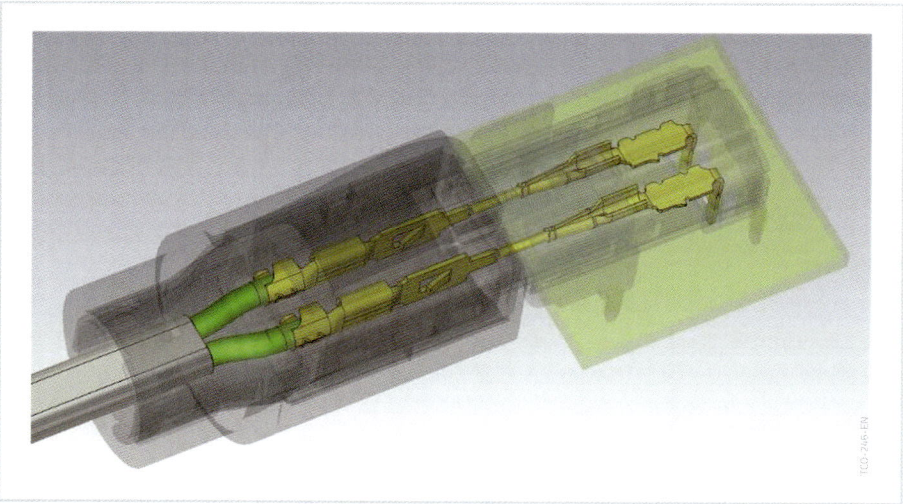

Fig. 1.289: Parallel guided and equal length contact paths to avoid runtime differences.

Further information on EMC filtering and ESD suppression is available at:
www.we-online.de/anp085.

2.2.4.4 ConFFC/FPC

FPC (Flexible printed circuit) connectors are used for connecting a FPCB (Flexible printed circuit board) or a FFC (Flexible flat cable).

They are a popular choice when for example a display or a keyboard needs to be connected. One reason is its flexibility and possibility to design compact and slim devices. The FPC connector can also be used as a space saving Board to Board connection when a low profile, secure, light weight connector with many conductors (4-50poles are available) are needed.

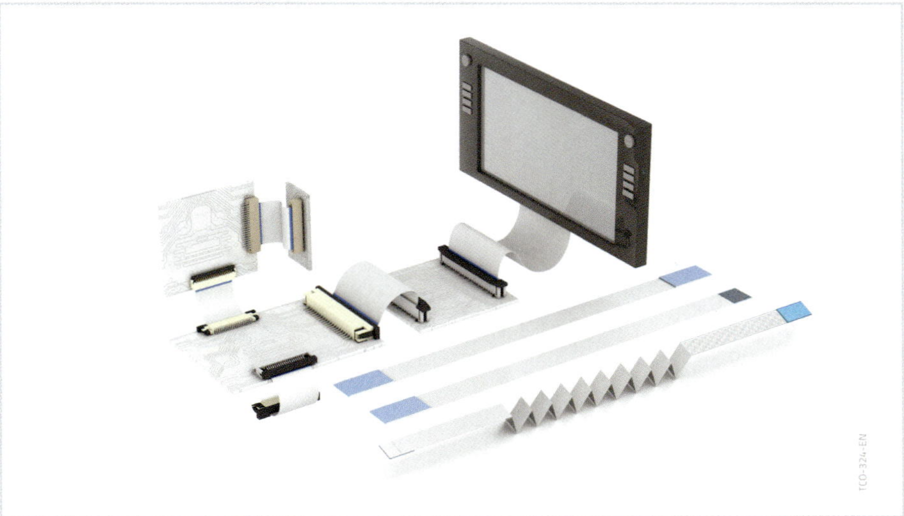

Fig. 1.290: WR-FFC & WR-FPC

FPC connectors can be divided into ZIF (zero insertion force) and LIF (Low insertion force)

When attaching the cable into the ZIF connector an Actuator must be locked to secure the cable with active locking. There are three different ways to lock the cable. Slider, Hinge and Back locking.

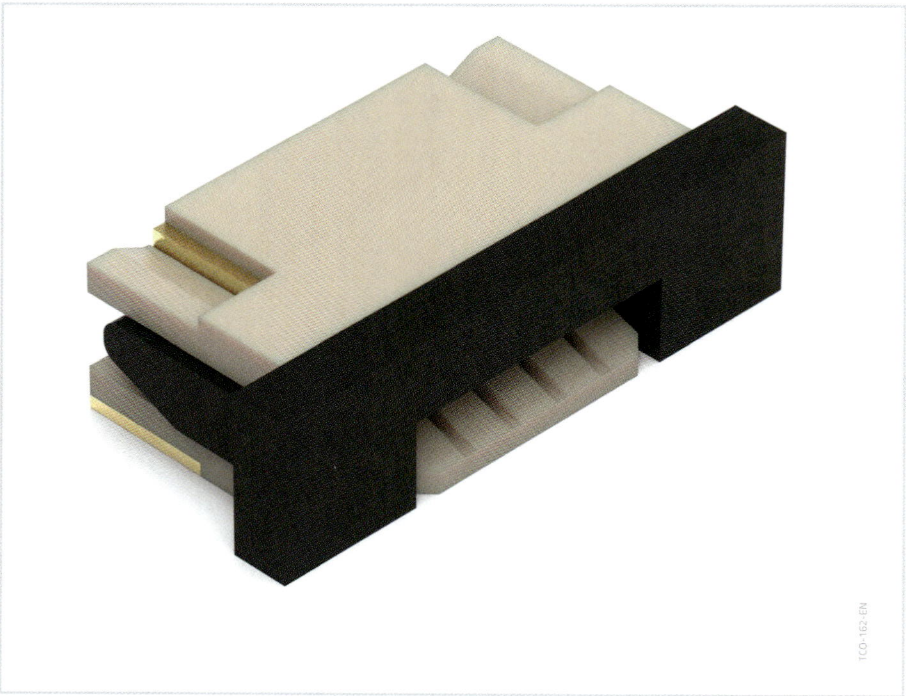

Fig. 1.291: ZIF slider type

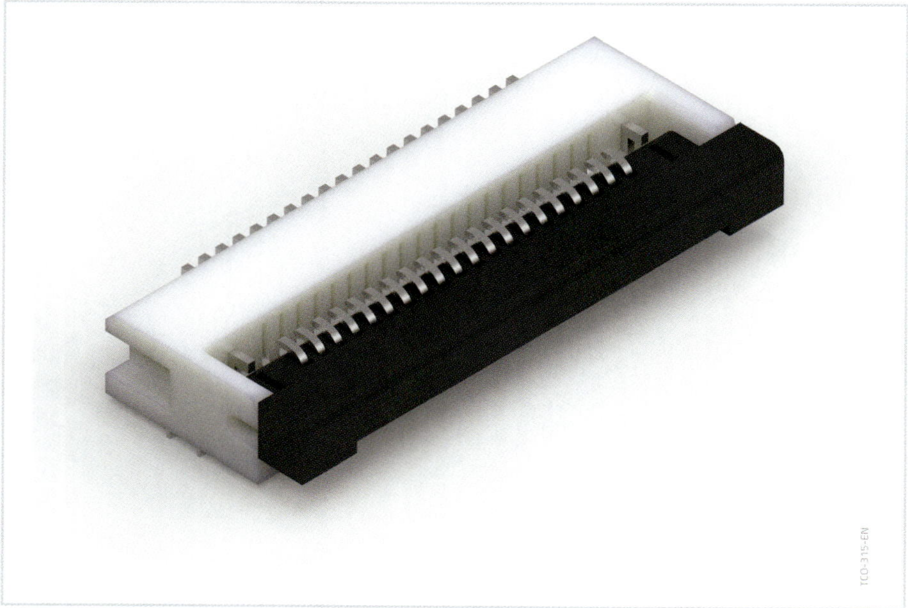

Fig. 1.292: ZIF hinge type

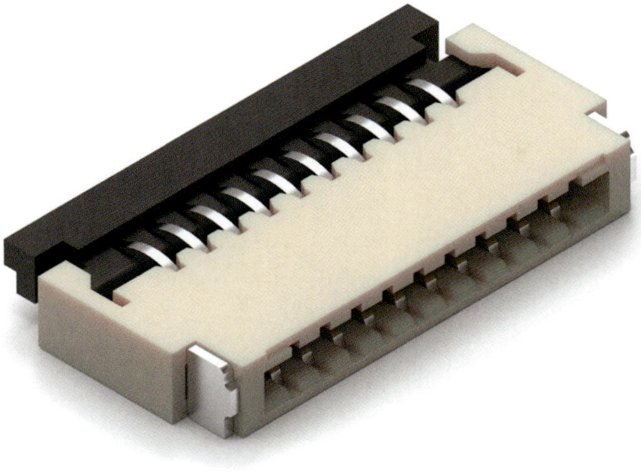

Fig. 1.293: ZIF Back locking type

When instead using the LIF connectors there is no actuator. The cable will be locked with passive locking when it is pressed into the connector house.

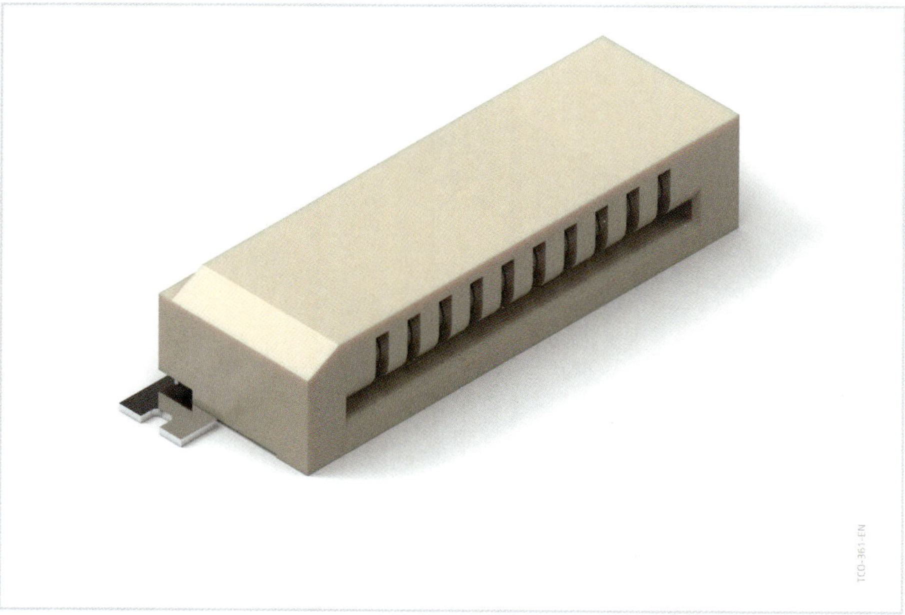

Fig. 1.294: LIF connector

FPC connectors are available with orientation Vertical/Horizontal, Top/bottom/dual contacts and 0,5/1 mm pitch.

FFC (Flat flexible cable) are light weight and reliable. As the name indicates, they are also very flexible and can be rolled and bent multiple times. They come in different standard lengths and requires no application tooling. Hence the assembly cost is low.

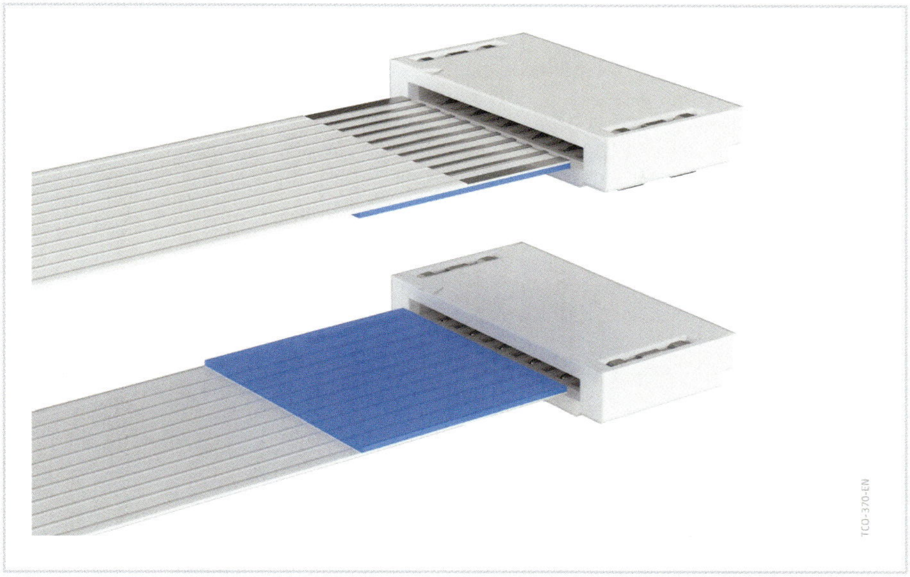

Fig. 1.295: Bottom & Top contact

Depending on if the FPC connector have top or bottom contacts and how it is used. There are two versions for the cable. Type 1 with the contacts on the same side and Type 2 with the contacts on the opposite side.

Fig. 1.296: Types of FFCs

2.2.4.5 ConCARD

SIM Card Connector

A Subscriber Identification Module (SIM) card is a type of Smart Card described by ISO/IEC 7810 and 7816 standards. Unlike the more generic Smart Cards, SIM cards are used to carry the mobile identification to access wireless phone networks.

Size and Form

SIM cards follow a lightly less strict set of rules than the two previous card standards. The golden contact plate can have different-looking contacts, as long as the wiring inside is consistent with the standard. The outer plastic card exists in 1FF, 2FF, 3FF and 4FF sizes, corresponding to SIM, MiniSIM, MicroSIM and NanoSIM (Figure 1.297).

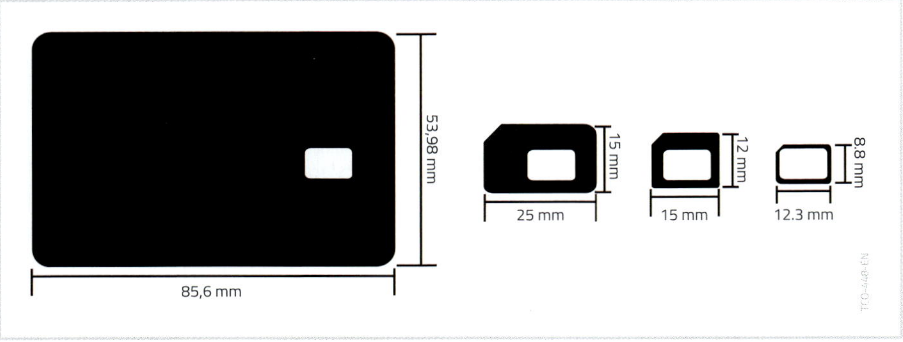

Fig. 1.297: SIM Cards Form Factors, from left to right : Standard SIM (1FF), MiniSIM (2FF), MicroSIM (3FF), NanoSIM (4FF)

SIM, MiniSIM and MicroSIM cards share the same thickness of 0,76 mm. NanoSIM is slimmer at 0,67.

SIM cards normally have 6 contacts, however it also exists a variant with 8 contacts called the MegaSIM pinout (Figure 1.298). This variant exists for standard SIM, MiniSIM and MicroSIM cards.

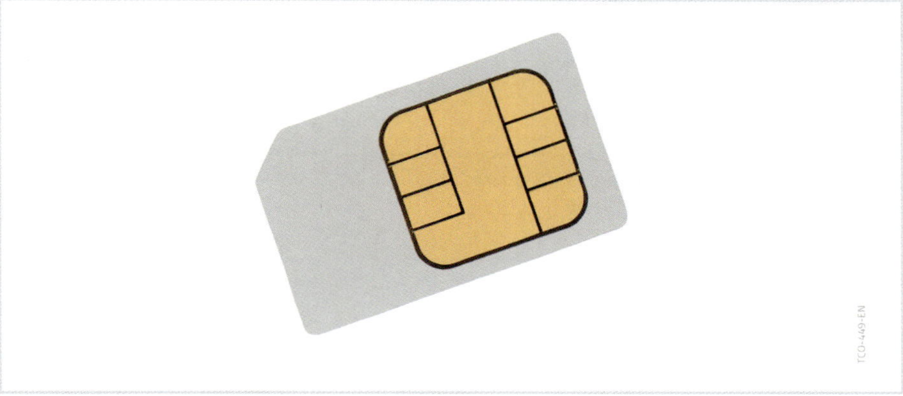

Fig. 1.298: A MicroSIM with 8 contacts (MegaSIM)

The connector itself uses spring-shaped contacts pushing the card on the shielding. These contacts ensure the mechanical retention and electrical connection at the same time.

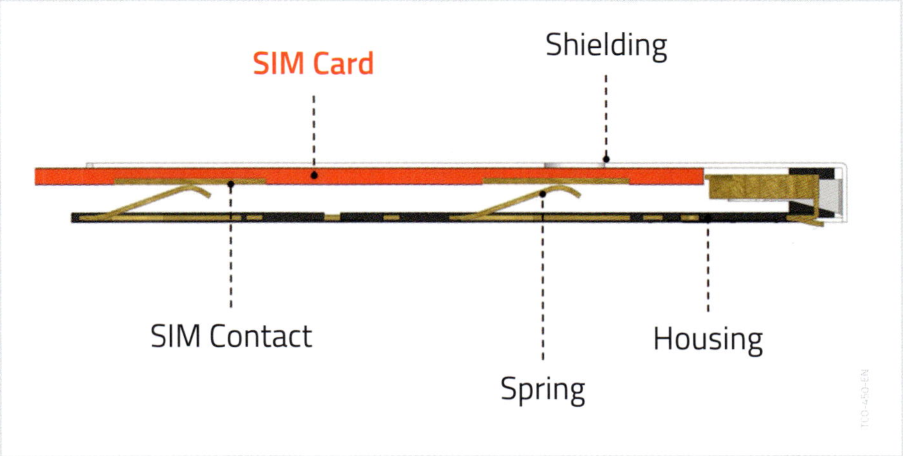

Fig. 1.299: SIM card and connector mating cross-section

Purely mechanical adapters exist from one form factor to another. One must be wary when using a NanoSIM to MiniSIM or MicroSIM adapter, as the NanoSIM is 0,09 mm thinner. Using a thinner card as intended may cause electrical discontinuity, lower retention force and weakness to vibrations

The structure of the card itself can be seen in Figure 1.300 UICC-style golden contacts on an insulator substrate are wire-bonded to the SIM chip itself (similar to the eSIM standalone chip). The chip-contacts module is then encapsulated and incorporated into the card plastic body.

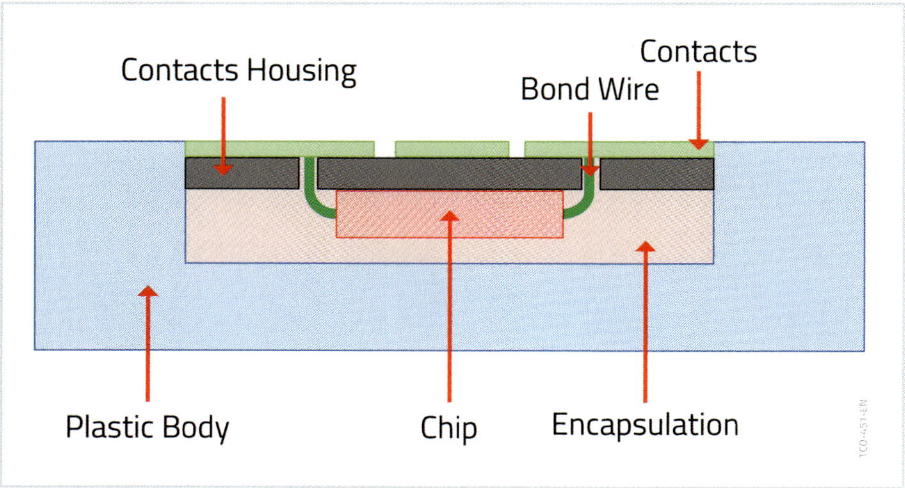

Fig. 1.300: SIM card cross-section schematic

Memory and protocol

Smart cards (or ICC, for integrated circuit cards) can provide identification, authentication, data storage, strong security authentication for Single Sign-On (SSO) and application processing. SIM cards, especially, store the IMEI and other access codes to the mobile phone network.

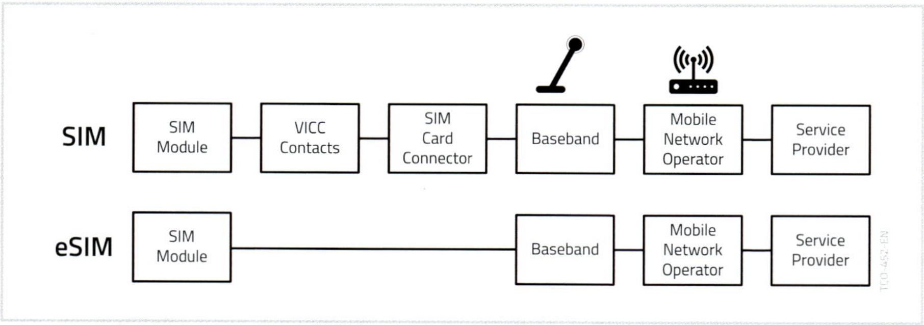

Fig. 1.301: SIM and eSIM architecture

Smart cards contain non-volatile memory storage components and perhaps dedicated security logic. Microprocessor cards can contain also volatile memory and microprocessor components that can be accessed with the additional MegaSIM contacts (Figure 1.302)

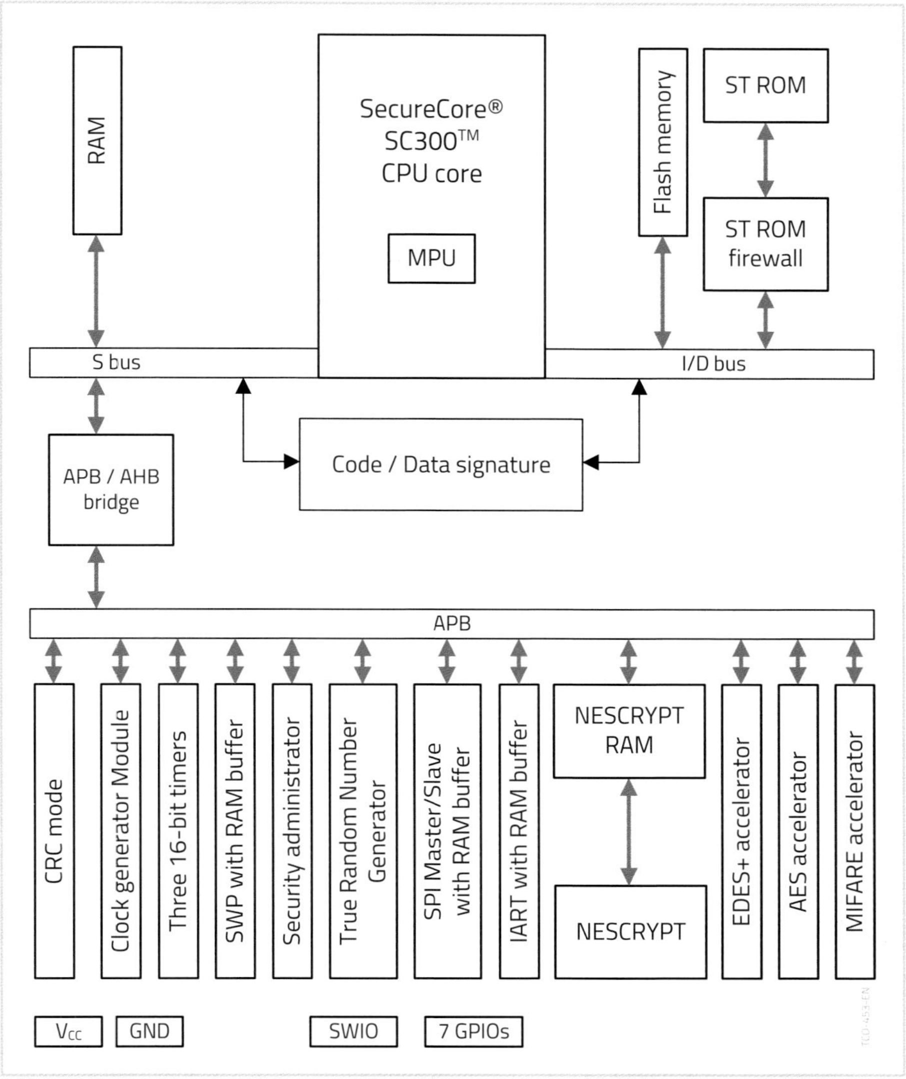

Fig. 1.302: ST33 ARM microprocessor architecture for ICC applications (STMicroelectronics). It includes 10 I/O channels, not all of them being used in Smart Cards.

The communication sequence between the SIM card and the connected device is as follows:

1. connection and activation of the contacts
2. reset of the card
3. answer to reset by the card
4. subsequent information exchange between the card and the interface device
5. deactivation of the contacts by the interface device

The use of the 4th and 8th contacts for MegaSIM cards does not follow this sequence.

Pin (SIM)	Pin (MegaSIM)	Name	Description
1	1	Vcc	+5 VDC power supply input (optional use by the card)
2	2	RESET	Reset signal, used to reset the card's communications.
3	3	CLOCK	Provides the card with a clock signal
–	4	RESERVED	AUX1, optionally used for USB interfaces and other uses
4	5	GND	Ground (reference voltage)
5	6	Vpp	Programing voltage input (optional).
6	7	I/O	Input or Output for serial data half-duplex
–	8	RESERVED	AUX2, optionally used for USB interfaces and other uses.

Tab. 1.20: SIM card and connector pinout

SD-Card Connectors

Secure Digital (SD) is a Non-Volatile Memory format developed by the SD Card Association (SDA) for use in portable devices.

Size and Form

The SD card exists in 3 form factors: Standard (2,1 mm thick), Mini (1,4 mm thick) and Micro (1,0 mm thick) with different dimensions Figure 1.303). The MiniSD card is obsolete today.

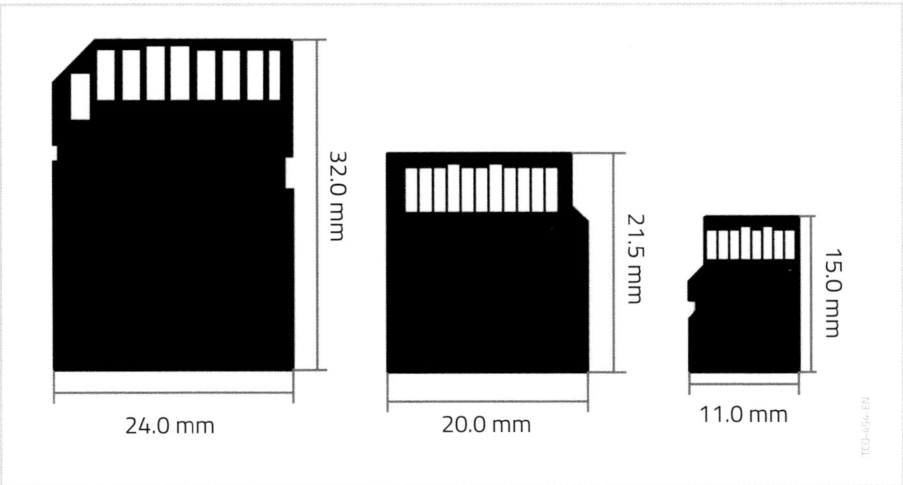

Fig. 1.303: Standard SD (left), MiniSD (center), MicroSD (right) form factor size comparison

The pin numbering for the cards and the connectors is as following:

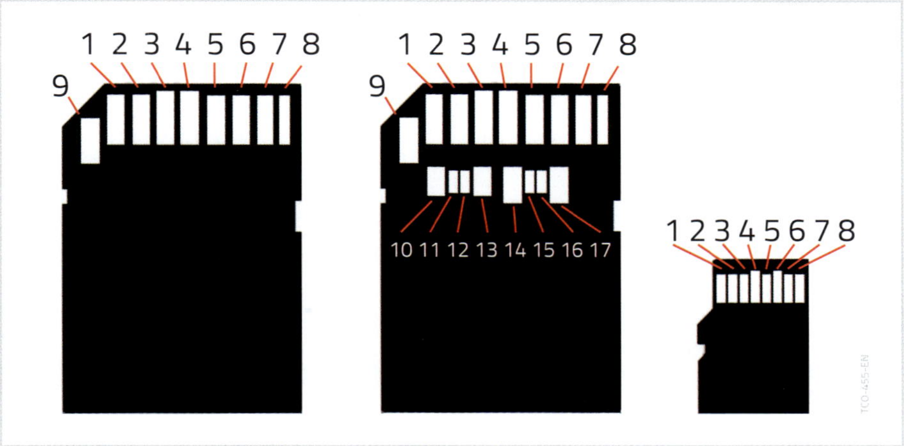

Fig. 1.304: Standard SD, SD 4.0 and MicroSD pin numbering

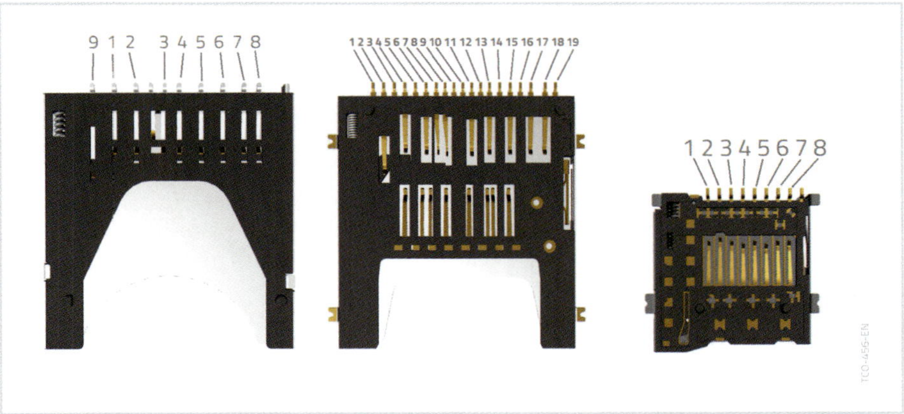

Fig. 1.305: Standard SD, SD 4.0 and MicroSD connector pin numbering. Pins without numbers are for additional card features

You can find the relation between pin number and signal description in Table. 1.21

Signal Name	Description	SD	SD 4.0	MicroSD
DAT2	Data Line 2	9	9	1
CD/DAT3	Card Detection/Data Line 3	1	1	2
CMD	Command/Answer	2	2	3
Vss 1	Ground	3	3	
Vdd	Power Supply	4	4	4
CLK	Clock	5	5	5
Vss 2	Ground	6	6	6
DAT0	Data Line 0	7	7	7
DAT1	Data Line 1	8	8	8
VSS3	Ground		10	
D0+	Lane 0 Differential Data		11	
D0-	Lane 0 Differential Data		12	
VSS4	Ground		13	
VDD2	Power Supply		14	
D1+	Lane 1 Differential Data		15	
D1-	Lane 1 Differential Data		16	
VSS5	Ground		17	

Tab. 1.21: SD card and connector pinout

It is to be noted that MicroSD to SD card adapters exist and are purely electromechanical adapters (Figure 1.306).

Fig. 1.306: A microSD to SD card adapter

Memory and transfer speeds

SD cards use flash NAND memory, which allows for high capacity.

Transfer speed incurs signal clocking up to 208 MHz. Power consumption during high-speed transfer may be up to 3 W. If different buses are used between cards and hosts, the transfer will be done at the lowest maximum speed. SD Express buses will revert to UHS-I speeds when not used with another SD Express bus.

How detect if a (micro)SD card is inserted in a card connector on the right way?

Before we go in deep how card detection can be implemented in to hardware designs, we are looking how the card detection in the card connector can look like. In Figure 1.307 you can see, how the mechanical switch, that is used for the mechanical card detection, is situated in one our microSD card connector. With this mechanical switch you detect if the (micro)SD card is inserted in the card connector. When the card is inserted the part within the red frame of figure 1 will be pushed against the shielding of the card connector. In this case the switch is closed and the microcontroller will receive a signal on his input pin.

Fig. 1.307: Cross section of microSD card connector with a mechanical card detection switch

In case that the card is inserted there will be no issue. But if the card physically is not inserted, and the switch in the card connector is not closed you could face some issues because of a floating line towards an input pin of the microcontroller. When the mechanical switch of the card connector is open, so the card is not inserted, the connected input pin is floating. So, neither a high nor a low logic state for the microcontroller. In this case the microcontroller might interpret the input value as a logical high or a logical low input. When the input pin of the microcontroller is floating a high impedance situation will occur. The high impedance state occurs when a pin is not pulled to a high or low level. In the floating situation you cannot predict how the microcontroller interpret the logic level.

To avoid this issue, you can add a pull-up or a pull-down resistor in your circuit. These resistors are used in logic circuits to ensure a defined logical level at a pin of a microcontroller. When you are using a pull-up resistor you will pull-up the logic level to a high logic level. For a pull-down resistor it is working in the opposite way, the logic level is pulled down to a low logic level. With this easy solution you can avoid a floating situation at an input pin of a microcontroller. Figure 1.308 and Figure 1.309 show both options.

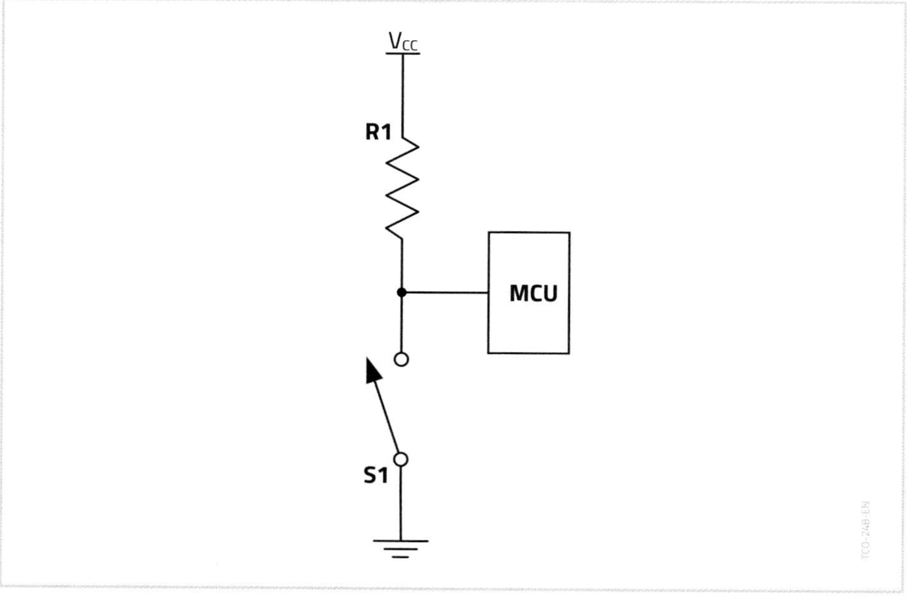

Fig. 1.308: Wiring diagram with a pull-up resistor

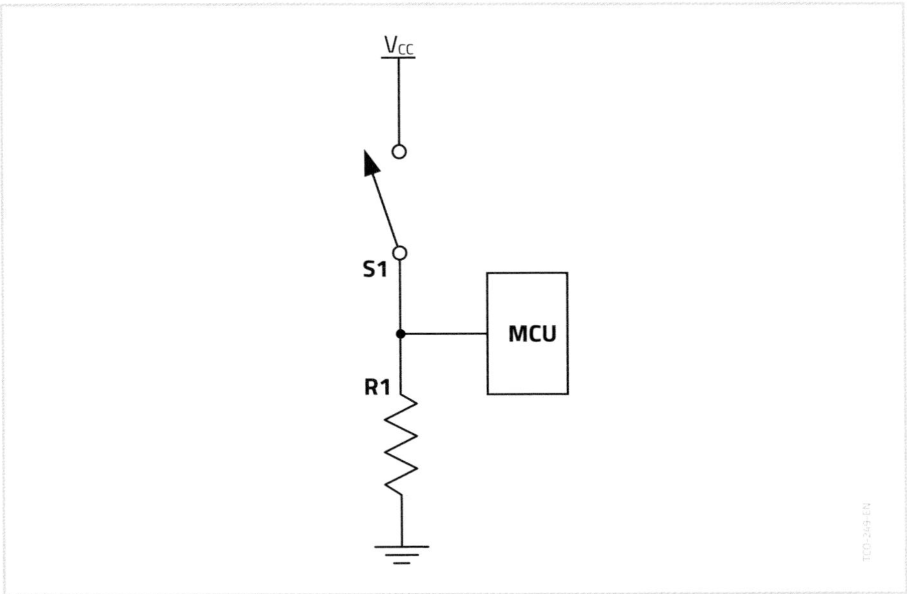

Fig. 1.309: Wiring diagram with a pull-down resistor

To choose the right value for the pull-up resistor you need to take two factors in consideration. The first factor is power dissipation. If you choose a resistor too low, a higher current will flow through the resistor when the switch is closed. However, when the chosen value of the resistor is too high, in combination of the leakage current at the input pin of the microcontroller, the input voltage at the input pin of the microcontroller could not be OK.

So we cannot give a fixed value, because not every microcontroller has the same input impedance. This is mentioned in the datasheet of the manufacturer of the microcontroller. A rule of thumb for a pull-up resistor, which is used in this sort of applications, is that the value is between 1 kΩ and 10 kΩ. A too high resistance value for the pull-up resistor could also give a disadvantage. The input pin can respond slowly on a voltage change. Because this is the result of coupling of the resistor and the total capacitance of the pin and wire, this forms a RC circuit. The higher the product of R and C, the more time you need to charge and discharge the capacitance.

For a pull-down is the approach to calculate the resistor value a little bit different. The value must be higher than the impedance of the logical circuit, otherwise it might be able that the voltage will be pulled down too much and the input voltage at the pin would remain at a constant logic low value. In this case we also cannot give fixed values of the resistor as it will depend on the microcontroller and the circuitry.

When the detection switch of the card connector is used, take in consideration to use a pull-up of pull-down resistor to avoid floating situations at the input pin of a microcontroller. When using a pull-up or pull-down resistor with the correct resistance value you will avoid strange behavior at the logic level of the input pin of the microcontroller. Below you can find at last two simulations of a pull-up circuit and a pull down circuit. This simulation is done with LTspice XVII. So you can have an idea how everything is done. In figure 4 you can see how the circuits are build up. In figure 5 you can see the result of the sumaltion for the pull up and in figure 6 you can see the result of the pull down resistor.

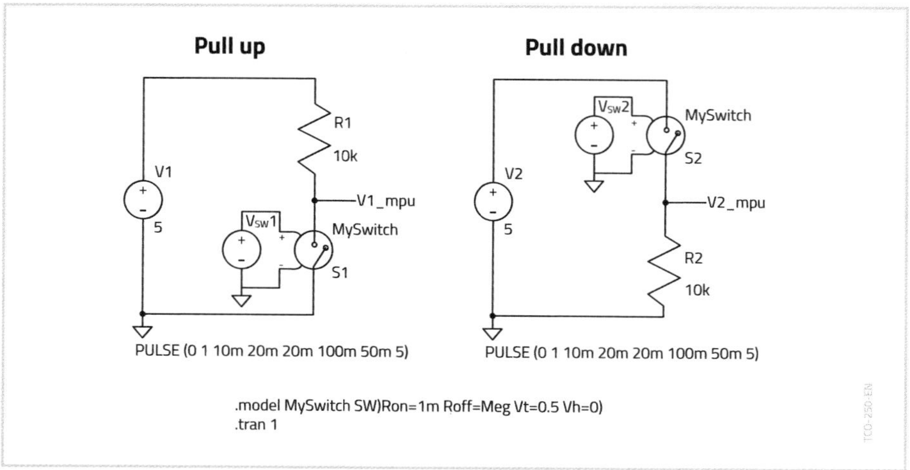

Fig. 1.310: Simulation circuit LTSpice XVII pull up and pull down resistor

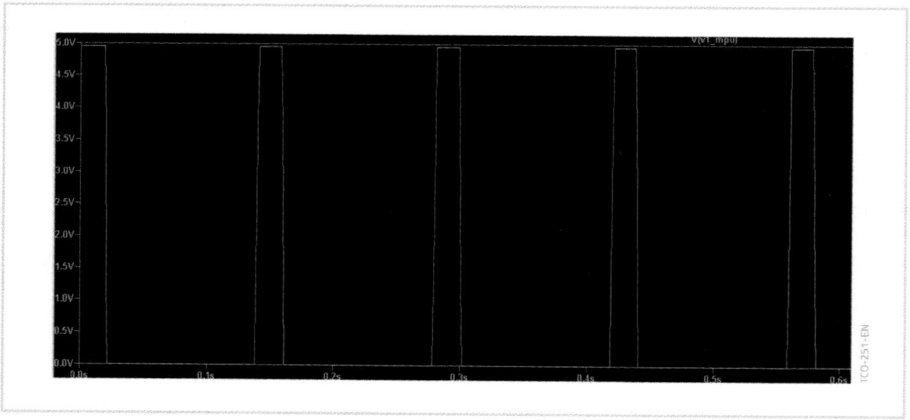

Fig. 1.311: Result of simulation Pull up resistor

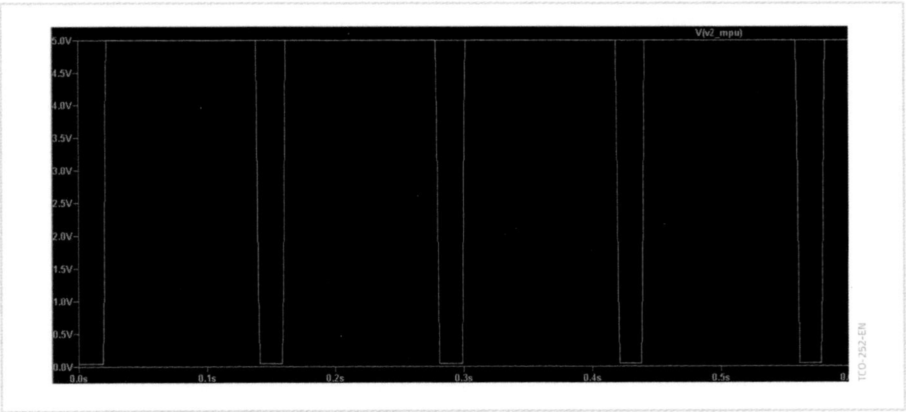

Fig. 1.312: Result of simulation Pull down resistor

CompactFlash Card Connectors

The CompactFlash card is a derivative of the PCMCIA card. Its 50 pins have equivalent in the 68-pins PCMCIA card standards and are small enough to be fitted in other computer card slots with an adapter

Types

CFexpress are clustered in three form factors shown in Table. 1.22.

	Type A	Type B	Type C
Contacts	13	21	64
PCIe Lanes	1	2	4
Width	20.0 mm	28.0 mm	2.8 mm
Length	38.5 mm	29.8 mm	3.8 mm
Thickness	54.0 mm	74.0 mm	4.8 mm

Tab. 1.22: CFexpress form factors

Size and Form

The CompactFlash card form factor is a 3.3 mm (Type I) or 5 mm (Type II) thick metal and plastic casing with 50 contacts.

Fig. 1.313: CompactFlash 50 contacts front panel

Both CompactFast and CompactFlash cards are 42.8 x 36.4 mm in lateral dimensions.

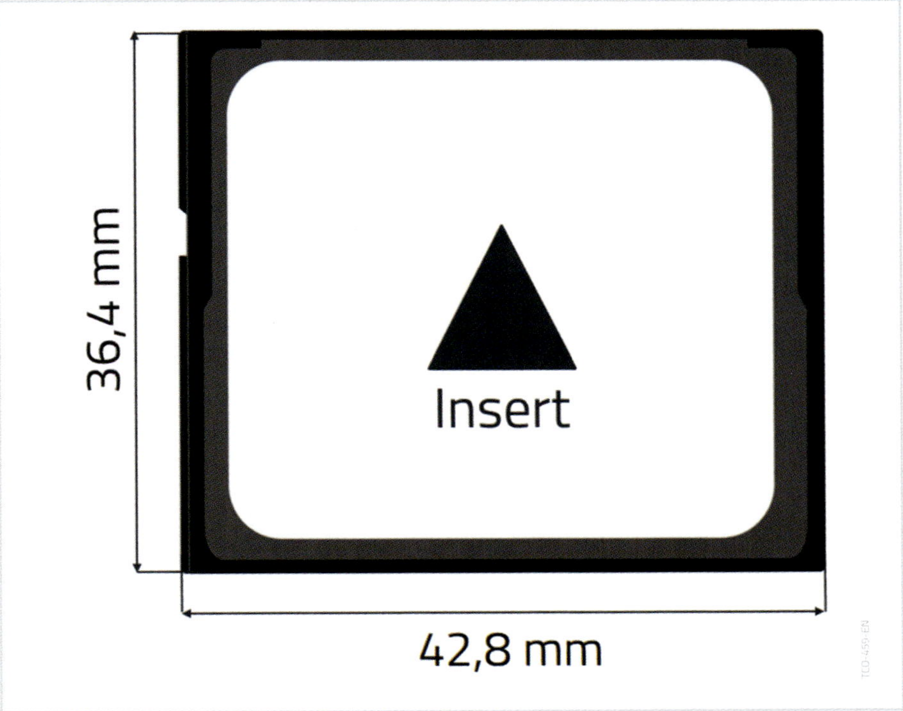

Fig. 1.314: CF card external dimensions

Likewise, the card connector for CompactFlash cards will feature 50 pins to match the card's contacts. Longer pins are power supply pins (Figure 1.315).

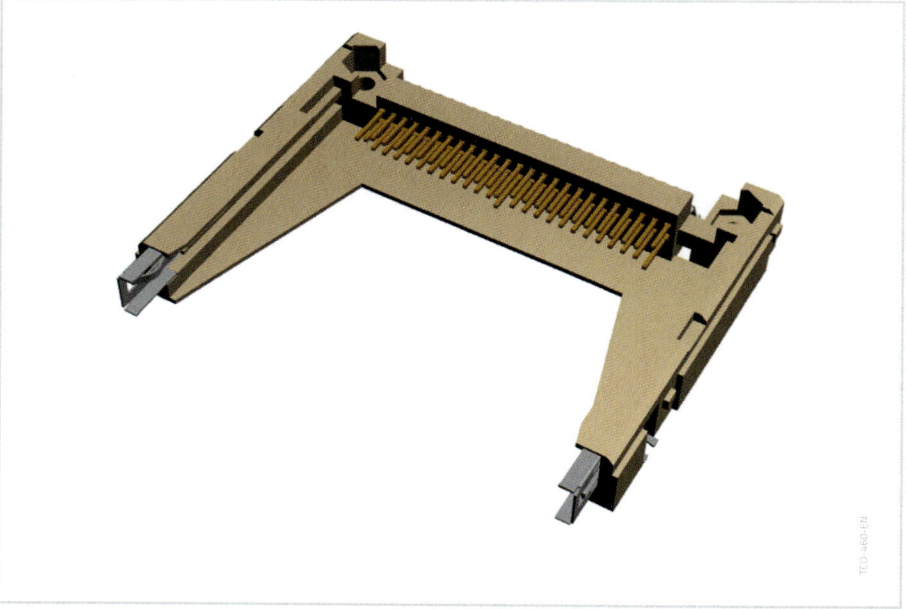

Fig. 1.315: SMT CompactFlash Card Connector (693120025011) with longer pins for power contacts

It is notable that such a connector can accommodate all the CompactFlash speed levels, because those are only defined by the active components of the system.

Depending on the configuration of the CompactFlash, the pinout may vary (Table. 1.23).

Memory			
Pin	Signal Name	Pin	Signal Name
1	GND	26	CD1
2	D03	27	D11
3	D04	28	D12
4	D05	29	D13
5	D06	30	D14
6	D07	31	D15
7	CE1	32	CE2
8	A10	33	VS1
9	OE	34	NU
10	A09	35	NU
11	A08	36	WE
12	A07	37	RDY/ BSY

I/O			
Pin	Signal Name	Pin	Signal Name
1	GND	26	CD1
2	D03	27	D11
3	D04	28	D12
4	D05	29	D13
5	D06	30	D14
6	D07	31	D15
7	CE1	32	CE2
8	A10	33	VS1
9	OE	34	IORD
10	A09	35	IOWR
11	A08	36	WE
12	A07	37	READY

TrueIDE			
Pin	Signal Name	Pin	Signal Name
1	GND	26	CD1
2	D03	27	D11
3	D04	28	D12
4	D05	29	D13
5	D06	30	D14
6	D07	31	D15
7	CS0	32	CS1
8	L	33	VS1
9	ATA_SEL	34	IORD
10	L	35	IOWR
11	L	36	WE
12	L	37	INTRQ

Memory				I/O				TrueIDE			
Pin	Signal Name	Pin	Signal Name	Pin	Signal Name	Pin	Signal Name	Pin	Signal Name	Pin	Signal Name
13	VCC	38	VCC	13	VCC	38	VCC	13	VCC	38	VCC
14	A06	39	CSEL	14	A06	39	CSEL	14	L	39	CSEL
15	A05	40	VS2	15	A05	40	VS2	15	L	40	VS2
16	A04	41	RESET	16	A04	41	RESET	16	L	41	RESET
17	A03	42	WAIT	17	A03	42	WAIT	17	L	42	IORDY
18	A02	43	NU	18	A02	43	INPACK	18	A02	43	NC
19	A01	44	REG	19	A01	44	REG	19	A01	44	H
20	A00	45	BVD2(H)	20	A00	45	SPKR	20	A00	45	DASP
21	D00	46	BVD1(H)	21	D00	46	STSCHG	21	D00	46	PDIAG
22	D01	47	D08	22	D01	47	D08	22	D01	47	D08
23	D02	48	D09	23	D02	48	D09	23	D02	48	D09
24	WP	49	D10	24	IOS16	49	D10	24	IOCS16	49	D10
25	CD2	50	GND	25	CD2	50	GND	25	CD2	50	GND

Tab. 1.23: Standard CompactFlash pinout tables for a memory configuration (left), a I/O configuration (center) and a TrueIDE configuration (right)

Memory

While originally based on Flash NOR memory, the CompactFlash standard is now on Flash NAND like MicroSD cards, as it was vertically scalable. This convergence has led to the NVMe standard, unifying most flash memory extensions technologies.

2.2.4.6 ConUSB

USB Connectors – The Universal Gate to Connectivity

The USB has evolved significantly since its creation, continually improving data transfer speeds and power efficiency. Compared to earlier versions, modern USB standards from 3.x family offer much higher bandwidth, making it easier to quickly transfer large files such as HD movies or databases. Alongside better power management, USB also maintains backward compatibility (for Type A and B), ensuring that older devices continue to work with the latest connectors.

History, Implementation and Main Purpose

The new communication technologies generate an enormous amount of content while storage is simultaneously becoming inexpensive. Therefore, the USB Implementers Forum (USB-IF) acknowledged the need for new generation of interconnection technologies and regularly updates USB specifications. This is mostly initiated by the promoter group made up of the following 6 companies: Intel, HP, Microsoft, NEC, Texas Instrument and NXP.

USB 3.2 aim to bring the following improvements compared to USB 2.0:

- Up to 40x higher data rate than USB 2.0 (20 Gbps vs. 480 Mbps)
- Separate data lanes for up & download enable to read and store files at the same time
- Thanks to higher current capacities, quicker charge of mobile devices through USB
- Better power efficiency thanks to the interrupt-driven protocol

USB Version	Current Designation	Data Rate*	Speed Name
USB 1.0	USB 1.0	1.5 Mbit/s	Low-Speed USB
USB 1.1	USB 1.1	12 Mbit/s	Full-Speed USB
USB 2.0	USB 2.0	480 Mbit/s	Hi-Speed USB
USB 3.0	USB 3.2 Gen 1x1	4.8 Gbit/s	SuperSpeed USB
USB 3.1	USB 3.2 Gen 2x1	10 Gbit/s	SuperSpeed USB 10 Gbps
USB 3.2	USB 3.2 Gen 2x2	20 Gbit/s	USB 20 Gbps
USB4	USB4 1.0	40 Gbit/s	USB 40 Gbps
USB4v2	USB4 2.0	80 Gbit/s	USB 80 Gbps
* Byte (1B) = 8 bit (8b)			

Tab. 1.24: Data rate evolution

Architecture, Protocol and Power Management Architecture: from High-Speed to Super-Speed

USB 3.0 consists of a physical super-speed bus in addition to the physical USB 2.0 bus (as illustrated by Figure 1.316). The protocol defines a dual simplex signalling mechanism with a rate of 5 Gbits/s. This enables the simultaneous transfer of data to and from the device, as opposed to the single duplex unidirectional USB 2.0 bus.

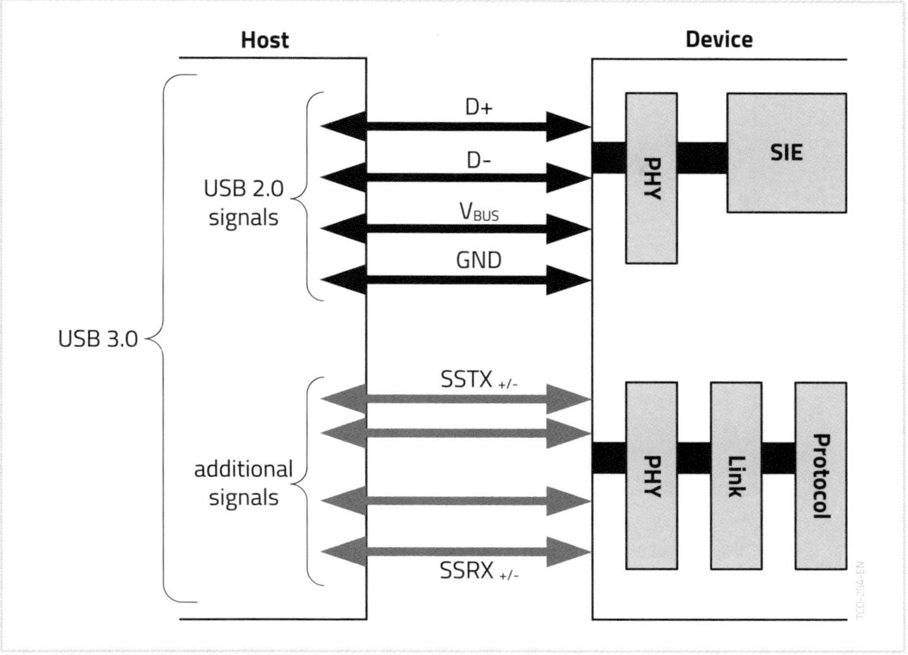

Fig. 1.316: USB 3.0 architecture

It is beyond the scope of this article to describe the full USB 3.0 architecture in detail. However, it is important to note that it was inspired by the layered PCIe architecture and the OSI model. Therefore, it includes a physical layer (PHY), a link layer and a protocol layer.

Power Management

Initially, USB battery charging was introduced, followed by a distinct standard known as USB Power Delivery that set guidelines for power management across all USB connectors. The first version supported up to 100 W (20 V, 5 A), while the latest revision (rev 3.1) supports up to 240 W (48 V, 5 A) under certain conditions.

There again, this article does not aim to describe how this works in detail, but it is worth noting that while USB 3.0 requires more power when active, it still consumes less power in the end. Indeed, its higher data rate and various power efficiency features let it remain far less active than in the High-Speed Mode. (See Figure 1.317 which compares power consumption during data transfer in High-Speed and Super-Speed Modes)

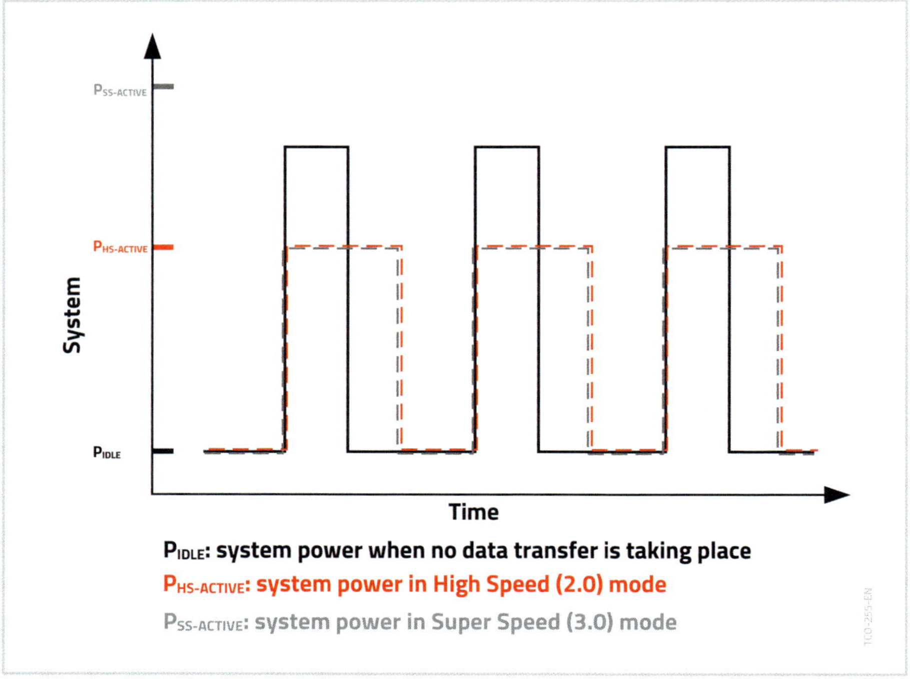

Fig. 1.317: Power consumption: comparison between High Speed mode & Super Speed mode

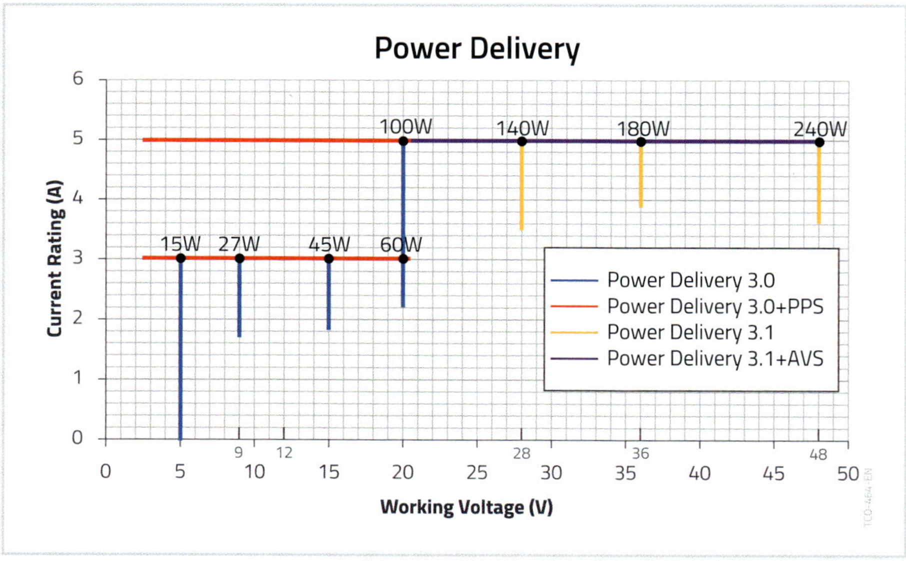

Fig. 1.318: Power delivery graph strategy

The graph shows how different versions of the USB Power Delivery (PD) standard power strategy by adjusting voltage and current to provide various power levels, ranging from 15 W to 240 W, based on device energy needs. Newer versions (PD 3.1) offer more flexible and powerful charging, suitable for energy-hungry devices like laptops and some other portable devices.

USB 3.0 vs. Other High Speed Protocols

Many other communication standards provide equivalent or even higher data rate than USB, namely: PCIe gen 7.0 (242 Gbits/s), Serial ATA (SATA) III (6 Gbits/s), Thunderbolt 3/4 (40 Gbits/s), HDMI 2.1 (48 Gb/s), DisplayPort 2.0 (up to 80 Gbits/s) ...

But PCIe and SATA III do not enable true plug & play application. Thunderbolt is a more complex technology, combination of PCI express and display port so the combination of monitor and data technologies. Therefore, they do not really fight for same applications. Same for HDMI which is rather related to video applications. Therefore, even if they look like direct competitors of the USB 3.0 technology, all this communication protocols rather focus on niche applications and do not compete on the exact same market.

Looking backwards, we should be reminded that the IEEE 1394 (Firewire) technology which was certainly the most reliable and effective serial interface (higher bandwidth and higher reliability ...) was only used on niche markets such as high-performance applications, while USB has been implemented in nearly every electronic equipment. As a consequence, technology is not always the key factor determining the success of a communication standard.

While talking about high data rate and improved power management, we must not forget that the main advantage of USB lies on its versability and ease of use. USB's ability to connect to a wide range of devices, from peripherals to storage and charging solutions, combined with plug-and-play functionality, has made it an excellent solution for consumers and manufacturers.

As these technologies evolve concurrently, it's clear that USB's low royalty fees and broad adoption will allow it to keep expanding, leaving plenty of room for further development.

USB 3.0 Connectors

Now, let's move on to the connector aspect in order to understand what this implies.

Designing a brand-new high-speed connector meeting the requirements is not an easy task, but adapting a more than 15-year-old existing one and making it high-speed capable while staying backward compatible is a really huge task.

The more specific requirements you add, like halogen-free plastic or special colour spec., the more complex it becomes to combine them together. Let's have a look at all these parameters, both individually and in combination with the others.

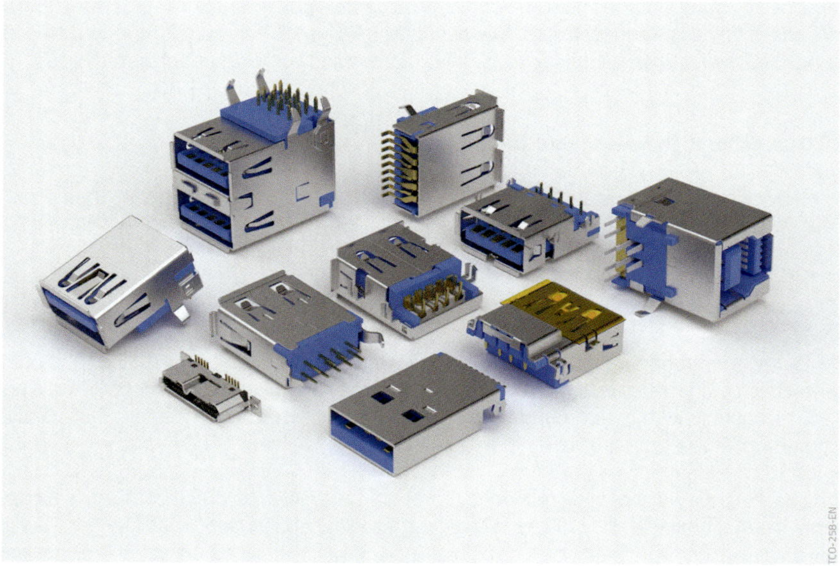

Fig. 1.319: Würth Elektronik USB 3.0 connector range

USB Organisation Basic Connector Specs

The connector section (n °5/ so called mechanical) of the USB specification contains over 60 pages of mechanical and electrical details regarding the supposed properties of connectors and what they should be able to stand. These pages do not even describe the PCB foot print... only the mating face.

The stacked USB is not even described. Therefore, a huge space remains for interpretation and connectors based on customer requirements.

Let's review some of the key criteria like pinning ... for the most important connectors. The USB organisation specification defines the following connectors:

- **USB 3.0 Standard A Plug and Receptacle**
- **USB 3.0 Standard B Plug and Receptacle**
- USB 3.0 Powered B Plug and Receptacle
- **USB 3.0 Micro B Plug and Receptacle**
- USB 3.0 Micro A Plug
- USB 3.0 Micro AB Receptacle

We will review the most important ones, mentioned above in bold.

USB 3.0 Connector: A Type

The USB 3.0 Standard A connector is defined as the host connector, supporting the Super-Speed mode. It has the same mating interface as the USB 2.0 standard A connector, but additional pins for two more differential pairs and a drain.

- Same interface as USB 2.0 standard A connector with 5 more pins
- Backwards compatible with the USB 2.0

Pin no	Name	Description	Mating seq.
1	VBUS	Power	Second
2	D–	USB 2.0 diff. Pair	Third
3	D+		
4	GND	Ground for power return	Second
5	StdA_SSRX–	Super Speed receiver differential pair	Last
6	StdA_SSRX+		
7	GND_DRAIN	Ground for signal return	
8	StdA_SSTX–	Super Speed transmitter differential pair	
9	StdA_SSTX+		
Shell	Shield	Connector metal shell	First

Note: Tx & Rx are defined from the host perspective

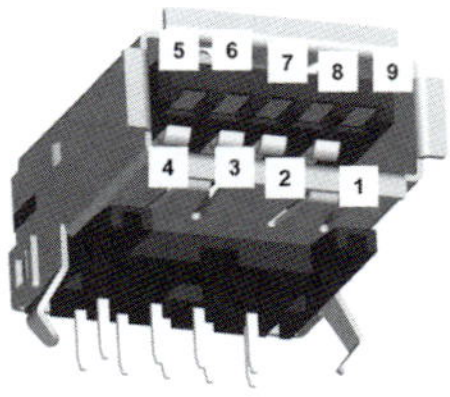

Tab. 1.25: USB 3.0 A type

USB 3.0 Connector: B Type

The USB 3.0 Standard B connector is intended for relatively large, stationary peripherals, such as external hard drives and printers... It's defined in such way that the USB 3.0 Standard B receptacle can accept either a USB 3.0 Standard B plug or a USB Standard B plug. Inserting a USB 3.0 Standard B plug into a USB 2.0 Standard B receptacle is physically impossible.

- Visually different from USB 2.0 standard but the receptacle USB 3.0 is backward compatible with the USB 2.0 standard plug (the opposite is not true)

Pin no	Name	Description	Mating seq.
1	VBUS	Power	Second
2	D–	USB 2.0 diff. Pair	Third or beyond
3	D+		
4	GND	Ground for power return	Second
5	StdB_SSTX–	Super Speed transmitter differential pair	Third or beyond
6	StdA_SSTX+		
7	GND_DRAIN	Ground for signal return	
8	StdA_SSRX–	Super Speed receiver differential pair	
9	StdA_SSRX+		
Shell	Shield	Connector metal shell	First

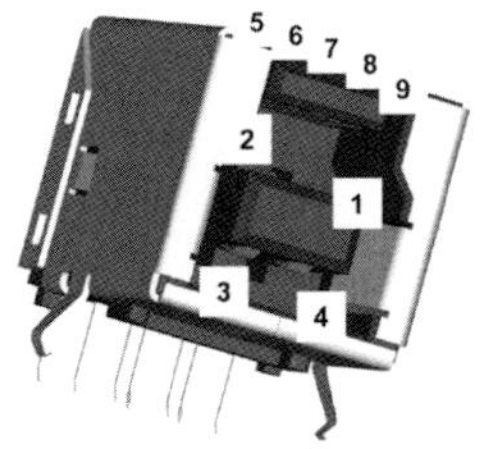

Note: Tx & Rx are defined from the device perspective. The USB org. also defined a potential powered B connector with 2 additional pins

Tab. 1.26: USB 3.0 B type

USB 3.0 Connector: Micro B Type

The USB 3.0 Micro B connector is defined for small handheld devices. It's compatible with the USB 2.0 Micro B connector, i.e. a USB 2.0 Micro B plug works in a USB 3.0 Micro B receptacle. The USB 3.0 Micro B plug cannot be inserted into a USB 2.0 micro B receptacle.

- Based on USB 2.0 connector with an additional part for the high speed transfer
- Backward compatible with the USB 2.0

Pin no	Name	Description	Mating seq.
1	VBUS	Power	Second
2	D–	USB 2.0 diff. Pair	Last
3	D+		
4	ID	OTG identification	
5	GND	Ground for power return	Second
6	MicB_SSTX–	Super Speed transmitter differential pair	Last
7	MicB_SSTX+		
8	GND_DRAIN	Ground for SS signal return	Second
9	MicB_SSRX–	Super Speed receiver differential pair	Last
10	MicB_SSRX+		
Shell	Shield	Connector metal shell	First
Note: Tx & Rx are defined from the device perspective			

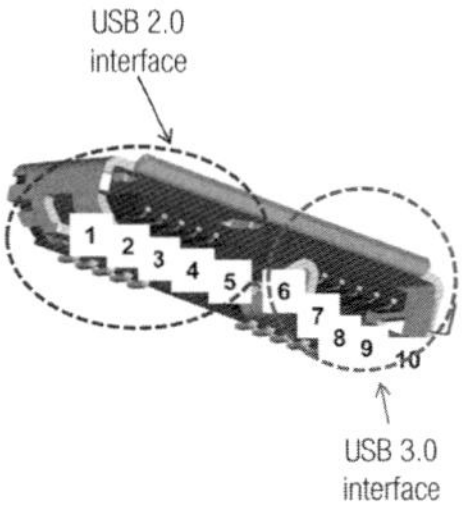

Tab. 1.27: USB 3.0 Micro B type

Material Specs

Component	Material	Comment	N °
Connector shell	Copper alloy or stainless steel, depending on durability requirements		1
Contact	Base material: copper alloy		2a
	Under plating: 2.0 µm Ni		2b
	Contact area plating: (Min) 0.05 µm Au + (Min) 0.75 µm Ni-Pd		2c
	Solder tail plating: (Min) 0.05 µm Au		2d
Housing	Thermoplastic capable of withstanding lead free soldering temperature	Pantone 300C (blue) is recommended for USB 3.0 standard A connector housings	3a
	Flammability rating: HB minimum		3b
	Halogen-free		3c

Tab. 1.28: USB org. material recommendation

1) Shell (Table 2.10)

Most important for the shell is to opt for a matt tin plating which enables much better solderability during the reflow process (compared to Ni plating). Only the plug should remain bright tin plated, since matt tin scratches easily (this could affect the visual aspect of USB sticks since the plug is not protected)

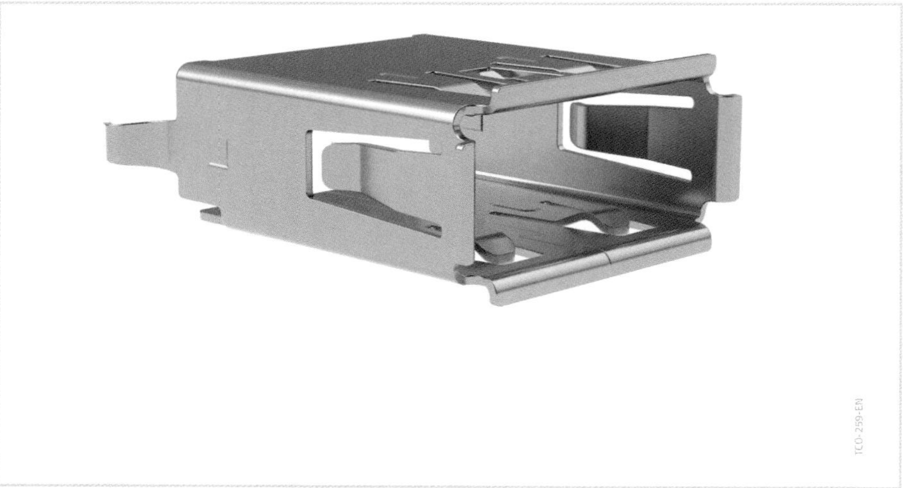

Fig. 1.320: USB 3.0 standard A type shell

2) Contact

As described earlier in the text, the contact material for connectors is always a copper alloy but there is a huge difference between a phosphor bronze and a basic brass in terms of quality (and price obviously, see section 1.5.2.4).

A Ni under-plating layer is a must to guarantee a good plating quality. This recommendation is therefore to be followed.

Here we have a case of critical recommendation which had better not been followed. First of all, suppliers looking for a low cost option will go for a simple flash gold able to stand a minimum of 1500 cycles. This will certainly be much more critical than in the past for USB 2.0. What was possible for USB 2.0 with a lower data rate will become more difficult to implement since the signal tends to flow at the surface when the data rate increases. (Therefore, the impact of a non-stable plating surface like flash gold could become critical).

In this case, a 0.76 µm gold would certainly be the best option to enable the 5000-mating cycle together with a smooth data transmission.

The USB organisation recommends a 0.05 µm Au + 0.75 µm Ni-Pd which tends to be equivalent and slightly cheaper than 0.76 µm gold. This plating was popular in the 80's when gold was really expensive but the Pd price has also increased a lot since that time. As a consequence, this plating is not so common anymore. Although well handled by Japanese companies, this plating is not mastered by Chinese plating companies. As all suppliers will manufacture these connectors in China, it could be critical to follow the USB recommendations. This really small saving may not be worth the high-quality risk that may follow.

A gold plating on the solder tail area is a nice option for whisker growth prevention but is not necessary since a good Ni under-plating with matt tin is sufficient for such a pitch (>1.5 µm).

Opting for gold on solder tail would lead to wettability issues since the gold-plated contacts are always more complex to tin solder.

Here, we would rather go a for a standard selective plating with 0.76 µm gold on the mating side and tin on the solder tail.

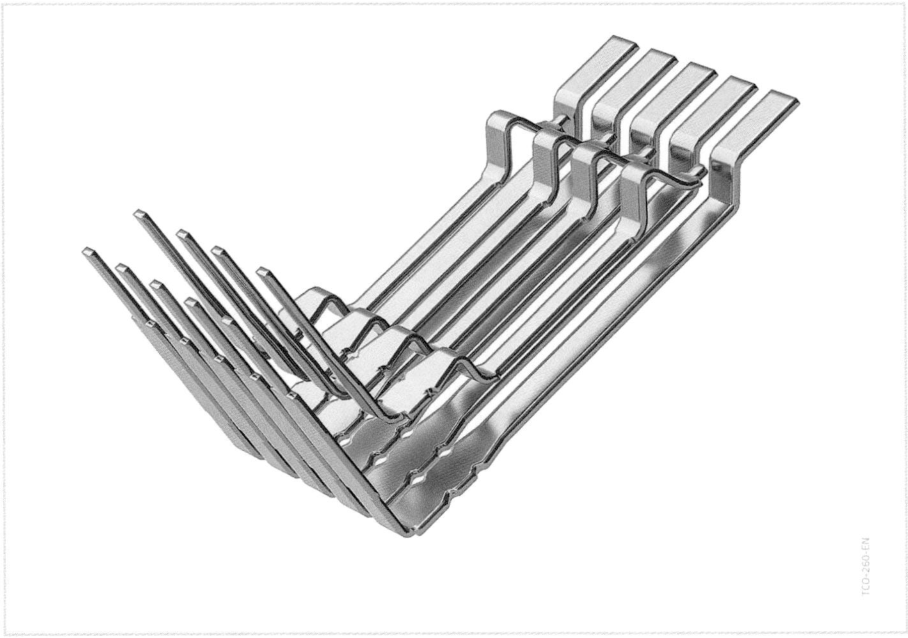

Fig. 1.321: USB 3.0 standard A type receptacle contacts

3) Housing

Here, not so many details even if this area is really critical.

Depending on the kind of lead-free we are talking about, whether wave or reflow, the impact on plastic selection is not the same. This will also impact the blue colour since the colour is likely to be affected during the reflow process (illustrated in Figure 1.322).

HB rating is the lowest flammability rating in the UL range and comes after V2 and V0 which is the clear standard used for 99 % of connectors sold on the electronic market.

It sounds really strange to state that the minimum rating required is HB rating, knowing that most customers request a V0 rating for safety reasons. Stating HB minimum is like mentioning that you expect a car with at least 4 wheels ... as if anyone would ever think of requesting only 3 ...!

Halogen-free is a trend in the electronic market and it's an eco-friendly decision to recommend its use. Nevertheless, there can be a strong impact on the plastic housing selection, since halogen is a key additive used to ensure a good flammability rating. Maybe one reason for this low HB flammability rating seen just above?!

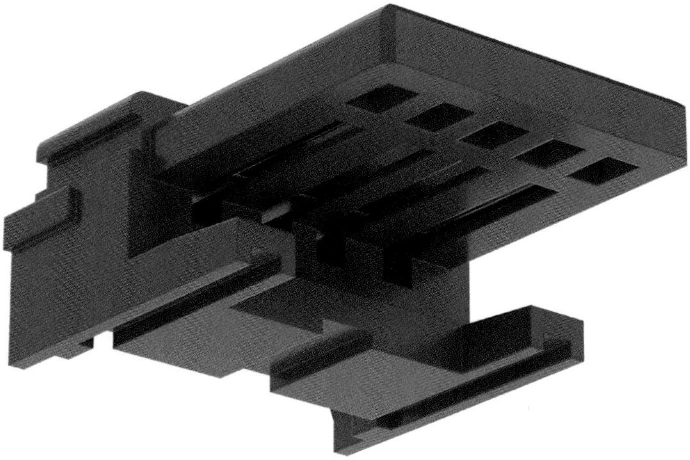

Fig. 1.322: USB 3.0 standard A type receptacle plastic housing

Combined Effects of these Material Requirements

Since both USB 2.0 and USB 3.0 may co-exist on a host, colour coding is recommended by the USB organisation to make it easier for users to differentiate the various ports.

Choosing to use the blue colour together with halogen-free plastic might seem trivial and the impact may not be obvious but this can have critical consequences.

We can already see that the USB organisation downgraded its recommendation to a low HB plastic flammability rating although this will hardly be accepted by end-product manufacturers who mainly request V0 plastic for their final products. Why going so low? Because halogen is the main additive used to make the plastic flame retardant. Without halogen, it is much more difficult to achieve a V0 rating.

As regards wave soldering, it remains possible to keep standard plastic like PBT or Nylon, to have them halogen-free and eventually even V0 rated. But when it comes to reflow soldering compatibility, everything becomes more complicated, especially because of the blue colour. Why?

Consider Figure 1.322 below. On the left we can see the plastic insulator in its original blue pantone 300C colour. On the right side, we can observe the result after reflow: the plastic turns into a horrible blue grey colour. Why? Because halogen-free flame retardant is not stable and react under heated conditions.

Combining UL 94V0 + blue colour + halogen-free + reflow is likely to produce such a result. Even if this doesn't directly impact the connector function, it will be hard to find customers who are willing to accept such a horrible blue/grey colour on the side of their computer...

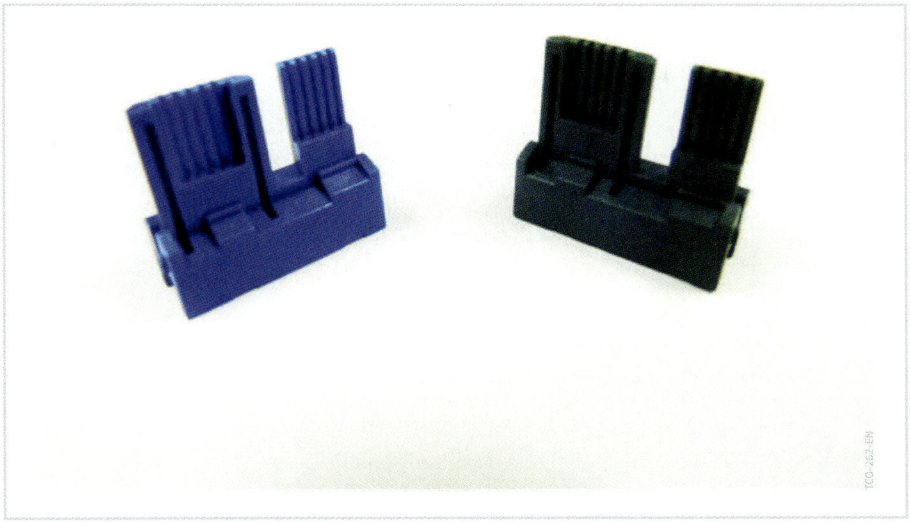

Fig. 1.323: UL 94 V0 halogen free blue PA9T before and after reflow soldering

There are not thousands of options to combine the four parameters together: either you go for a PA46/6T or 9T plastic and decide to downgrade one of the parameters, or you go for **LCP which is the only originally halogen-free plastic**.

Knowing that the UL flammability rating is the only parameter you can downgrade to remain compliant with the USB recommendation (down to HB), there is a high risk that suppliers will offer this option to save costs.

LCP is definitively the best choice if you aim to maintain the best quality possible.

USB Organisation: Connector Certification

While this chapter is written, only 3 laboratories worldwide are able to certify USB connectors. Two of them are located in the USA and one is located in Taiwan.

Whoever wants to have their connectors certified has to deal with one of these 3 labs to obtain one of the most wanted certificates in the field of connectors: the TID, Test Identifier Number which certifies that your connectors are officially approved.

Here, you must consider a cost ranging from US$ 20K to US$ 30 to have your connectors certified (depending on the generation 2.0 or 3.0). And even before that, you must be a member of the USB organisation (yearly membership of US$4K). This is quite a price to pay but this is certainly the only way to differentiate yourself from your competitors and to prove that your connectors comply with the USB IF spec.

This is the choice that we made at Würth Elektronik. Therefore, we decided to have our USB connectors progressively USB-IF certified.

A certified list of connectors can be found under **https://www.usb.org/products**

Conclusion

USB 3.0 targets first the PC market and all consumer devices requiring high data rate and volumes of data transfer such external storage drives ... but like the former USB protocol versions, time will also come for other applications whether they are industrial, medical...

At the moment USB 3.0 chips are still significantly more expensive than the 2.0 one, therefore it limits a mass USB 3.0 spread but we can guess that this time will come.

Nevertheless, it's important to consider that the connector choice for USB 3.0 application is crucial. The connector is the gate from which every signal enters or exits a system. A wrong choice can lead to a lower speed, damaged signal or other critical issues. Therefore, it should not be forgotten that, to maintain backward compatibility and new expectation as halogen free... brought much more stress on connector side. Raw material used to make the connector should carefully be monitored in order to avoid bad influence on the complete application.

To come to the end, **the connector is the gate to high speed, select it carefully!**

For further information, visit **www.we-online.com\usb**

USB 3.2 Connectors – Redefining Data Transfer and Power Delivery

On July 2013, a new version of USB specification was released and officially published in August 2014: USB 3.1 or Super Speed +.

What's different in this new specification? Data rate, power delivery features and new connector type but still backward compatible with USB 3.0 & 2.0 standards. Therefore, it is a huge step forward and with this improvement, a universal communication protocol may be established.

When USB 2.0 was able to transfer data at 480 Mbps and USB 3.0 at 5 Gbps, USB 3.1 can achieve 10 Gbps data rate. For comparison purpose, a 25 GB HD movie could now be transferred in just more than 30 seconds against 70 seconds with USB 3.0 protocol.

Standard USB 3.1 connectors will be able to supply more power than USB 3.0 (9 W for A, B and Micro B and 15 W for Type C vs 4.5 W before).

Type A connector remains retro compatible but design is slightly changed with shielding reinforcement. On Power-Delivery compatible connector, two detection pins are added for insertion and power:

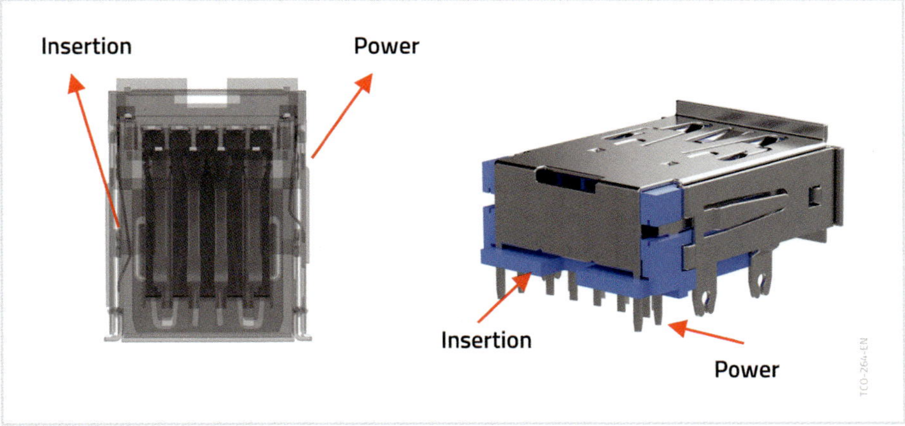

Fig. 1.324: USB 3.1 Type A with Power-Delivery feature supports 5 A over 20 V

Type B & Micro Type designs have some extra-shielding but are really close to USB 3.0 (EMI-RFI performance increase).

Besides high transfer rate aspects, one of the main reasons why USB 3.1 is highly awaited is the new reversible connector: Type C.

Type-C design was introduced to reduce USB diversity and solve many flaws from his predecessors. This modern connector features a reversible design, allowing users to plug it in either way, which simplifies connectivity and reduces wear and tear. Its introduction marked a shift towards a more universal and efficient interface, capable of meeting the demands of contemporary electronic devices.

Mechanical aspects of Type C

It features a compact, symmetrical design that allows for reversible insertion. It includes 24 pins arranged symmetrically, robust materials for up to 10,000 insertion cycles, and metal shielding to protect against electromagnetic interference. The connector also incorporates a secure retention mechanism and durable plastic housing, ensuring reliable and long-lasting performance. Additionally, it can be translated into multiple variants of orientations, pin configurations on PCBs, different heights, and more, making it suitable for a wide range of devices.

Fig. 1.325: USB 3.1 Type C receptacle views

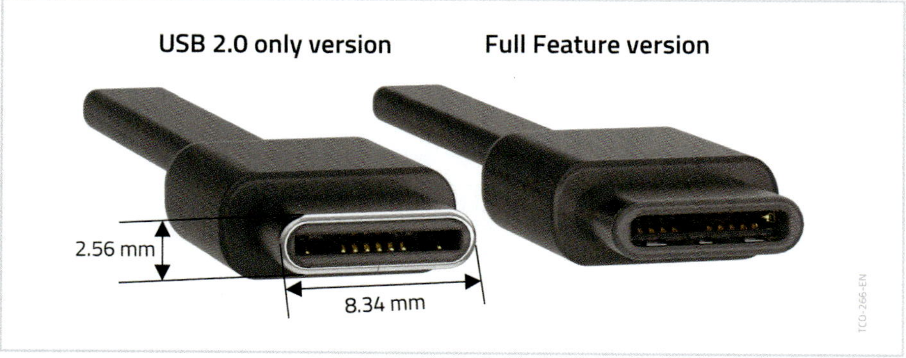

Fig. 1.326: USB 3.1 Type C plug views

More than this feature, Type C connector is able to handle 15 W (3 A under 5 V) and with a connector size close to Mini USB. For supporting reversibility, 24 pins are needed as shown in the figure next:

A1	A2	A3	A4	A5	A6	A7	A8	A9	A10	A11	A12
GND	TX1+	TX1-	Vbus	CC1	D+	D-	SBU1	Vbus	RX2-	RX2+	GND
GND	RX1+	RX1-	Vbus	SBU2	D-	D+	CC2	Vbus	TX2-	TX2+	GND
B12	B11	B10	B9	B8	B7	B6	B5	B4	B3	B2	B1

Fig. 1.327: USB 3.1 Type C receptacle pin-out

A1	A2	A3	A4	A5	A6	A7	A8	A9	A10	A11	A12
GND	TX1+	TX1-	Vbus	CC1	D+	D-	SBU1	Vbus	RX2-	RX2+	GND
GND	RX1+	RX1-	Vbus	SBU2			CC2	Vbus	TX2-	TX2+	GND
B12	B11	B10	B9	B8	B7	B6	B5	B4	B3	B2	B1

Fig. 1.328: USB 3.1 Type C full-feature plug pin-out

The full-feature plug has "only" 22 pins because B6 & B7 contacts are not implemented for avoiding cross-talk issues and extra-stubs.

USB-C cables come in various types including passive and actives cables, with active cables offering extended lengths and enhanced performance.

Electrical and Data aspects of Type C

Type-C connector supports high data transfer rates, up to 10 Gbps with USB 3.1 and up to 40 Gbps with USB4, ensuring fast and efficient data transmission. It also supports USB Power Delivery (USB PD) specifications, allowing for power delivery up to 100 watts (20 V at 5 A) or 240 W in Extended Power Range mode, which is sufficient to charge larger devices like laptops.

Additionally, the connector can carry various protocols, including USB, HDMI, DisplayPort, and Thunderbolt, providing versatile connectivity options.

USB Type-C connector offers numerous benefits making it future-proof solution for modern electronic devices. The evolving legislation, such as the European Union's mandate for USB Type-C as a common charging port, further underscores the importance of standardization in enhancing user convenience and reducing electronic waste.

These regulatory actions drive Wurth Elektronik to promote the extensive adoption of USB Type-C, guaranteeing a superior and more sustainable connectivity experience for our customers.

2.2.4.7 ConHDMI

HDMI (High Definition Multimedia Interface) connectors are dedicated for full or compressed Audio and Video signals. The ConHDMI series offered by Würth Elektronik are HDMI connectors fully compliant with the HDMI 1.4b standard, which means:

- Maximum data rate: 10.2 Gb/s
- Max Resolution: 4K (3840 x 2160) pixel
- Durability : up to 10.000 insertions for type A and 5000 for type C and type D.

The range of products proposed includes three different form factors:

- SMT Horizontal Type A or Standard HDMI (with and without M3 panel flange)
- SMT Horizontal Type C or mini-HDMI
- SMT Horizontal Type D or micro-HDMI.

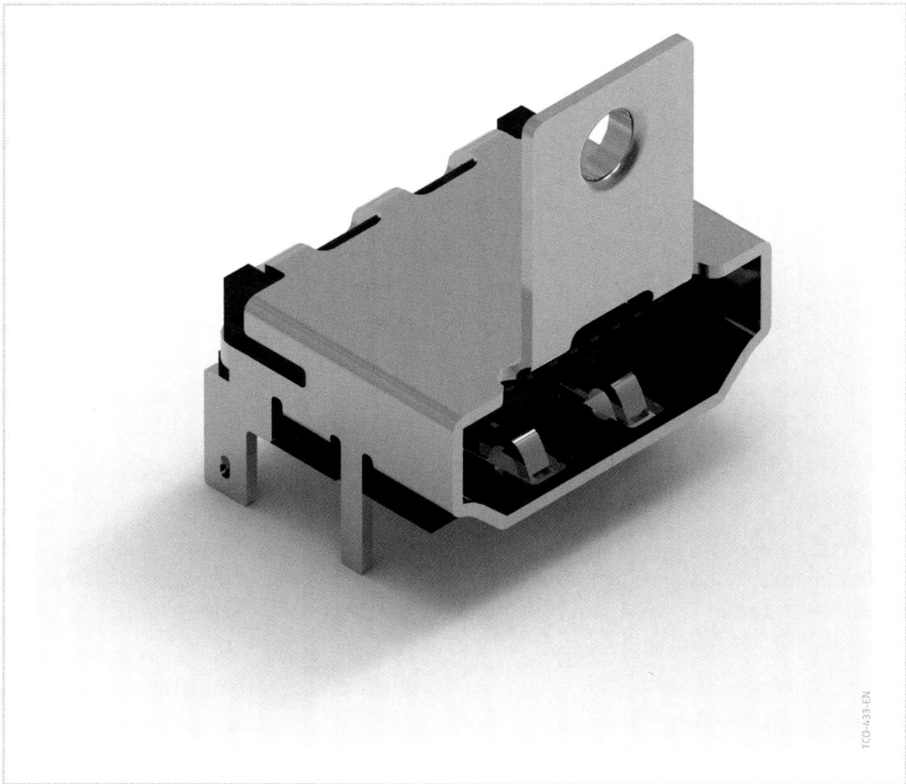

Fig. 1.329: HDMI Type A with Flange

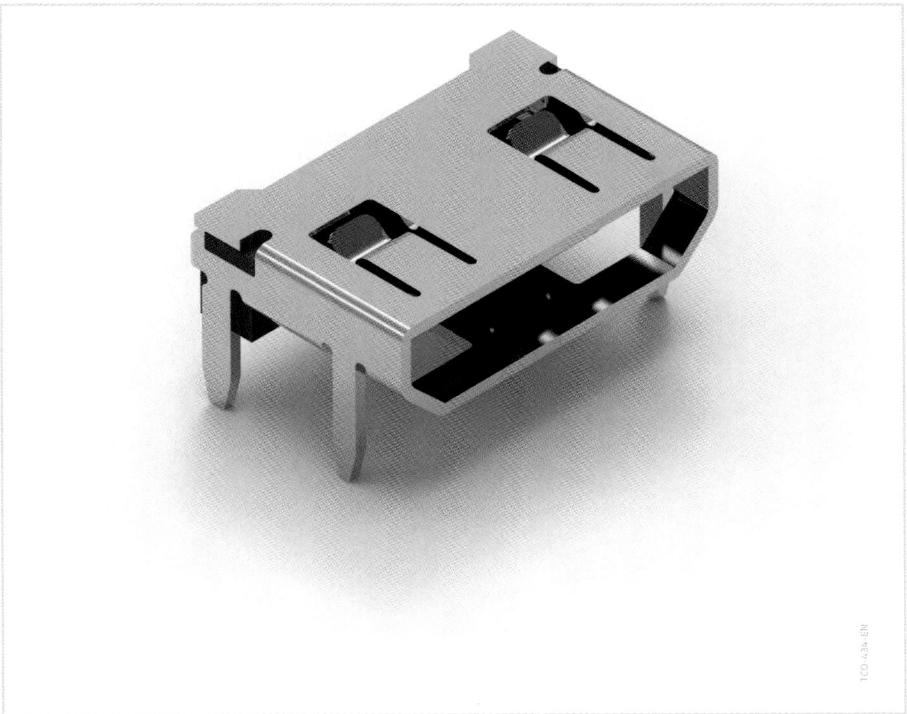

Fig. 1.330: HDMI Type C

Fig. 1.331: HDMI Type D

The standard pinout is made up of 19 pins which support:

- 3 TMDS Channels for signals
- 1 TMDS Channel for Clock
- DDC Channel
- CEC Channel
- Hot Plug Detect
- +5 V
- Ground

The standard pin assignment is the same for Type A and Type D, but is slightly different for the Type C. The main differences are that the positive signals of the differential pairs are swapped with their relative shield, the Ground is assigned to the pin 13 instead of pin 17 , the CEC is assigned to the pin 14 instead of pin 13 and HEC- is assigned to the pin 17 instead of 14

Pin Number	Signal Name	Pin Number	Signal Name
1	TMDS 2+	11	TMDS Clock Shield
2	TMDS 2 Shield	12	TMDS Clock-
3	TMDS 2-	13	CEC
4	TMDS 1+	14	HEC-
5	TMDS 1 Shield	15	SCL
6	TMDS 1-	16	SDA
7	TMDS 0+	17	Ground
8	TMDS 0 Shield	18	Power
9	TMDS 0-	19	Hot Plug/HEC+
10	TMDS Clock+	Shell	Shield

Tab. 1.29: HDMI Standard Pin Assignment for Type A

Pin Number	Signal Name	Pin Number	Signal Name
1	TMDS 2 Shield	11	TMDS Clock+
2	TMDS 2+	12	TMDS Clock-
3	TMDS 2-	13	Ground
4	TMDS 1 Shield	14	CEC
5	TMDS 1+	15	SCL
6	TMDS 1-	16	SDA
7	TMDS 0 Shield	17	HEC-
8	TMDS 0+	18	Power
9	TMDS 0-	19	Hot Plug/HEC+
10	TMDS Clock Shield	Shell	Shield

Tab. 1.30: HDMI Standard Pin Assignment for Type C

Pin Number	Signal Name	Pin Number	Signal Name
1	Hot Plug Detect/HEAC-	11	TMDS 0-
2	Reserved/Utility/HEAC+	12	TMDS Clock+
3	TMDS 2+	13	TMDS Clock Shield
4	TMDS 2 Shield	14	TMDS Clock-
5	TMDS 2-	15	CEC
6	TMDS 1+	16	DDC/HEAC/CEC Ground
7	TMDS 1 Shield	17	SCL
8	TMDS 1-	18	SDA
9	TMDS 0+	19	Power
10	TMDS 0 Shield	Shell	Shield

Tab. 1.31: HDMI Standard Pin Assignment for Type D

TMDS: Transition Minimized Differential Signal
CEC: Consumer Electronic Control: Device control function between HDMI Video/audio devices
Hot Plug: Allows to detect the plug of a device.
HEC: HDMI Ethernet Channel: allow to support bidirectional Ethernet communication up to 100 Mb/s
SCL/SDA: I2C serial clock for Display Data Channel (DDC Channel)

2.2.4.8 RF Connectors

Applications and characteristics of coaxial connectors

Line-bound high-frequency signals are susceptible on the one hand to the coupling in of interference signals, and on the other hand to decoupling into the environment. Without appropriate shielding, high-frequency lines therefore act like antennas and emit or receive undesirable electromagnetic waves. For this reason, coaxial lines and connectors have become established in wire-based high-frequency technology. Thanks to their shielding (outer conductor), they attenuate both unwanted signal coupling into and decoupling from the inner conductor to such an extent that they can usually be neglected in applications.

In coaxial connectors, single-ended signals are always transmitted with referenced grounding. The signal-carrying inner conductor is located in the center of a coaxial cable. An insulating dielectric separates it from the outer conductor. This shielding establishes the connection with the ground potential. The inner conductor, dielectric and outer conductor are arranged coaxially, thus giving this type of cable its name (Figure 1.331check number).

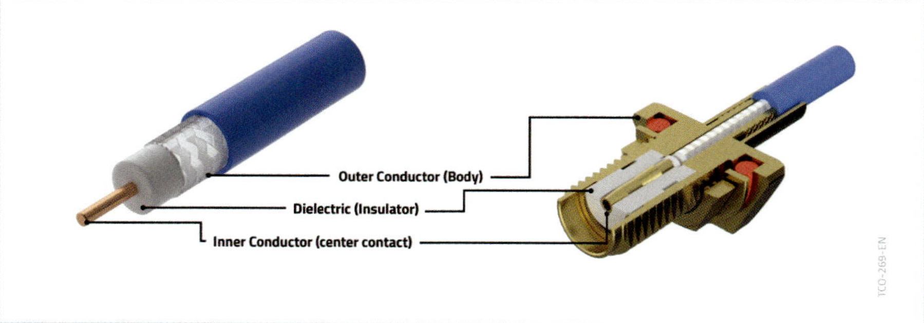

Fig. 1.332: Typical structure of a coaxial cable (left) and termination of a coaxial connector (right).

In the case of coaxial cables, it is precisely the product structure as critical points, since it is important to maintain effective shielding there. While the shielding structure of coaxial cables is usually homogeneous, a discontinuous transition is almost unavoidable when the cable is mostly connected to a coaxial connector. By means of suitable design measures, connectors must also ensure the shielding effect over the entire plug-in connection from cable to plug and socket to another cable or printed circuit board. In addition, changes in the diameter of the inner conductor lead to impedance differences and thus cause signal reflections.

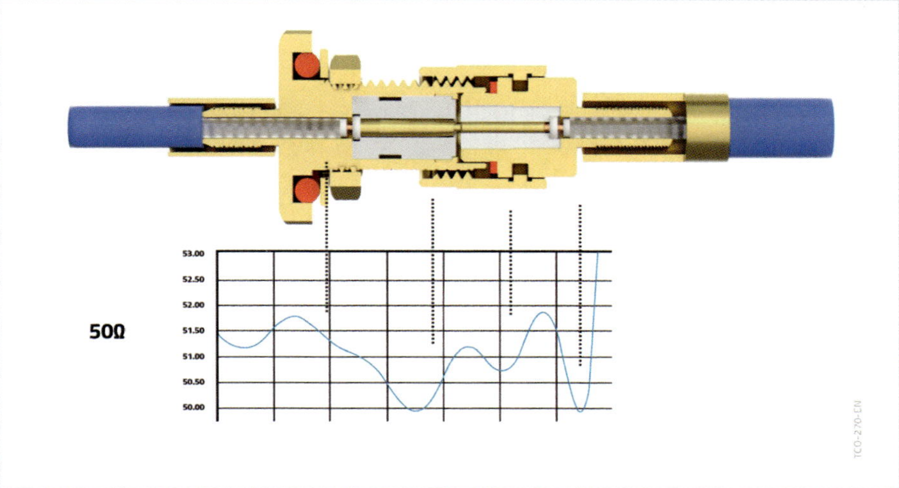

Fig. 1.333: Deviations from the ideal wave impedance of 50 Ω occur at the junctions of the coaxial line and the coaxial connector.

Characteristic impedance and wave propagation

An important parameter for the optimum transmission of high-frequency signals is the characteristic impedance of the RF waveguide, which should have the same value over the entire transmission distance. The characteristic impedance Z_L results from the geometry of the coaxial cable and the impedance of free space Z_0 (376,62 Ω):

$$Z_L = \frac{Z_0}{2\pi\sqrt{\epsilon_r}} \ln\left(\frac{D}{d}\right) \quad \text{(EQ 24)}$$

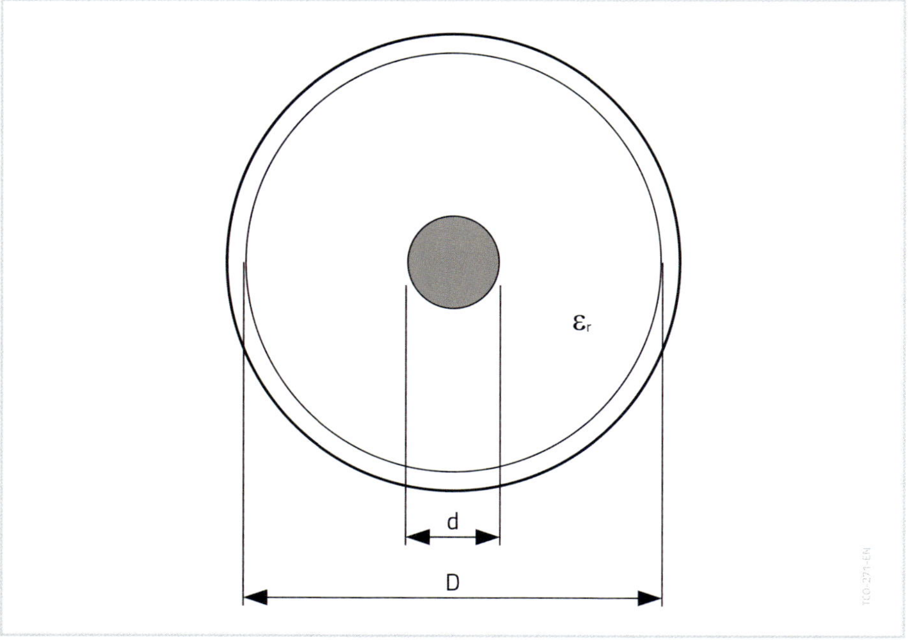

Fig. 1.334: Coaxial line in cross-section. The ratio of the diameters of the inner conductor (d) and outer conductor (D) determines the characteristic impedance.

Typical values for the characteristic impedance of an RF line are 50 and 75 Ω. The diameter of the inner conductor is smaller in the 75 Ω version than in the 50 Ω version.

Typically, transverse electromagnetic waves (TEM) propagate over a coaxial line, where the electric field in the insulating dielectric is radially directed and the magnetic field is concentrically directed. The signal propagation is subject to higher losses with a usually relative permittivity $\varepsilon_r > 1$ of the dielectric compared to air.

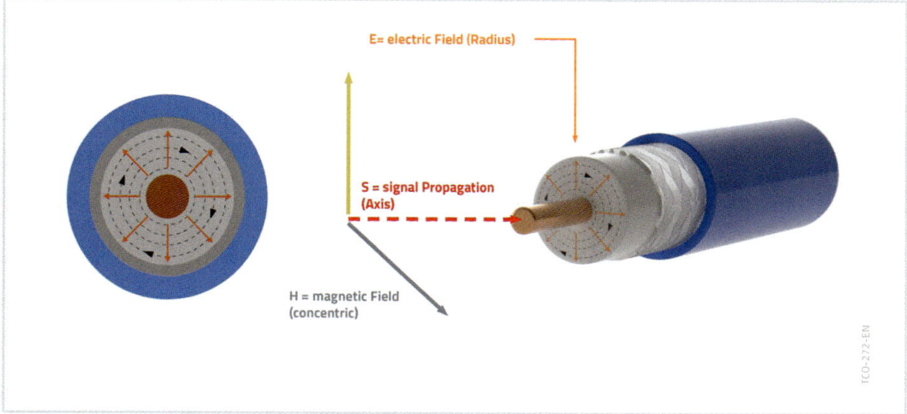

Fig. 1.335: Propagation of transverse electromagnetic waves in a coaxial line.

Impedance mismatches in an RF transmission line cause power losses and reflected energy. Voltage Standing Wave Ratio (VSWR) is a method of measuring such impedance mismatches. VSWR is defined as the ratio between the transmitted and reflected voltage standing waves and is considered an ideal match when VSWR = 1. A common example is an RF power amplifier connected to an antenna via a transmission line. Influencing factors here are the electronics, connectors, jacks, cables, i.e. all RF transmitting components of the system. A high SWR indicates poor efficiency of the transmission line and reflected energy.

Types of coaxial connectors and frequency response

There are numerous types of coaxial connectors, for example N-Type, BNC, TNC, Belling-Lee, F, MCX, MMCX, SMA, SMB, SMP, K, V, W to name just a few de facto standards. Depending on the connector type, the frequency spectrum extends into the one- and two-digit gigahertz range, with the widespread SMA types often specified up to 18 GHz, the more advanced K connectors up to 40 GHz, V connectors up to 70 GHz and W types even up to 110 GHz. The following section explains the coaxial RF connector types SMA, SMB, SMP, UMRF, MCX,MMCX, BNC, TNC and N Type from Würth Elektronik (www.we-online.com/coax) as examples.

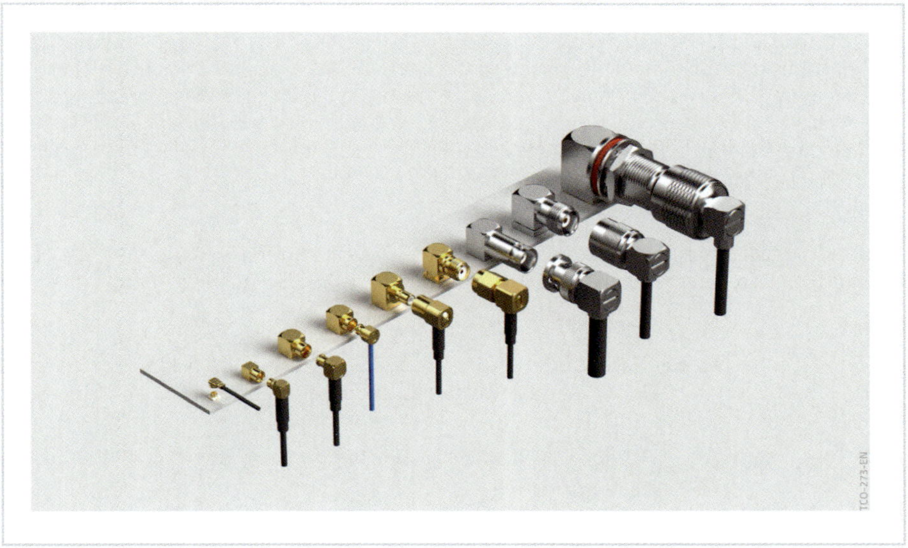

Fig. 1.336: The portfolio of Würth Elektronik eiSos includes the coaxial RF connector types UMRF, MMCX, MCX, SMP, SMB, SMA, SMB, BNC, TNC and N-TYPE

Update for below table with TNC ad BNC included

	N Type	BNC	TNC	SMA/ RPSMA	SMB	SMP	MCX	MMXC	UMRF
Order code	604...	607...	607...	603.../630...	616...	601...	606...	660...	636...
Interface	MIL-STD-348						IEC 61169-36	IEC 61169-52	–
Interface Mechanism	Thread	Bayonet	Thread	Thread	Snap on	Snap on	Snap on	Lock Snap	Lock Snap
Nominal Freq. Range	DC~11 GHz	DC~4 GHz	DC~11 GHz	DC ~ 18 GHz	DC~4 GHz	DC~18 GHz	DC~6 GHz	DC~6 GHz	DC~6 GHz
Impedance	50 Ω								
Tool	–	–	600 633 0101	600 633 0101	–	600 601 0201	–	–	600 636 001
Design Kit	603301 (PCB Connector)	603001 (Cable Connector)	603301 (PCB Connector)	603001 (Cable Connector)	616001 (PCB & Cable Connectors)	601001 (PCB & Cable Connectors))	606001 (PCB & Cable Connectors)	660001 (PCB & Cable Connectors)	636001 (PCB & Cable Connectors)

Tab. 1.32: Overview of the product range of coaxial RF connectors from Würth Elektronik and their technical specifications.

SMA and RPSMA

SMA (Sub Miniature A) connectors are widely used in RF technology, where the coupling nut and plug or thread and socket belong together respectively (Fig. 2.X.6). The SMA is specified in the MIL-STD-348 standard. A modified version of these are RPSMA connectors (Reverse Polarity Sub Miniature A), in which the coupling nut is combined with the socket and the thread with the plug. A reliable electrical connection – especially in case of occurring vibrations or humidity – is ensured by screwing the SMA connectors (or RPSMA) with a defined torque of 57 Ncm. Typically, the frequency response is up to 18 GHz or higher.

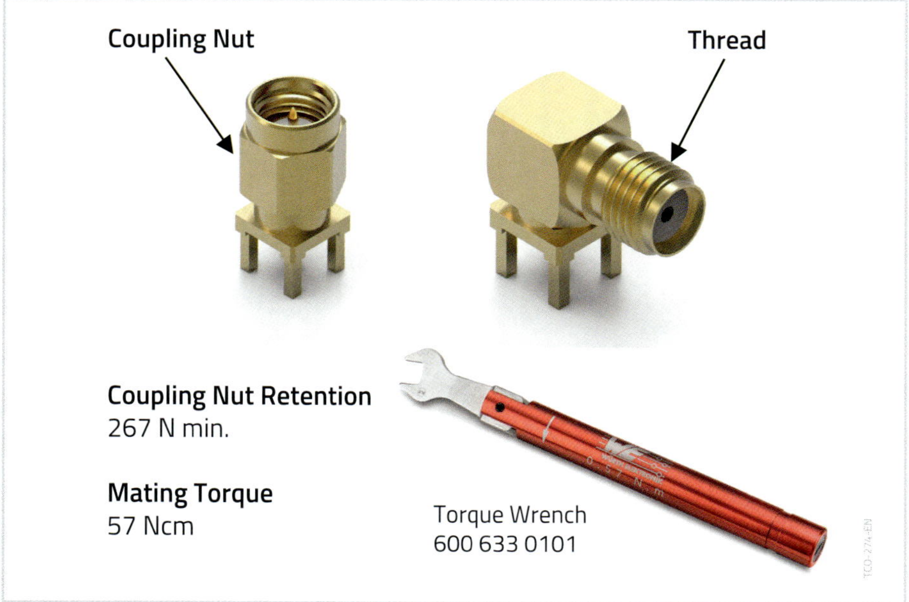

Fig. 1.337: SMA connectors can be screwed with a defined torque and thus ensure a reliable electrical connection.

N-Type

The N-type is a proven screw-thread RF connector developed by Paul Neill of Bell Labs in the 1940s. The N-type is specified in the MIL-STD-348 standard. N-type connectors operate from DC to 11 GHz at an impedance of $50\ \Omega$. Its rugged structure is shock resistant and suitable for outdoor application. N-type connectors are available as Plug and Jack, THT, 4-hole-flange and cable connectors. Versions include straight, right angle and rear mounted bulkhead supports panel mount igreess protection up to IP67.

Fig. 1.338: N-type connectors feature a screw-on-coupling and a rugged structure.

SMB, MCX and MMCX

In contrast to SMA connectors, the SMB (Sub Miniature B), MCX (Miniature CoaX) and MMCX (Micro Miniature CoaX) types use snap-on locking. This allows them to be plugged and unplugged quickly and rotated through 360°. While SMB types operate from DC to 4 GHz and require mating forces of 62.3 N, MCX and MMCX types are specified from DC to 6 GHz and require mating forces of 63 N and 18 N, respectively. Mating disconnect forces are typically less than this (SMB 8.9-62.3 N, MCX 8-20 N and MMCX 6 N).

Fig. 1.339: SMB, MCX and MMCX connectors rely on a snap-on locking mechanism and can thus establish an electrical connection easily and quickly.

SMP

SMP (Sub Miniature P) connectors also rely on a snap-on locking mechanism. They can also be rotated through 360°, but allow a frequency response from DC to 18 GHz. A special feature are three different connector types (plug) with different degrees of engagement and thus mating forces and permissible mating cycles:

1. FD (Full Detent) requires a mating force of 68 N and provides a holding force of 22 N for 100 mating cycles.
2. LD (Limited Detent) requires a mating force of 45 N and provides a holding force of 9 N at 500 mating cycles.
3. SB (Smooth Bore) requires a mating force of only 9 N and provides a holding force of 2.2 N at 1000 mating cycles.

The socket (jack) always remains the same. Thanks to these three versions, SMP connectors can be used to establish both quick mate and unmate connections as well as stable board-to-board connections, e.g. in the FD adapter-SB combination. The FD adapter connection offers high holding forces, while a board with SB adapter can be plugged and unplugged up to 1000 times with low forces.

Three different Interfaces designed for Plug:

→ „Full Detent" (FD) / „Limited Detent" (LD) / „Smooth Bore" (SB)

Three level of retention forces are created

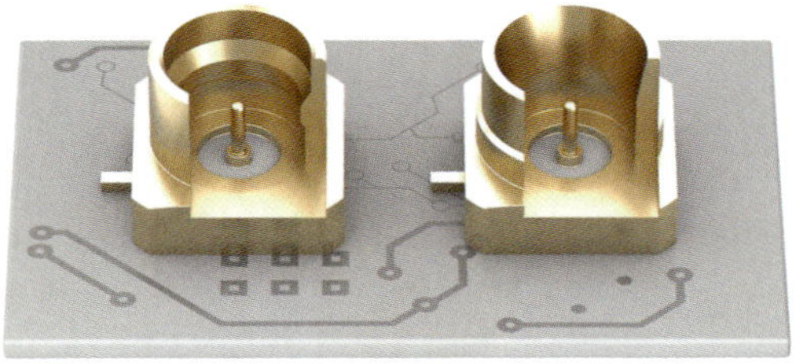

Mechanism \ Interface	Full Detent	Limited Detent	Smooth Bore
Force to Engage	68N max.	45 N max.	9 N max
Force to Disengage	22 N max.	9 N max.	2.2 N max.
Mating Cycles	100	500	1000

Fig. 1.340: SMP connectors rely on a snap-on locking mechanism and offer three versions on the plug side with different mating forces and permissible mating cycles. The socket (jack) is the same for all three versions.

BNC and TNC

The BNC (Bayonet Neill–Concelman) connector is a miniature quick-connect/disconnect RF connector used for coaxial cable connections. The BNC connector features a bayonet coupling mechanism, which allows for quick and secure connections. The bayonet system involves two small pins on the female connector that engage with the slots on the male connector, enabling a quarter-turn twist to lock the connector in place. This design ensures a firm connection while also allowing for easy disconnection when needed. This has made BNC connectors a staple in test and measurement environments, where equipment is often reconfigured. They are particularly common in television broadcasting, test equipment, and amateur radio setups. BNC connectors are also used in networking, where they were used to connect coaxial cables.

The TNC (Threaded Neill–Concelman) connector is used widely in various communication systems. TNC connector was developed as an enhancement of the BNC connector, primarily to provide better performance at higher frequencies.

The TNC connector is similar in design to the BNC connector but with one critical difference: the TNC connector has a threaded coupling mechanism instead of a bayonet locking system. This threaded interface ensures a more secure and stable connection, especially under conditions where vibration or movement could cause a BNC connector to come loose. The standard TNC connector is designed to operate in frequencies up to 11 GHz.

One significant advantage of TNC connectors is their resilience to environmental factors. The threaded design helps maintain a tight connection, reducing the likelihood of signal degradation due to moisture or dust ingress. For both BNC and TNC connectors, their size is between N Type and SMA connectors.

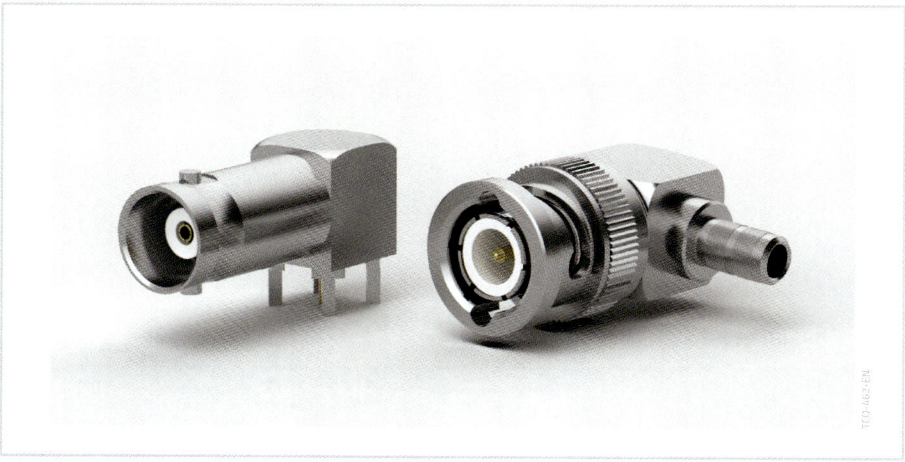

Fig. 1.341: BNC PCB receptacle and cable plug

Fig. 1.342: TNC cable plug and PCB receptacle

UMRF

The very compact UMRF (Ultra Miniature RF) connectors can be used from DC to 6 GHz and typically consist of a socket for PCB mounting with an area of 3.1 x 3.0 mm2 and a 90° angled crimp plug for coaxial cables with a diameter of 0.81 to 1.37 mm. When mated, the connector has a height of only 2.5 mm from the PCB. Mating and unmating require a special tool, which is available from WE. Up to 30 mating cycles are possible.

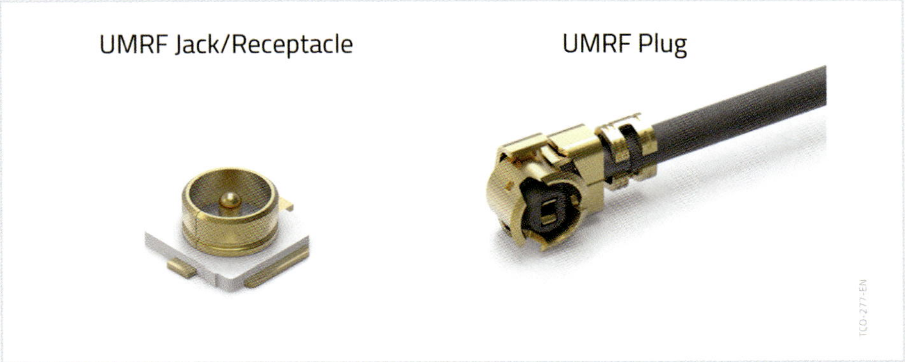

Fig. 1.343: UMRF Cable Plug and PCB Receptacle

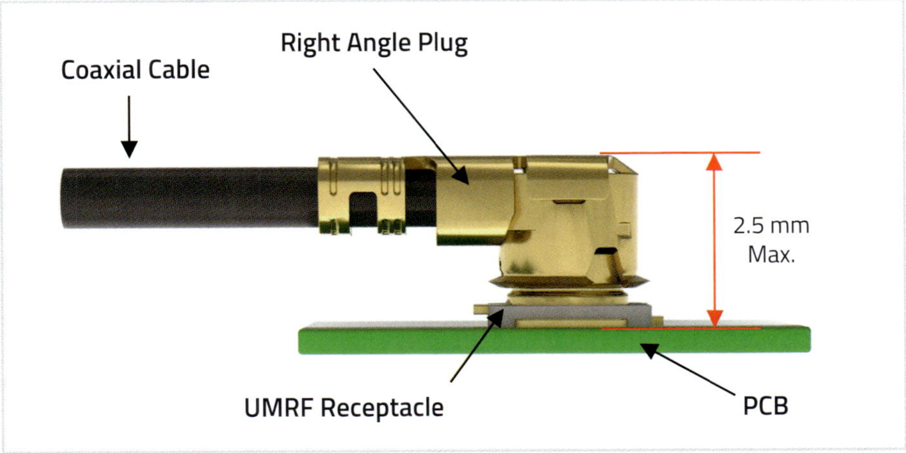

Fig. 1.344: UMRF connectors rely on a snap-on locking mechanism and manage up to 30 mating cycles. Only 2.5 mm bears a UMRF connector on a printed circuit board (bottom).

Adapters and assembled cables

Coaxial adaptors are essential components used to connect different types of coaxial cables or connectors in RF systems. WE offers a variety of adapters to bridge the gap between various connector types or genders, allowing for seamless integration and communication between devices with different interfaces. Coaxial adaptors are designed to maintain the electrical characteristics of the RF system, including impedance matching (usually 50 or 75 ohms) and low VSWR (Voltage Standing Wave Ratio). Coaxial adaptor ensures minimal signal reflection and loss, especially in high-frequency applications. Please look up Würth Elektronik (www.we-online.com/coax) for different coaxial adaptor options.

JCh(PM): The picture expression too much of coaxial adaptor product range, I prefer to revised the wording and remove the picture to avoid misunderstanding.

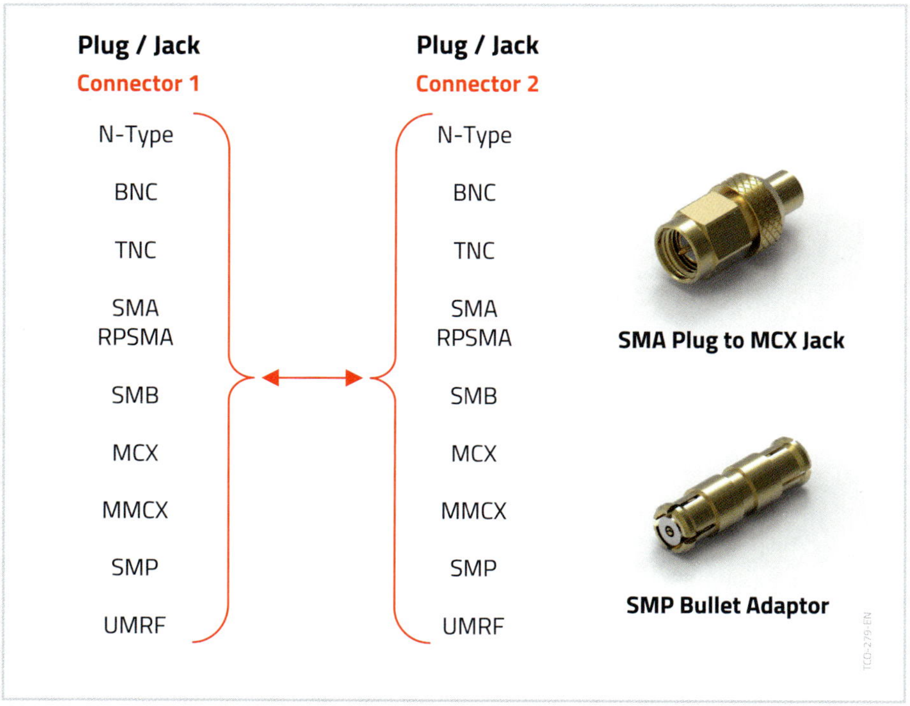

Fig. 1.345: WE offers a wide range of adapters for connecting different connector types.

For cable assembly products, WE offers variety of pre-assembled coaxial cables in length 152.4 mm (6 inch) and 304.8 mm (12 inch) length. Plugs and sockets of interface r types threaded, snap-on, bayonet , which can be mounted to cables of both flexible and hand-formable types. Please look up Würth Elektronik (www.we-online.com/coax) for different coaxial cable assembly production options. Fig. 2.X.11 provides an overview of WE standard coaxial cable type option.

Cable	Type	O.D.	UL	Jacket	Picture
RG58 RG174	Flexible	4.95 mm 2.8 mm	1354	PVC	
1.13 mm 1.32 mm 1.37 mm	Flexible	1.13 mm 1.32 mm 1.37 mm		FEP	
Low Loss 195	Flexible	4.95 mm	-	PE	
RG178 RG316	Flexible	1.8 mm 2.48 mm	1971	FEP	
RG142 RG316DB	Flexible (Double Braid)	4.95 mm 2.9 mm	1971	FEP	
RG402 (.141") RG405 (.085")	Hand- formable	3.58 mm 2.15 mm	1375	-	
RG405 (.085") RG047 (.047")	Flexible	2.64 mm 1.42 mm	1971 1354	FEP	

Fig. 1.346: WE offers a wide range of cable assembly with different connector types.

2.3 Würth Elektronik Packaging

Today, packaging is not only for protection, but also the first physical contact the customer gets before using the product. Packaging has to protect, inform and be suitable for customer´s processes (manual, semiautomatic or automatic).

2.3.1 Why Different Packaging?

Different types of packaging are available for connectors. The choice of the packaging depends on the assembly process.

When placing a connector on a Printed Circuit Board (PCB), the automatic system is the most common process in SMT (Surface Mount Technology). This system is called the "Pick and Place", during which robotic machines are used to place surface-mount devices (SMD) or through-hole technology connectors (THT) onto a PCB. They are used for high speed, high precision placing. This system normally uses pneumatic suction caps. The "Pick and Place" system is mostly used with reels and trays and in some cases with tubes. In order to use a reel with this automatic system, the manufacturer must have a feeder. Once the PCB is clamped onto the machine, the nozzle picks up an individual connector from the feeder or tray, rotates them to the correct orientation and then place them on the appropriate pad on the PCB.

Tube packaging can be used for THT and SMT connectors. The inside shape of the tube (different exterior and interior designs exist depending on the connector's shape) allows for holding the connector tightly thus avoiding the shock during transport or handling.

Carton boxes and plastic bags are commonly used for connectors with no risk of damage during handling or transport, like terminal blocks (TBL), jumpers, pin headers (PHD) and are adapted to the size of connectors.

Fig. 1.347: Tape & reel

Fig. 1.348: Bag

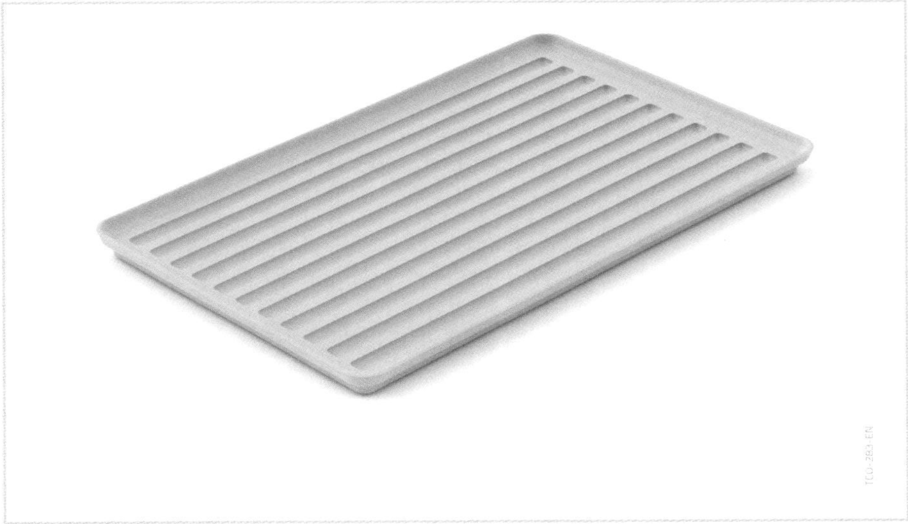

Fig. 1.349: Tray

Fig. 1.350: Tube

2.3.2 Packaging Types

2.3.2.1 Tape & Reel

Tape & reel packaging is the best packing solution for SMT connectors.

Würth Elektronik uses only one reel type for connectors. The diameter of this reel is 13" (30.02 cm). Würth Elektronik follows the requirements of the IEC 60286-3 standard for packaging.

To assess the peeling force of the cover tape, we follow these parameters on Table. 1.35 with peel speed of 300 mm/minute.

Peel force of Cover Tape	Limits	in average
Pulling force [N] Max	0.74 ±0.34	0.74 ±0.15
Pulling weight [g] Max	75 ±35	75 ±15

Tab. 1.33: Pulling force of cover tape

To allow customers to use the reel, it's necessary to have a post-carriage and a leading feed zone. Concerning the feeding direction, the tape orientation, the sprocket, and the cavity holes, all the information is available on the data sheet.

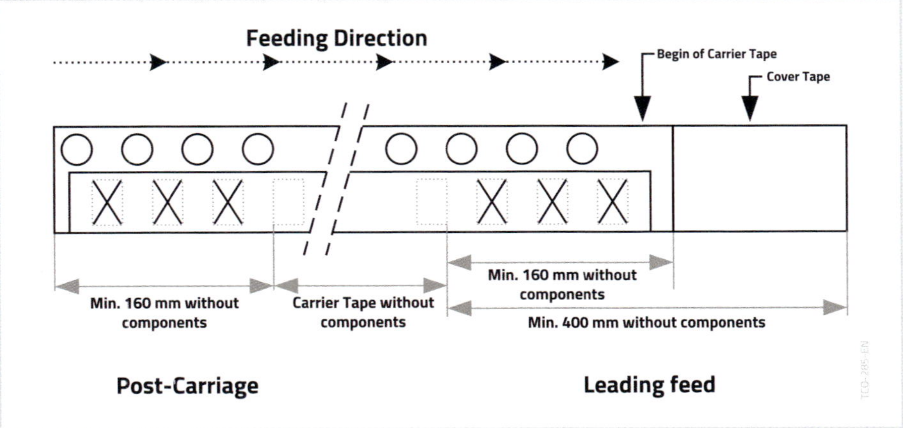

Fig. 1.351: Post-carriage and leading feed

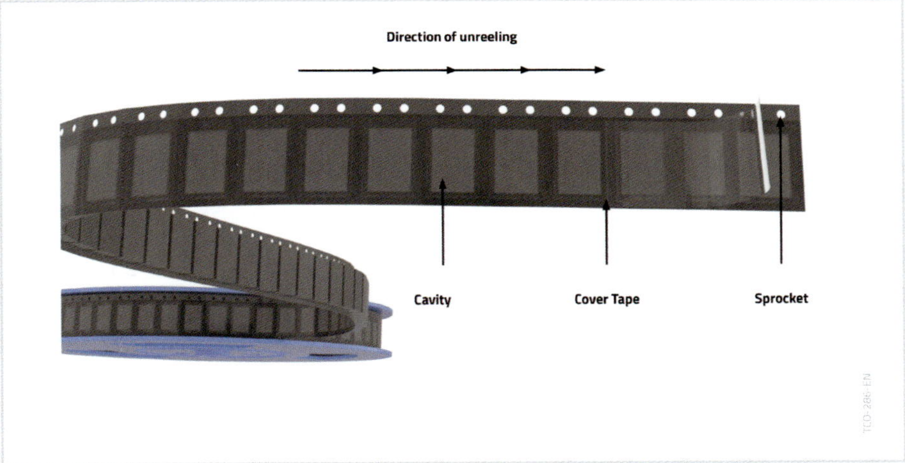

Fig. 1.352: Unreeling direction

2.3.2.2 Tray

Tray packaging is popular because it can be used for the automatic processing of SMT devices (if the exposed surface of the connector is big and flat enough) or THT and even for manual processing.

Trays are a very good way to protect connectors because you can design a lot of different shapes that fit perfectly with the connectors that will be placed inside to avoid damage during handling or transport.

There are two main types of tray packaging:

Separate Cavity Holes: For connectors prone to handling damage, the first type of tray features individual molded cavities. This design keeps each connector isolated, eliminating the risk of shocks and bent contact pins (USB, HDMI ...)

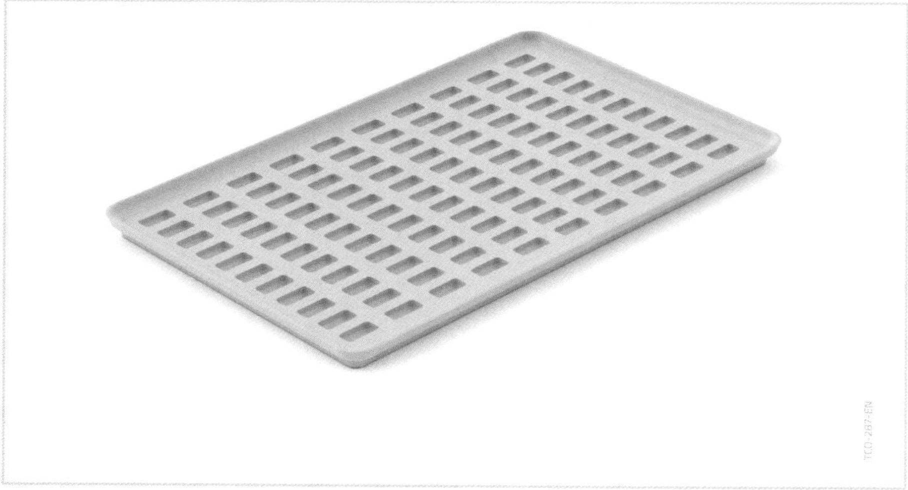

Fig. 1.353: Separate cavity holes

Rows within the tray: This second type of packaging utilizes internal rows to house the connectors. While this design protects them from external shocks, the connectors may experience slight contact with each other during transport. Therefore, only robust connectors with no risk of damage are suited for this type of tray (like BHD, D-SUB, PHD, Socket Header ...).

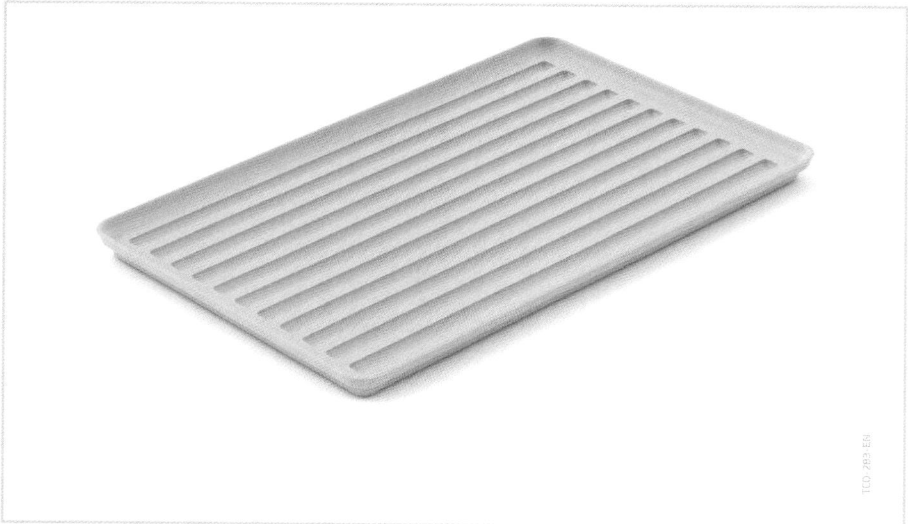

Fig. 1.354: Rows inside the tray

2.3.2.3 Tube

Tube packaging can be used for THT and SMT connectors.

The plastic material is very strong and protects connectors from external shock. The plastic shape of the tube allows the connector to be kept inside tightly and to prevent contact pins from being bent.

Below you will find a few examples of tube shapes that will illustrate how tubes can be adapted to all types of connectors:

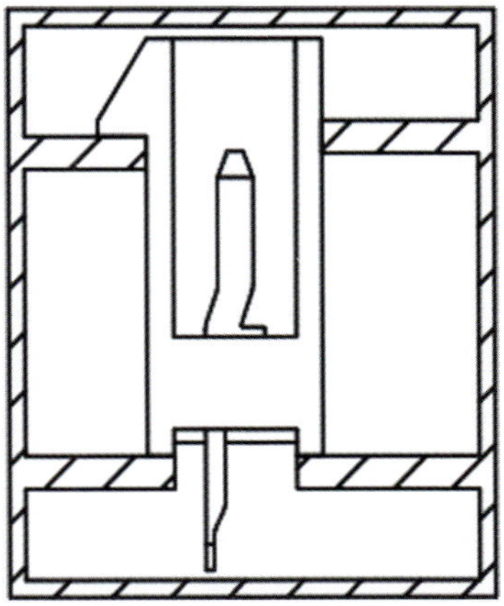

Fig. 1.355: Tube shape example for straight single row connector

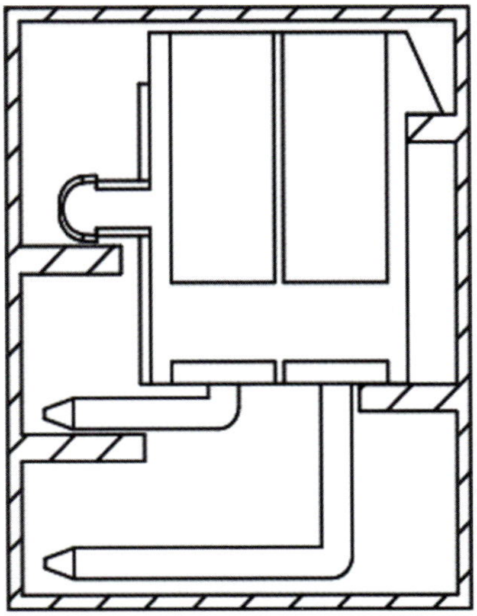

Fig. 1.356: Tube shape example for dual row angled connector

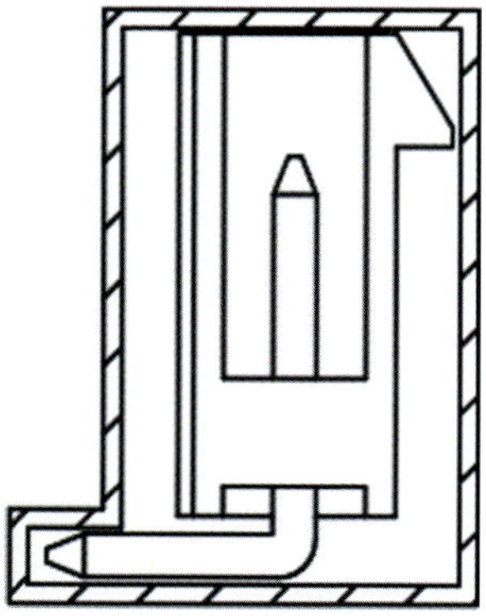

Fig. 1.357: Tube shape example for single row angled connector

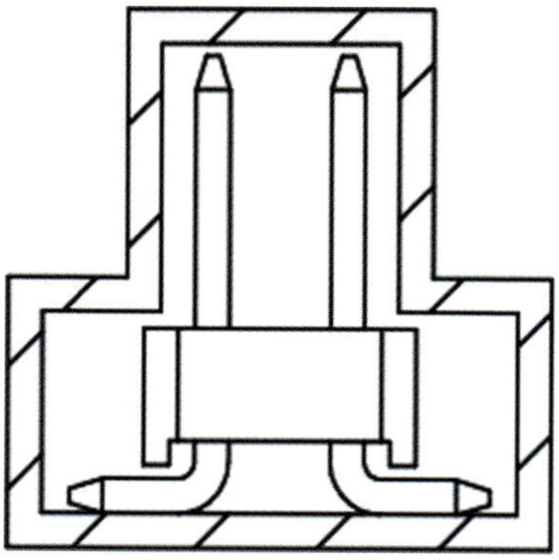

Fig. 1.358: Tube shape example for dual row SMT header

2.3.2.4 Bag/Carton Box

Bag packaging is commonly used for connectors having no risk of damage during transport or handling. The bags are sealed to ensure the best possible floor life of the components but also to create a micro-climate around the component in order to protect it from environmental influences like high humidity.

2.3.3 Labelling

2.3.3.1 Standard Label

Our standard labels are made to provide all necessary information to the customer while ensuring Wurth Electronik has comprehensive traceability for monitoring of the products sent to customers.

Würth Elektronik standard layout: 2D Data Matrix + 1D Code 128 + human readable text + logos

Fig. 1.359: Würth Elektronik standard label

The barcodes use the IEC EN 62090:2002 standards:

- Intervals between the "lines"
- Constitution of the "2D-Barcode"
- Identifier for the "Code 128"

All information mentioned in "Code 128" barcodes are combined in one data matrix.

Fig. 1.360: Würth Elektronik standard label layout

Part number, quantity, article description and logos correspond to the information for customers using Würth Elektronik components.

Lot number and date code are important for monitoring in the event of a complaint. It allows Würth Elektronik to find other samples from the same production batch. This way Würth Elektronik can investigate and test the quality of a nearly identical product.

2.3.3.2 Tube Label

Due to the lack of space on some packaging types like tubes, Würth Elektronik developed a smaller label to identify the products.

"Code 128" barcodes were removed and only the combined data matrix remains. This way if the human readable information is partially destroyed, it is still possible to scan the data matrix to collect the rest of the information.

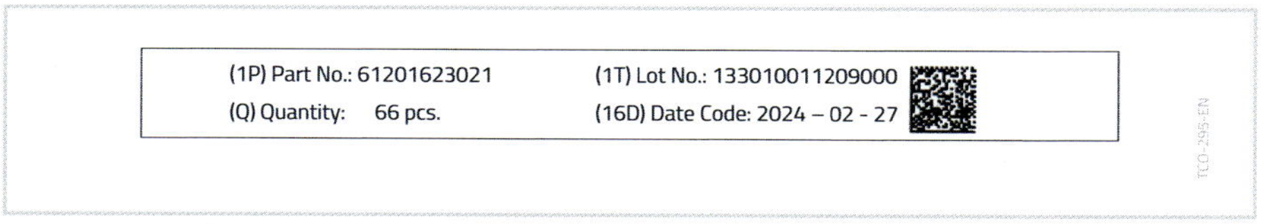

Fig. 1.361: Würth Elektronik tube label

2.3.3.3 MSL

MSL is the Moisture Sensitive Level of the component. The handling of moisture sensitive components needs to be done according to **JEDEC J-STD033** and **JEDEC J-STD020**. The MSL is categorized into six levels, and each describes the sensitivity of different components to moisture.

These levels range from 1, indicating the lowest sensitivity to moisture, all the way up to 6, signifying the highest sensitivity. Crucially, the packaging a component receives must directly correspond to its MSL rating. This ensures the component is shielded from harmful moisture exposure during processing, which can lead to a variety of potential issues and defects. By tailoring packaging to a component's specific MSL, we guarantee its safe arrival and optimal performance.

Würth Elektronik differentiates packaging between MSL=1 and MSL>1.

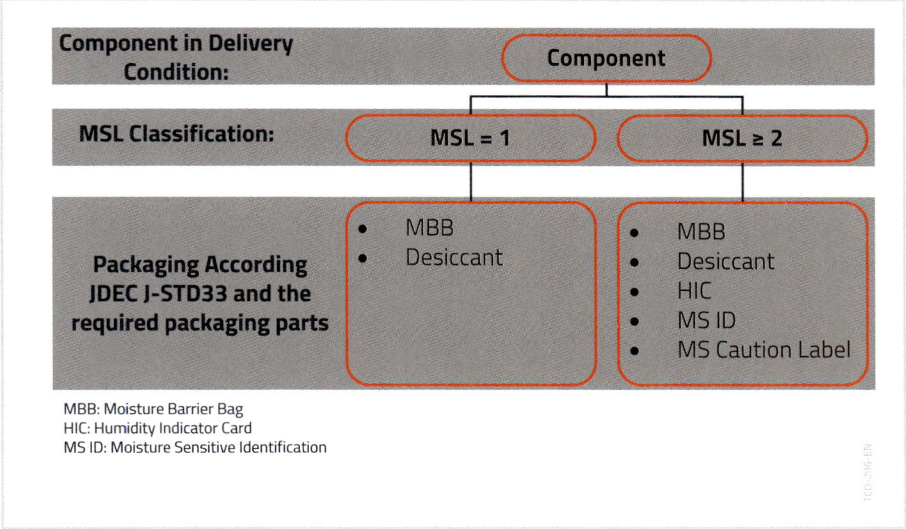

Fig. 1.362: MSL classification

2.3.4 Recommended Storage Conditions

WERI connectors that are considered MSL1 and packed in their original closed packaging are not subject to storage time limits (according to the IPC/JEDEC J-STD-033A). For unpacked connectors Würth Elektronik highly recommends adhering to "controlled storage room" (CSR) conditions.

In case of non-compliant exposure and to prevent any defect due to moisture, Würth Elektronik suggests predrying and solderability tests (per IPC/EIA/JEDEC J-STD-002B Category 1) prior to solder processes.

CSR conditions are defined as 20 °C ±10 °C @ maximum 60 % relative humidity.

Predrying baking: The moisture desorption is a function of the plastic thickness, time and temperature.

Bigger plastic part thickness	Baking temperature	Baking time
0.5 mm	120 °C	1 hour
1 mm	120 °C	2 hours
2 mm	120 °C	4 hours

Tab. 1.34: Baking suggestions according to IPC/JEDEC J-STD-020B

It should be noted that the above information applies to PCB connectors only, which are soldered by wave or reflow soldering processes.

Trilogy of Connectors

Applications

Part 3: Applications

3.1 Emergency-Call System for Urban Traffic (Audio and Video)

The emergency-call system is integrated in ticket vending machines on bus stops, train stations, tramways etc. In case of emergency (i. e. violence against people or medical emergency) the passenger can call the police by pushing an emergency-call button.

The big advantage is, that there is not only an audio connection, but as well a real time video stream. So police will have immediately a live overview about the situation and can initiate suitable actions.

As all vending machines are already secured against vandalism, there were no special housing requirements to consider. Another point is, that those machines are already available on most of the urban traffic stations, so it is possible to implement this system in existing terminals.

Fig. 1.363: Ticket vending machine

3.1 Emergency-Call System for Urban Traffic (Audio and Video)

Fig. 1.364: Control unit for ticket vending machine with and without housing

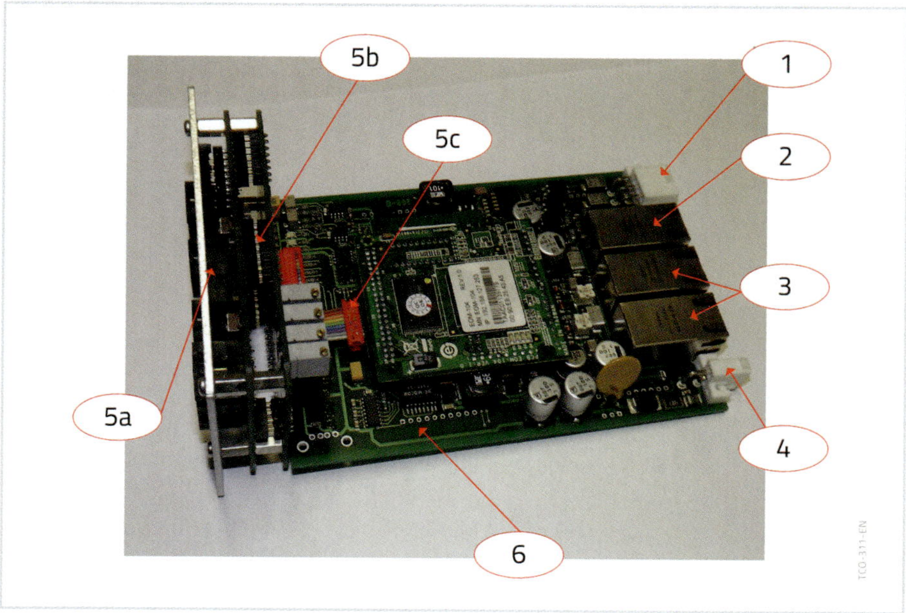

Fig. 1.365: Control unit for vending machine with different interfaces

1) Control connection to vending machine

This interface allows the system to check the status of the ticket vending machine. Additionally, it gives the possibility to display any pre-programmed content on the terminal monitor after pushing the emergency button.

Fig. 1.366: Emergency call solution with display for pre-programmed content

Fig. 1.367: 6 pin dual row Mini Power Connector (WR-MPC4) with 4.20 mm pitch with flanges

Connector solution:

- PCB-Side: Würth Elektronik order code 64900621022
- Wire-Side: Würth Elektronik order code 649006113322 (housing)
- Würth Elektronik order code 649006113322 (crimp contact AWG24-18)

Key features for the decision for MPC4 series:

The connector shows an active locking. This ensures a secure fitting of the mating halves. Second point is the coding. When assembling the device the wire connector cannot be mismatched or turned. Furthermore it is a flange type which is possible to be additionally screw-mounted to the PCB for further increase of mechanical stability.

2) I/O interface – Ethernet

The main communication to the emergency call center is performed through a standard 10/100 Base-T Ethernet connection – in this application directly linked to the emergency call 110 (police)

Fig. 1.368: Modular Jack 8P8C with integrated magnetics

Connector solution:

Würth Elektronik order code 7499411122

3) 2x I/O interface – Ethernet PoE+

Enables to equip the terminal with max. 2 cameras as sometimes one camera cannot cover the whole environment. Due to the PoE+ version the cameras can be powered directly via Ethernet cable system.

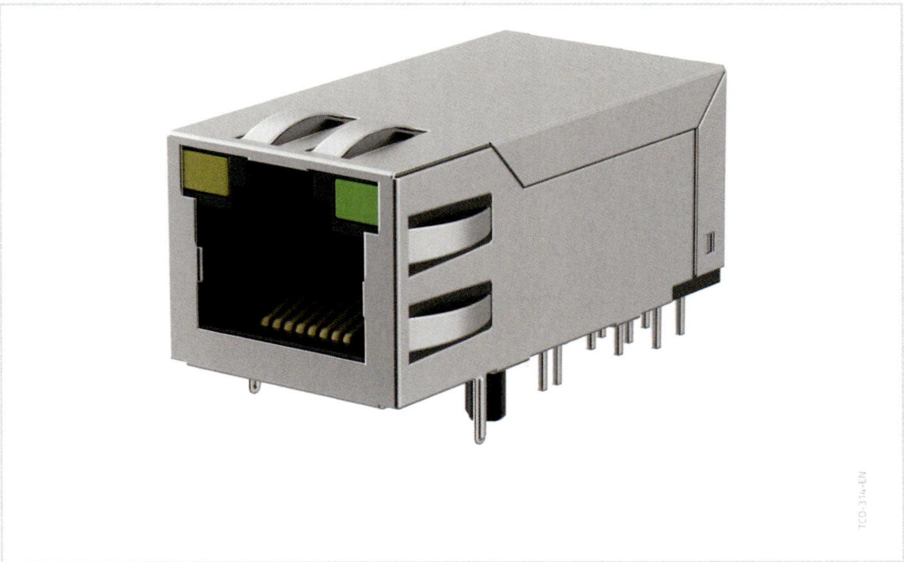

Fig. 1.369: Modular Jack 8P8C with integrated magnetics and PoE feature

Connector solution:

Würth Elektronik order code 7499411122

4) DC Power Supply

Besides the electrical performance the main focus for the power supply connection was set on robustness and easy handling in the field. The Mini Power Connector series 4.20 mm pitch with flange as 2 pin version has been chosen. Again the key features like active locking, coding and flange mount ability were the main criteria.

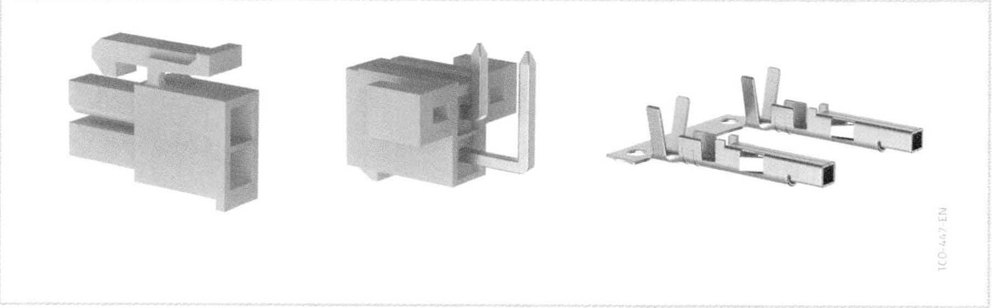

Fig. 1.370: 2 pin dual row Mini Power Connector (WR-MPC4) with 4.20 mm pitch with flanges

Connector solution:

- PCB-Side: Würth Elektronik order code 64900221022
- Wire-Side: Würth Elektronik order code 649002113322 (housing)
- Würth Elektronik order code 64900613722 (crimp contact AWG24-18)

5) Diagnostic and Programming Interface

The diagnostic and programming interface is needed for installation and basic adjustments (volume, equalizer, IP-address configuration etc.). It contains 2 serial interface ports (4 pins and ground each) and one NF test signal (2 pins).

5a)

Fig. 1.371: 12pin Micro Power Connector (WR-MPC3)

External connection:

- PCB-Side: Würth Elektronik order code 66201221122 Wire-Side: Würth Elektronik order code 662012113322
- Würth Elektronik order code 66200113722

Key features:

- active locking
- keying

5b)

Fig. 1.372: 10 pin 2.54 mm Pin- & Socket Header (WR-PHD)

Internal connection:

- PCB1 (incoming board):
- SMD Socket Header with through hole option to realize BTB height
- Würth Elektronik order code 610010243021

PCB2 (input protection board):

- THT Pin Header
- Würth Elektronik order code 61301021121

5c)

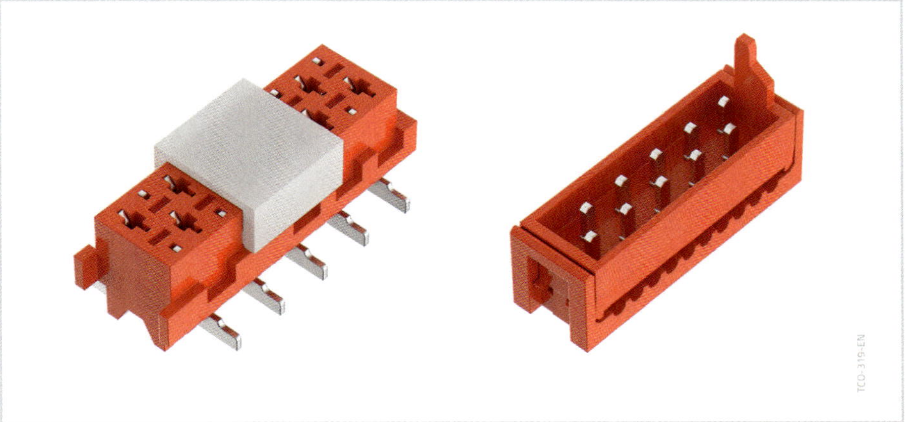

Fig. 1.373: MiniModule (WR-MM) IDC Connector & SMT socket with flat ribbon cable 10pin

Internal Connection:

- PCB2 (input protection board): Female SMT Connector
- Würth Elektronik order code 690367281076

Flat ribbon cable: Male IDC connector

- Würth Elektronik order code 690157001072
- PCB3 (processing unit):
- Female SMT Connector
- Würth Elektronik order code 690367281076

Key features:

- Cable solution as on PCB edge no space left for a BTB connection.

6) Optional Layout for Board To Board Connection

Additional vias allow to implement an ad of automatic information service board but this is currently not in use.

3.2 USB Panel mount adaptor[1]

Fig. 1.374: Panel mount adaptor for USB 2.0 and USB 3.0

The idea was, to create a possibility for USB connection in any housing that is small enough, easy to mount and has no need for long routings on PCB.

The reason is, that routing USB on a PCB is becoming more difficult if longer data-lines are needed. It becomes even more critical if we start to talk about USB 3.0 instead of USB 2.0 as the speed is increasing. For 10 cm of routing on PCB the loss will be around 5 dB at 7.5 GHz!

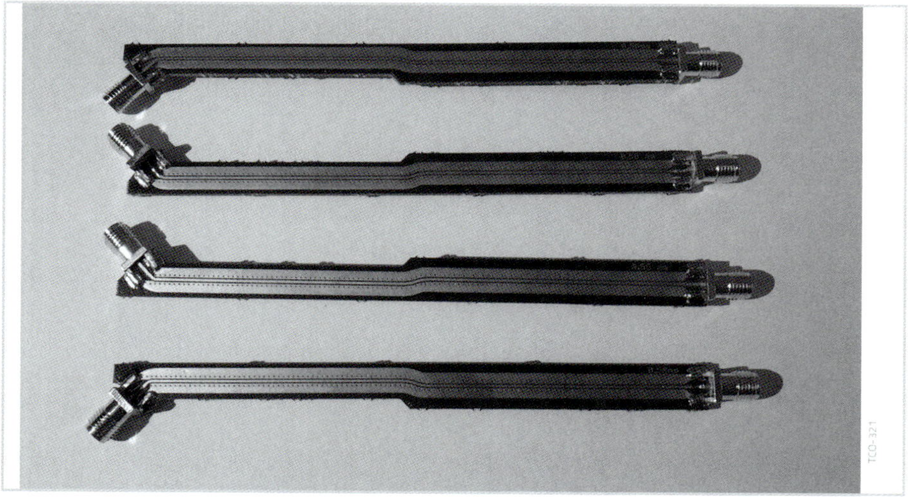

Fig. 1.375: Testee for measuring loss at 7.5 GHz on PCB

So we need to think about how to avoid this problem and find an easier solution. The basic thought is that usually in a housing some wires need to be put in place. So if this is the case anyway, we can use standard USB cables which are already designed for our signal specification

[1] by K&K Prime Engineering

and we use them to connect the adaptor to the board close to the point where the USB signal is received to avoid long routings.

The adaptor is available in two versions: one for USB 2.0 and one for USB 3.0, which are basically built up in the same way, but differs in some points.

The USB 2.0 version adapts a micro USB 2.0 connector (device input, order code 629 105 136 821) to an USB 2.0 Type A connector (output to board, order code 629 004 160 21). Additionally, it has a status LED and a 470 Ω resistor for the LED.

For panel mounting, two brass spacers M3x10 (WA-SBRH Brass Spacer Stud order code 970 100 354) I/I are soldered on separately designed pads. They are also used for ground connection to the housing. All parts are SMT for easier production. Of course, instead of micro USB you could also use MiniUSB or USB Type B.

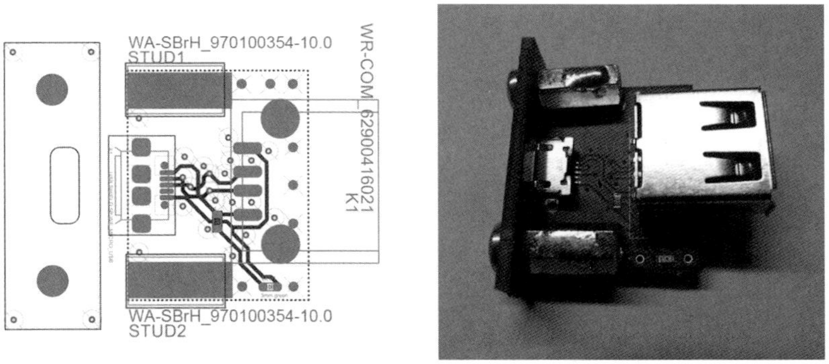

Fig. 1.376: Schematic and PCB for USB 2.0 panel mount adaptor

The USB 3.0 version adapts an USB 3.0 Type B connector (order code 692 221 030 100) to an USB 3.0 Type A connector (order code 692 121 030 100). Of course, it is also equipped with status LED, resistor and soldered spacer studs (WA-SBRH Brass Spacer Stud 970 100 354) for mounting and ground connection. Additionally, it contains a TVS diode. The main focus is the signal routing in this case, because the signal lines need to be done in equal length and low resistant ground connection is relevant.

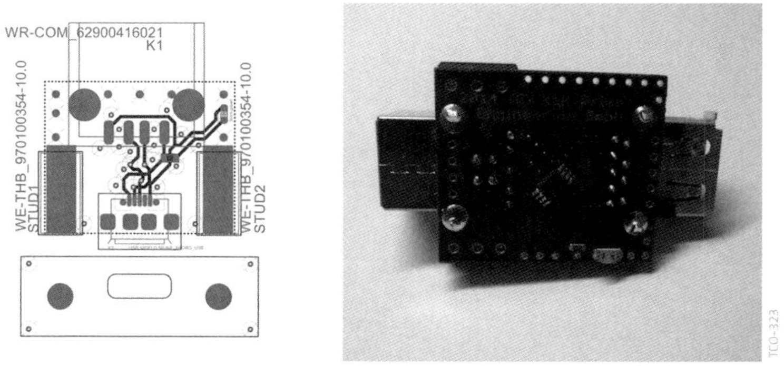

Fig. 1.377: Schematic and PCB for USB 2.0 panel mount adaptor

3.3 Measurement of the Characteristic Impedance of a Flat Flexible Cable (FFC)

Flat flexible cables are usually used in LCD screens, board-to-board solutions for medical devices, test and measurement equipment, industrial applications, consumer electronics and any other kind of equipment requiring a small, strong and flexible cable for data transmission purposes.

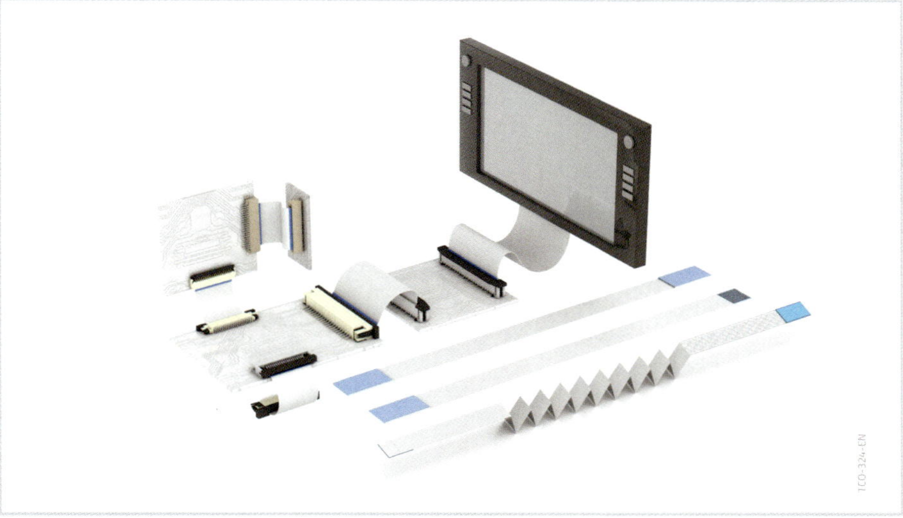

Fig. 1.378: Possible use of ZIF connectors and FFC cables

In order to meet customers' requirements, Würth Elektronik offers 0.5 mm and 1 mm – pitch FFCs – the most common pitches – in various lengths.

Würth Elektronik components used:

- WR-FPC SMT ZIF Horizontal 1.0 mm (686 1xx 144 22)
- WR-FFC Flat Flexible Cable Type 1, 1.0 mm (686 6xx xxx 001)
- WR-FPC SMT ZIF Horizontal 0.5 mm (687 1xx 145 22)
- WR-FFC Flat Flexible Cable Type 1, 0.5 mm (687 6xx xxx 002)

Since it is a compact, thin and light product, it is suitable for low-space applications. Moreover, these cables can be bent without impacting the transmission performance.

Today's technologies require ever higher data rates with efficient transmission. The characteristic (or line) impedance is a key indicator to assess the performance of cables or data lines in terms of data transmission. This fundamental point deserves to be highlighted. The present article therefore summarizes two different methods to measure the impedance: time domain reflectometry and frequency analysis. These two complementary processes enable customers to measure both the characteristic impedance and signal losses through the cable. Most of the time, customers want to know the maximum frequency that can be transmitted by the cable according to a certain losses value.

The study covers FFCs with a 0.5 mm pitch and a 1 mm pitch.

3.3 Measurement of the Characteristic Impedance of a Flat Flexible Cable (FFC)

Fig. 1.379: Flat flexible cables

In practice, measurements were carried out by characterizing the cable & FPC couple. It provides realistic test conditions, just like in customers' applications.

Basics on characteristic impedance

The characteristic impedance is an important parameter for signal transmission. If the impedances of a source, a cable and a receiver do not match, a reflection phenomenon will prevent signals from being fully transmitted. Therefore, the cable impedance must be as close as possible to the source and receiver impedances.

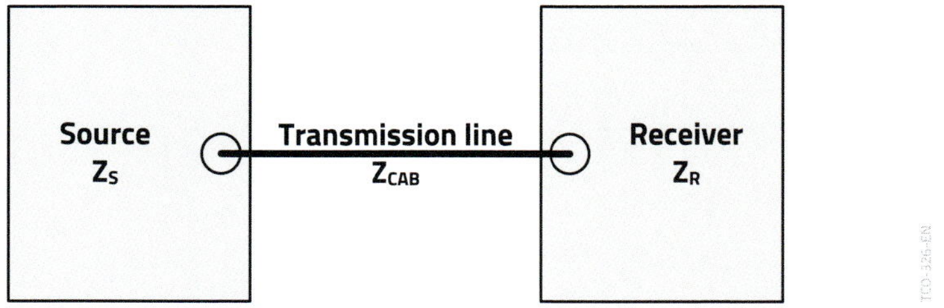

Fig. 1.380: Characteristic impedance block diagram

If a cable is used to provide high power signals and part of it is lost, this may disturb the source and/or the receiver and impact the power efficiency of the whole system. If a cable is used to transmit data, this is less important because signals are usually powerful enough to be decoded at receiver level.

The key indicators to quantify losses are transmission and reflection coefficients. Let's assume that the receiver (Z_R) is adapted to the transmission line. Then, equations are as follows:

$$\text{Reflection coefficient } \rho = \frac{Z_{CAB} - Z_S}{Z_{CAB} + Z_S} \text{ and transmission coefficient } \tau = 1 - \rho \qquad \text{(EQ 25)}$$

The reflection phenomenon is not linear. The table and curves below quantify the part of the signal that is transmitted/reflected according to the cable characteristic impedance (for Z_S = 50 Ω):

ZCAB	40 Ω	50 Ω	60 Ω	75 Ω	100 Ω	120 Ω	150 Ω	200 Ω
	11.1 %	0 %	9.1 %	20 %	33.3 %	41.2 %	50 %	60 %
	88.9 %	100 %	90.9 %	80 %	66.7 %	58.8 %	50 %	40 %

Tab. 1.35: Reflection and transmission coefficients according to the transmission line characteristic impedance (for Z_S = 50 Ω)

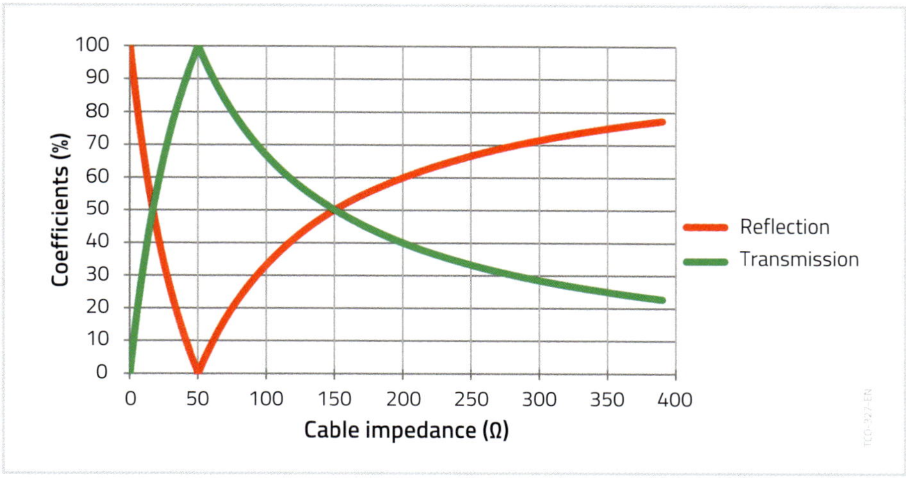

Fig. 1.381: Reflection and transmission coefficients according to the transmission line characteristic impedance (for Z_S = 50 Ω)

According to the previous table and curves, the more the impedances vary, the more important the reflection. This is why impedance matching is important in order to avoid losses during transmission.

It must be noted that the characteristic impedance depends on geometrical parameters of the transmission line as well as on the materials electrical permittivity. Further investigation needs to be conducted to determine the exact physical properties of the materials used and the correct formula to predict the impedance value.

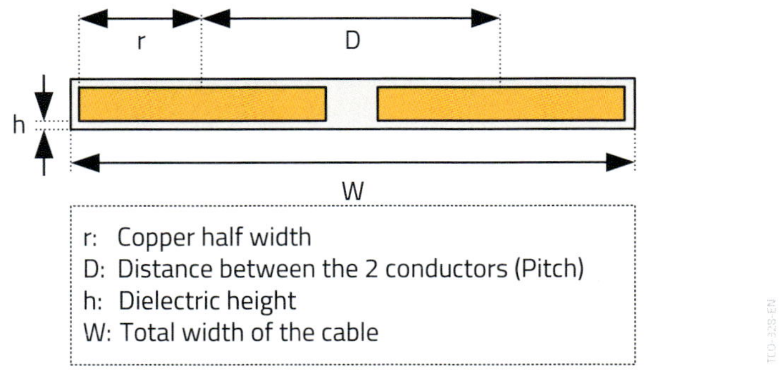

Fig. 1.382: FFC sizing description

According to the literature, the characteristic impedance depends on dimensions (r, D and h) and on the dielectric permittivity ε_R:

$$Z_{CAB} = \frac{276}{\sqrt{\varepsilon_R}} \ln(f\{r;D;h\}) \qquad \text{(EQ 26)}$$

In order to have a good estimation, the dielectric permittivity and dimensions must be precisely known. Since many different materials are used to manufacture FFC products (blue PET, white PET, glue), it is difficult to assess the dielectric permittivity. This is why measurements are taken to quantify the characteristic impedance.

Measurement methodology

The characteristic impedance measurement can be performed according to two methods:

- Unbalanced measurement (single):
  One conductor transmits the signal and the other one ensures the grounding return (classical assymmetric transmission). Boards have been designed to measure it (as per the following block diagram).

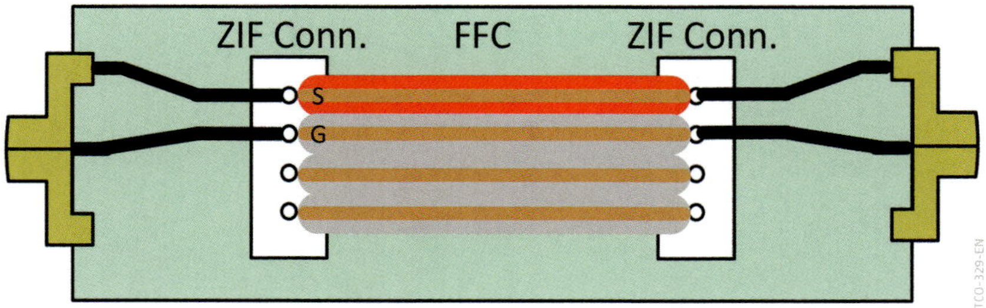

Fig. 1.383: Signal ground measurement board block diagram

- Balanced measurement (differential):
  Multiple grounding acts like a shielding and stabilizes the GND with a lower impedance to the reference GND in order to obtain a behaviour close to coaxial transmission. Such measurements with FFCs can be taken in two ways:
  - Using a FFC where the signal is transmitted in one line surrounded by one ground line on each side (ground – signal – ground):

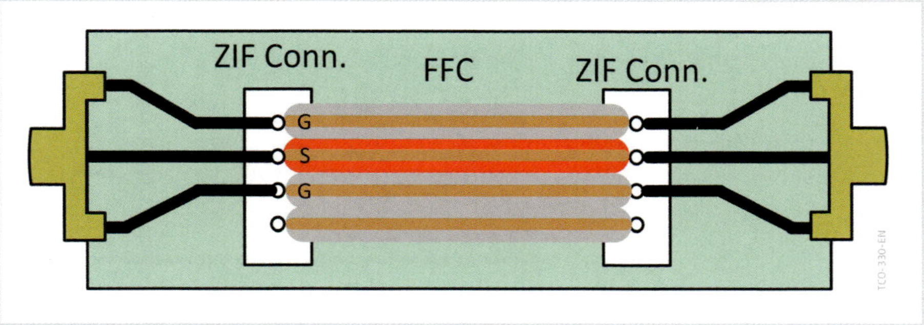

Fig. 1.384: Ground-signal-ground measurement board block diagram

  - Or using a shielded FFC that can be intrinsically used for unbalanced measurement:

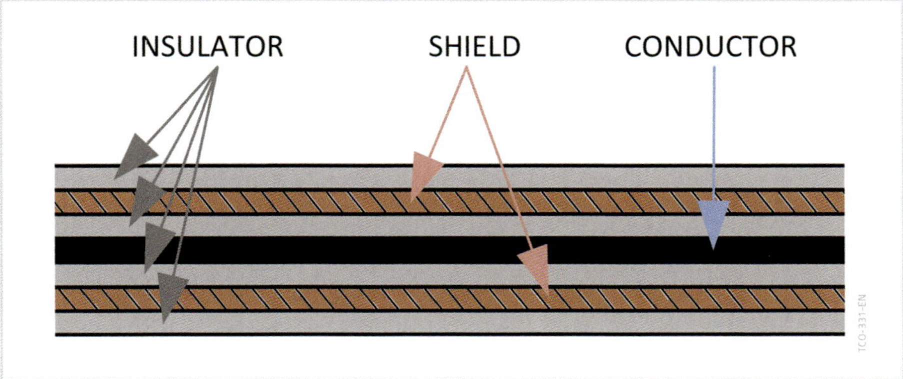

Fig. 1.385: Slice view of a shielded FFC

This product, which is not part of the Würth Elektronik catalogue, is only relevant for applications where shielding requirements are tough to achieve. In this product, the shielding tape is wrapped over the copper.

Among the previously explained alternatives, the following methods were selected:

- Unbalanced: Signal-ground measurement (Figure 1.382)
- Balanced: Ground-signal-ground measurement (Figure 1.383)

3.3 Measurement of the Characteristic Impedance of a Flat Flexible Cable (FFC)

Two types of equipment can be used to measure impedance:

- Time Domain Reflectometry (TDR) mode oscilloscope – measurement of impedance at each point of the FFC.
- Network analyzer – measurement according to the frequency

The equipment that was used is the E5071C by Agilent. It is both a network analyzer and a TDR Mode Oscilloscope.

The measurement process is described in the following block diagram:

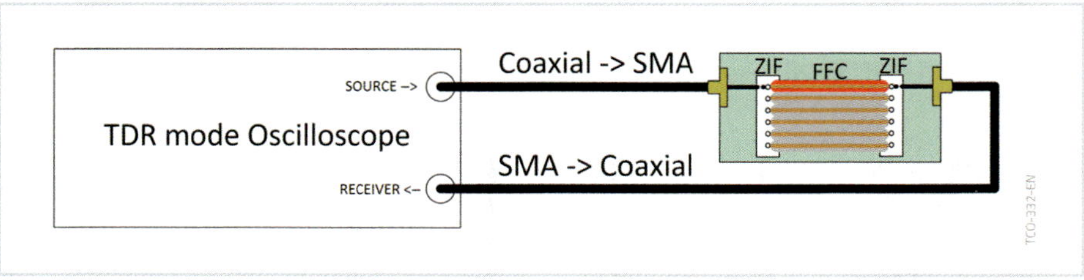

Fig. 1.386: Measurement block diagram

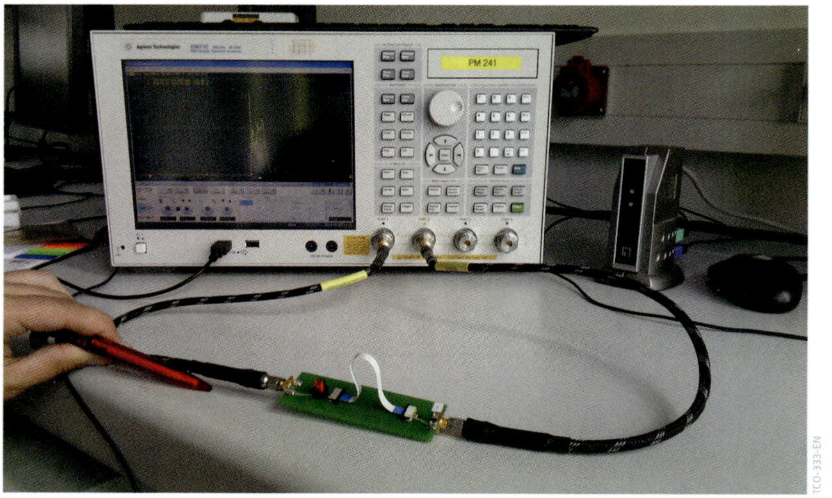

Fig. 1.387: Measurement bench

Impedance values and discontinuities are measured by sending the signal (step) from the source to the receiver through the device under test. Source and receiver impedances are 50 Ω (fixed setting of the equipment). For the time domain reflectometry mode, the rise time of the step (from 10 % to 90 % of the maximum step value) must be pre-set and the equivalent frequency bandwidth test is:

$$BW = \frac{0.22}{\text{rise time}_{10\%\text{-}90\%}} \qquad \text{(EQ 27)}$$

Measurement results – Time Domain Reflectometry:

According to competitors and to the literature, the impedance of this type of cable should be close to 100 Ω.

Figure 3.31 shows a typical curve obtained by using the time domain reflectometry configuration (with a 200 mm FFC – 1 mm pitch – rise time = 32ps):

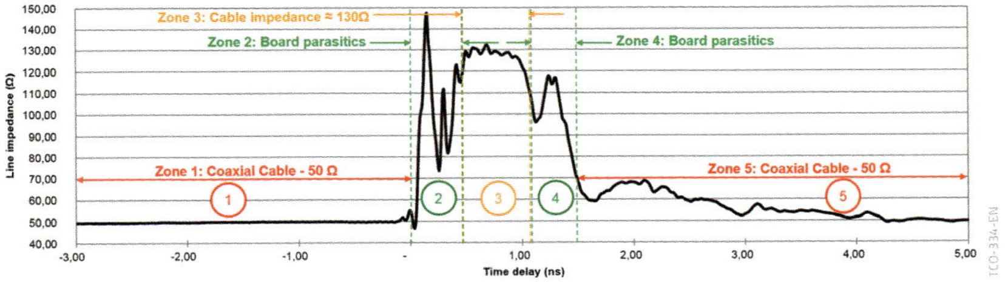

Fig. 1.388: Measurement results for a 100 mm FFC – 1 mm pitch – unbalanced mode

The main advantage of this technique is the possibility to measure the impedance at each point of the board:

- Zone 1: Impedance of the coaxial cable (matched at 50 Ω)
- Zone 2: Parasitic introduced by SMA termination, soldering and FPC connector
- Zone 3: Impedance of the FFC (≈ 130 Ω)
- Zone 4: Parasitic introduced by SMA termination, soldering and FPC connector (≈ Zone 2)
- Zone 5: Impedance of the second coaxial cable (matched at 50 Ω)

In Zone 3, the impedance is unstable and this is due to:

- SMA transitions and ZIF connectors that cause disturbances and instability of the measurement,
- the support tape of the measured cable (made of PET material – different thickness and colour, therefore a different characteristic impedance).

For instance, the below table summarizes the measurements based on reflectometry mode method (rise time is set to 32ps for a better measurement accuracy – equivalent frequency is 6.88 GHz).

	0.5 mm pitch	1 mm pitch
Unbalanced	125 Ω	130 Ω
Balanced	85 Ω	85 Ω

Tab. 1.36: Characteristic impedance of FFCs according to the length

According to the reflection formula (for 50 Ω input and output network at equipment level), the reflection percentage and equivalent losses in dB at this frequency are as follows:

	0.5 mm pitch	1 mm pitch		
	Reflection	Losses	Reflection	Losses
Unbalanced	42.9 %	2.43 dB	44.4 %	2.55 dB
Balanced	25.9 %	1.30 dB	25.9 %	1.30 dB

Tab. 1.37: Reflection coefficient and equivalent losses of cables at 6.88 GHz

The balanced mode (ground-signal-ground) should be used for better efficiency during transmission. This method is an indicator of the cable losses. Therefore, it's not the most realistic way to quantify these losses. It's more relevant to use a frequency depending chart to determine the losses of the "PCB + transitions + ZIF connector + FFC" set, even if it's a little pessimistic.

Measurement results – Network analyzer

As explained in the last paragraph, the frequency characteristic is a key point. For many applications, losses must be predicted according to a full frequency range and not only according to a single one.

The network analyzer was used to quantify the losses of the cable up to 3 GHz (which covers most applications).

For comparison purposes, the below curve highlights the differences between balanced and unbalanced modes.

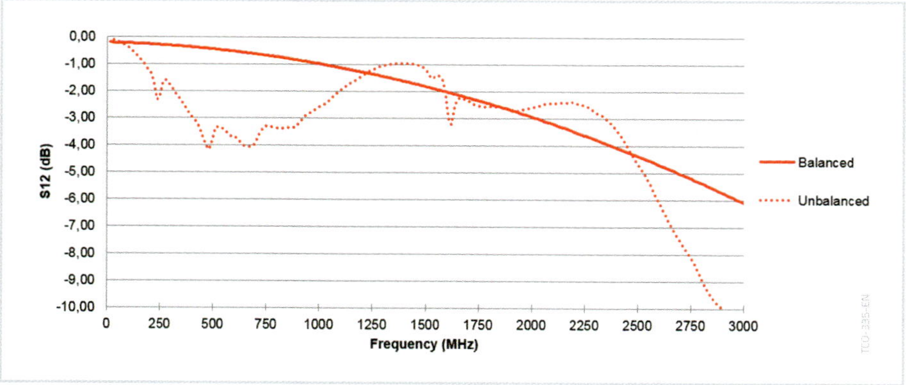

Fig. 1.389: Losses according to the frequency for a 50 mm FFC – 1 mm pitch

3.3 Measurement of the Characteristic Impedance of a Flat Flexible Cable (FFC)

It's clear that losses are more important in unbalanced mode. Moreover, losses vary more according to the frequency. The cable attenuation is more constant and more stable if the cable is used in balanced mode (coaxial like).

Based on those considerations, the next graphs show the balanced mode behaviour for 0.5 mm and 1 mm pitch FFCs:

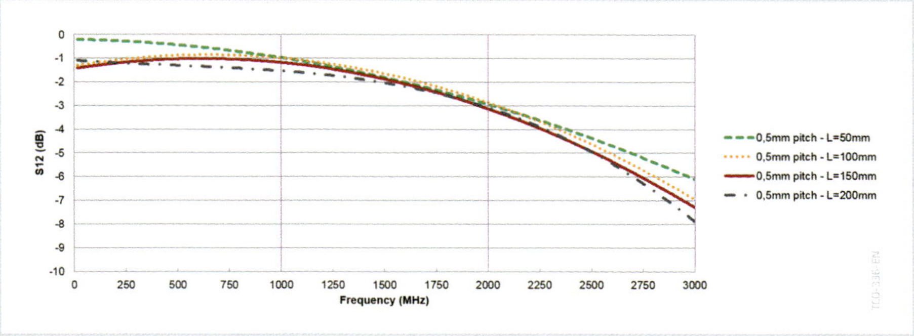

Fig. 1.390: Losses according to the frequency for 0.5 mm pitch FFCs

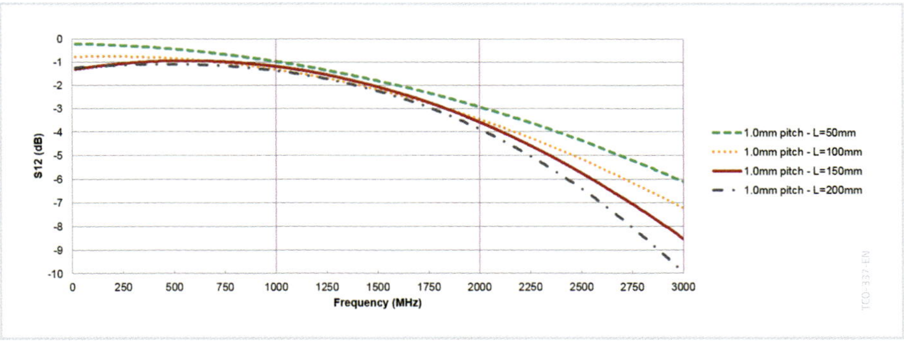

Fig. 1.391: Losses according to the frequency for 1 mm pitch FFCs

As expected, with a 200 mm length cable, the attenuation is more important than 50 mm length cable. Indicative losses including parasitic of the board and SMA terminations are below 4 dB (60 %) at 2 GHz and between 6 dB (75 %) and 10 dB (90 %) at 3 GHz (depending on the length).

Conclusion

This paper describes a functional method to determine the characteristic impedance of a Flat Flexible Cable. The typical curves included in this document illustrate the process that was used.

The characteristic impedance is measured in balanced mode (coaxial like) and unbalanced mode. As demonstrated, the balanced mode is recommended to obtain a better system stability and to avoid extra-losses.

TDR mode and network analyser measurements were performed. With the TDR mode, only the FFC impedance is measured for a defined frequency. This setup has shown that the cable length does not influence the characteristic impedance.

The network analyser covers a full frequency range and measurements are performed in conditions that are similar to customers' applications (PCB, transitions, connectors and cable). It is more representative to measure the impedance value which is a key information.

The results obtained with this second method highlight that a "Flat Flexible Cable + FPC" set (0.5 mm and 1 mm pitch – Type 1 and 2 – all lengths) can be used for frequencies under 2 GHz (signal loss around 4 dB).

Depending on the final application and the maximum acceptable losses, the frequency range can be wider (see Figure 1.389 and Figure 1.390).

No impedance changes or extra-losses were observed when bending the FFC.

3.4 Impedance matched for high-speed data communication

Development of new devices goes often together with contrary requirements: high speed data communication and low budget BOM (bill of material). This chapter tries to bridge these requirements by enhancing capabilities of common available part for higher speed data in this case Würth Elektronik Ribbon Flat Cable WR-CAB.

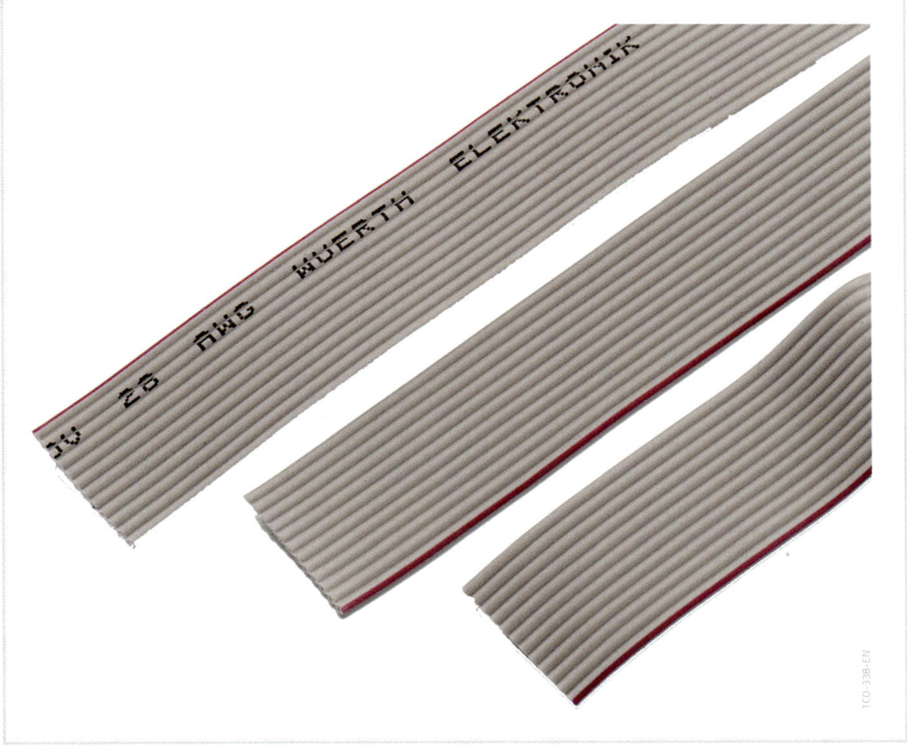

Fig. 1.392: WR-CAB Ribbon Flat Cable 1 mm (639 xxx 156 21)

3.4.1 Differential signal on a two-wire connection

A common way to transfer high speed signals is differential mode. This mode has its advantages in good immunity against interference and is therefore used for USB, HDMI and other communications. High speed signals need to be impedance matched in order to avoid reflections and resulting distortion. Matching a two wire connection was already investigated in 1888 by Austrian physicist Ernst Lecher. The impedance of a two wire line depends on its geometric and material used:

$$Z_1 = \frac{1200}{\sqrt{\varepsilon_R}} \text{acrosh} \left(\frac{a}{b}\right) \text{ (see[2])} \qquad \text{(EQ 28)}$$

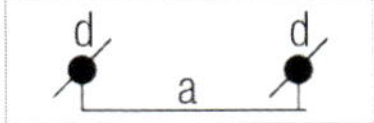

Another way to calculate the impedance provides the datasheet. The inductivity and capacity per meter can be used in below formula:

$$Z_1=\sqrt{\frac{L}{C}} \qquad \text{(EQ 29)}$$

With:

L: Inductance [H/m]
C: Capacitance [F/m]

In using "1.00 mm Ribbon Flat Cable WR-CAB" datasheet this results in:

$$Z_1=\sqrt{\frac{L}{C}}=\sqrt{\frac{1.45\ \mu H/m}{10\ pF/m}}=190\Omega \qquad \text{(EQ 30)}$$

Such a communication line can be easy simulated using LT spice:

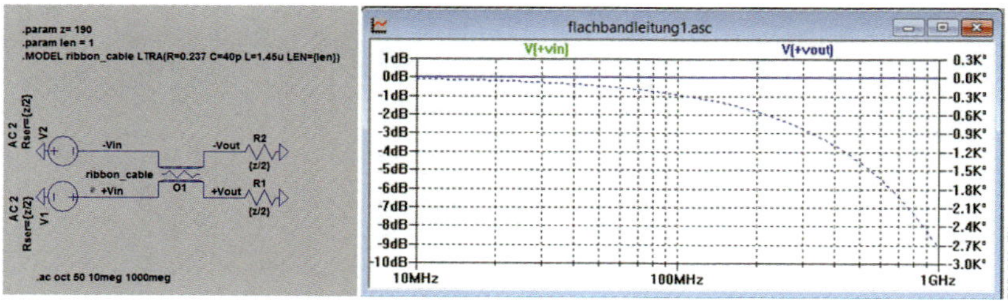

Fig. 1.393: Simulation of communication line in LT Spice

3.4.2 Differential signals in a ribbon cable

As soon as you have multiple wires lying flat side by side in a ribbon cable impedance is impacted as well as they couple between each other. This can also be simulated using LTspice. As a simple approach each wire is seen as split in two times a two wire connection, one to the left and a one to the right. Resistance and inductance are adjusted by a multiplication of 2.

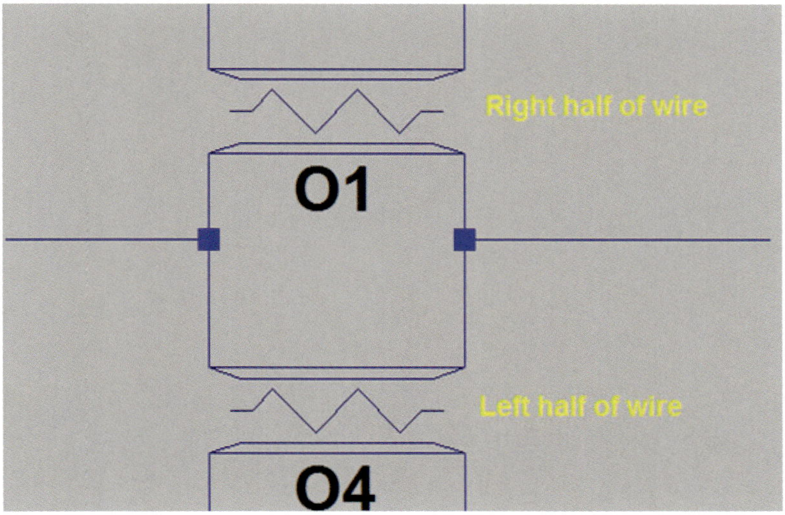

Fig. 1.394: Simulation of a single wire in LT Spice

The simulation of an 8 wire ribbon cable looks like this:

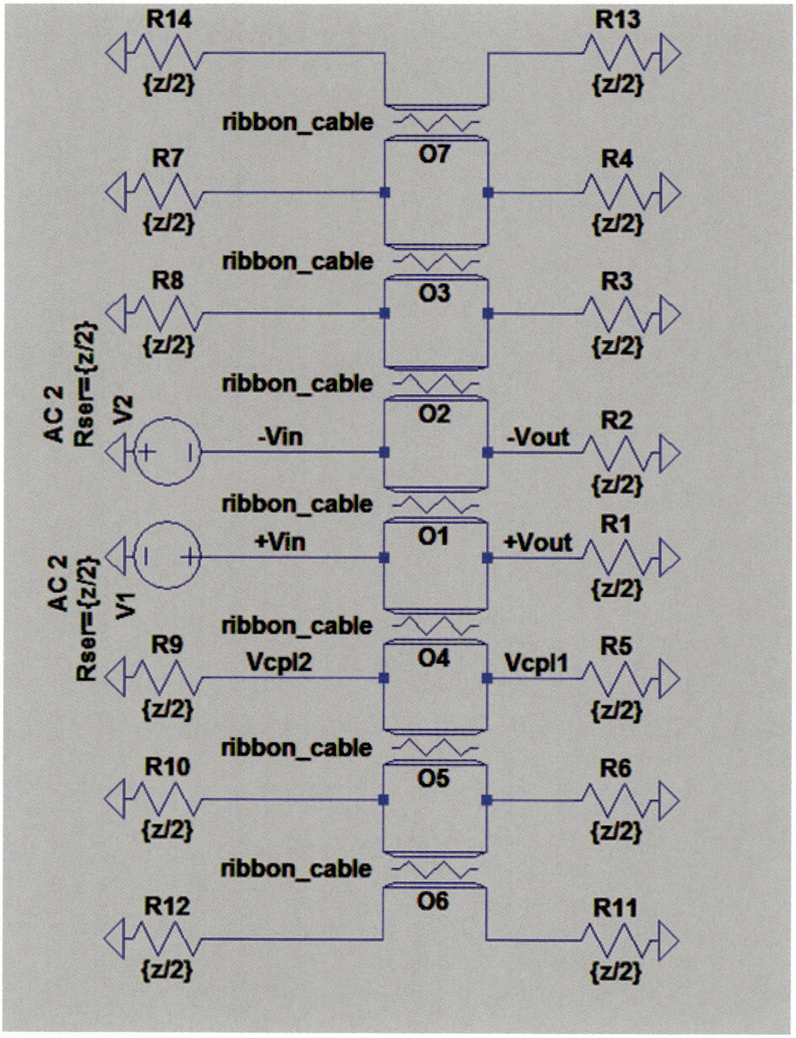

Fig. 1.395: Simulation of an 8 wire ribbon cable

Using above simulation we see standing waves on the transmission as a result of impedance mismatch and coupling.

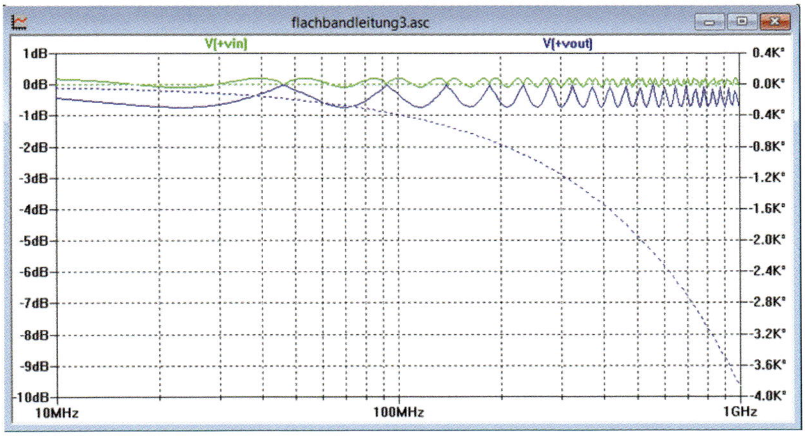

Fig. 1.396: Impedance mismatch and adaption to reduce ripple

By adapting the impedance on output and input to 88 Ω (= 176 Ω differential) the ripple can slightly be reduced.

The coupling to neighbour transmission is shown below: V(vcpl1), V(vcpl2):

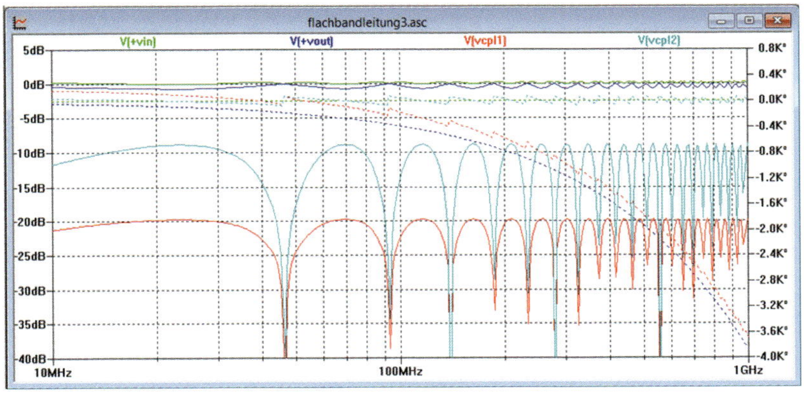

Fig. 1.397: Coupling of neighbour transmission lines

A ribbon cable acts like a directional coupler and couples to more to reverse port Vcpl2 then to Vcpl1 by about 10 dB. Such a coupling can impact data transmission especially when bidirectional communication is used.

3.4.3 Differential signals in ribbon cable with reduced coupling

As Würth Elektronik offers beside 1.27 mm ribbon cable also 1 mm ribbon cable a viable approach is to add a grounded wire on each side of a transmission pair without increasing the space requirement too much. A try in simulation showed excellent results.

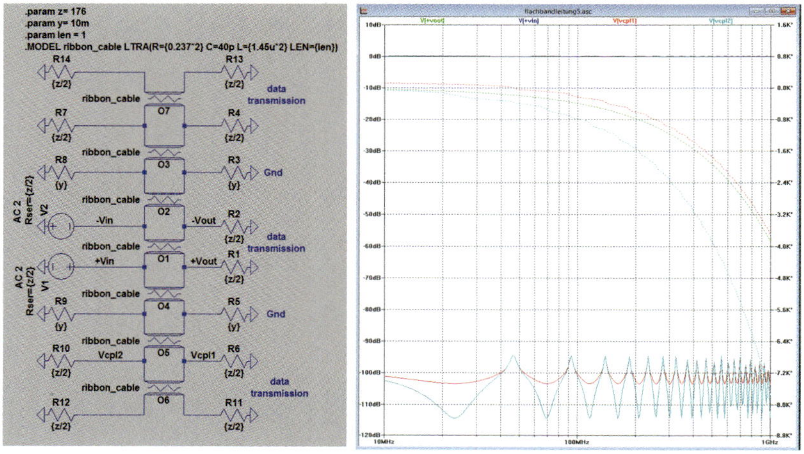

Fig. 1.398: Simulation of transmission lines separated by grounded lines

The coupling between different transmission pairs (Vcpl) is significant reduced but also the frequency response looks much better (Vout+).

3.4.4 Summary

Using "1.00 mm Ribbon Flat Cable WR-CAB" for high-speed data transmission is a viable approach. Taking in concern impedance matching and spending a grounded wire beside each transmission pair high data rates can be achieved with reduced interference between transmission pairs. The total width per pair is then 3 mm (= two transmission lines and a ground line). But this is only a 20 % increase to 2.54 mm when using a standard 1.27 mm ribbon cable with transmission pairs side by side.

This simulation does not consider electromagnetic coupling as it is no electromagnetic simulation. However, it shows a potential way to go when high transmission rates and budget constraints affect you in the development.

3.5 A Bridge to connect PCBs

3.5.1 Scope

Many devices are still made of multiple PCB modules. One reason is the mix of different layer requirement like FPGA require 8 to 12 layers, Analog just 4 layers and Power only 2 layers. Doing all PCBs in high order layer technique would make the process longer and device expensive. Another reason is modularity and interfacing. The analog team can work independent from power supply and digital control team. However, the boards need to be connected together in an easy and cost effective way to allow fast and secure manufacture of the device as well as simple exchange of broken boards at service.

Such a connection has to carry various loads: Power line, digital signals and analog signals. Taken all this into concern we developed a bridge to connect PCBs with a target: PCBs placed aside each other (same level) with a defined space (1 mm), connected with each other (by 22 connections), secure fastened (with screws M3) and protected against misplacement.

3.5.2 The Approach

Low price, high frequency digital and analog signals and size constraints has driven us to choose Würth Elektronik WR-PHD Dual SMT Pin Header 1.27 mm spacing (order code 623 235 521). With a total height of 4.2 mm the SMT Dual Pin Header and SMT Dual Socket Header Bottom Entry 1.27 mm (order code 623 326 235 321) was the preferred choice. The new WA-SMSI SMT Steel Spacer with 4 mm (order code 977 404 036 0R) may allow a tight mounting and prevent misplacing. Our idea was to put two socket headers on a small PCB and connect the pins with each other. The four pins at the corner of each header as well as the screws are connected to ground prepare the base for impedance controlled wiring of digital and analog signals. The counterpart at the modules consists of one spacer and one pin header. Having the corner pins at ground reduces the risk of unintended electrical shortcut by tools, probes or others.

We choose the 26 pin version as a balance of signals to carry, size and cost.

3.5.3 The Design

The schematic is quite simple, but the layout was tricky.

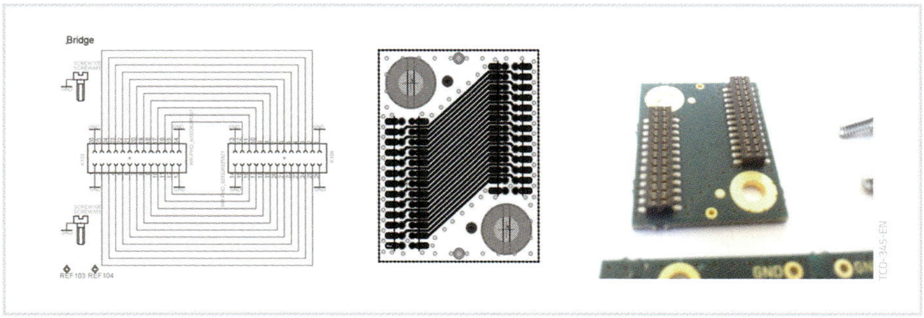

Fig. 1.399: Schematic, layout and picture of the bridge to connect

Each connection from one module to the other module has to have the same length (and therefore group delay). Independent if the connection is made at center or side, independent is first or second row is selected. Especially reaching the same length on different rows but not compensating the row spacing required a new approach. As a result all connections are made on the bottom side with vias at pin position. These vias can now be used as test points free of charge. The width of wires is calculated for 50 Ω impedance which is standard on most communication systems. To reach this low impedance a four layer PCB was used. This number of layer ensures also a good ground connection between modules.

3.5.4 Test Setup

Doing a design in best intention is one thing, verifying performance a second thing. For verification we designed a test bench to test power line, analogue connection and digital transmission. The test bench consists of a PCB which is separated in two independent electrical parts like two modules. To test power capabilities, solder points for high current cables are prepared. Analog connection can be tested with RF connectors type SMA and digital communication is done with a USB 3.0 extension.

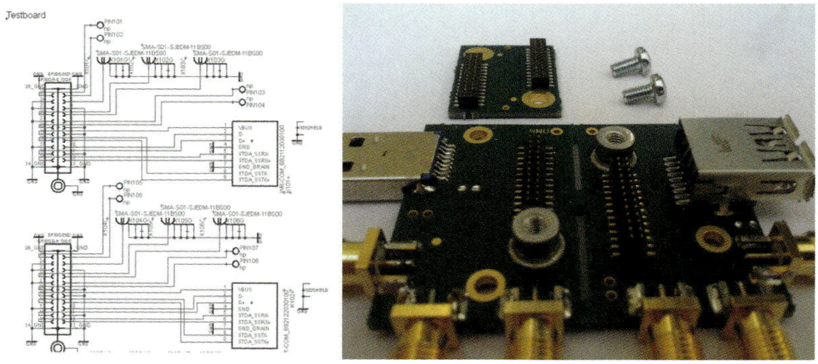

TCO-346-EN

Fig. 1.400: Schematic and picture of the two parts of the test board of the bridge to connect

On the picture the two halves are easy to recognize. They are only connected with a bridge and fastened by two M3 screws.

TCO-347-EN

Fig. 1.401: The test board for the bridge to connect mounted completely

3.5.5 Test results Power transmission

The power transmission was tested up to 3 A. The limitation was our lab power supply. Although the PHD connection is only specified for 1 A we have not recognized any issue up to 3 A.

The resistance from pin pad of one module to pin pad on the other side differs slightly if inner or outer row was used.

- Resistance for pins at row away from PCB edge: 37 mΩ
- Resistance for pins at row closer to PCB edge: 32.5 mΩ

Resistance measured at module pad and includes headers/sockets on both sides as well as bridge PCB. There was no degradation recognized on any current up to 3 A. There was no temperature increase detected. The isolation exceeded our measurement capabilities of 100 MΩ.

Analog Connection

In using a vector network analyser, we checked performance of a bridge connection having a second test line close by. This allows verifying inner row path and outer row path as well as isolation and crossing isolation between each other.

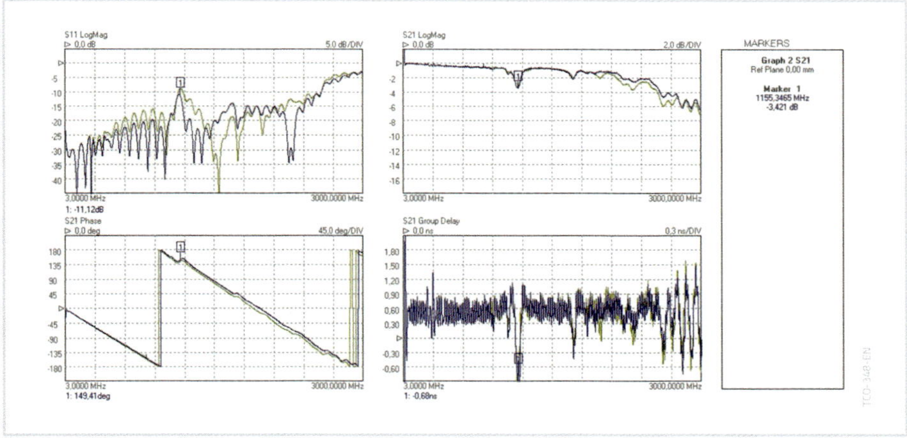

Fig. 1.402: Measurement results for the test board

The blue line is inner path (rows closer to PCB edge). The olive coloured line shows the outer path (rows away from PCB edge). Both paths show good results up to 1 GHz. A resonant effect can be detected at marker frequency of 1.155 GHz. However, both lines show similar behaviour and a group delay difference in only a few pico seconds.

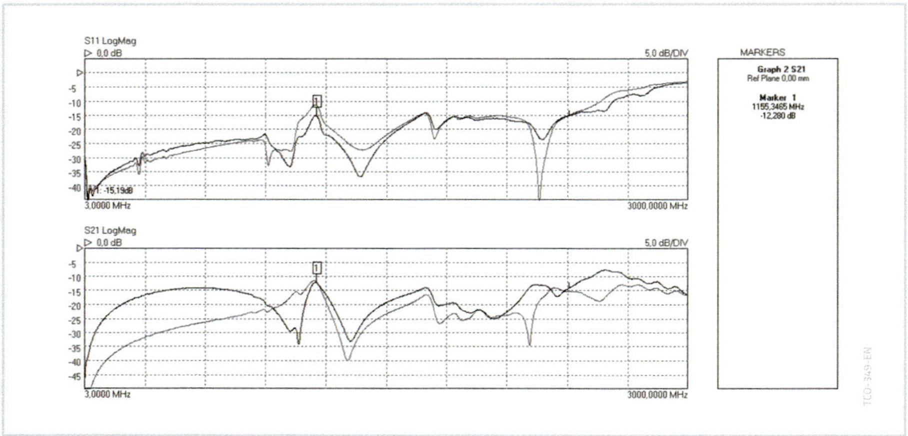

Fig. 1.403: Coupling of the test board

The isolation above shows as grey line coupling in same direction and as black line coupling in opposite direction. It is easy to see the coupling on two adjacent lines is mainly in opposite direction like a directional coupler. The coupling is below 14 dB (= 1/25) up to 1 GHz.

Digital transmission

Unfortunately, we are not yet equipped with USB 3.0 devices. Therefore we tested USB 2.0 high speed. We have not recognized any error or speed reduction in copying large files over the bridge.

3.5.6 Summary

The bridge allows connecting PCB modules inside a device in a reliable way. It carries power supply, control lines and analog signals. It is easy to mount and to disassemble.

3.6 REDFIT IDC – Skedd technology in use

3.6.1 Debugging

The advantages of SKEDD technology make it very suitable for debugging applications. Often plug connections are only needed briefly to upload or debug programs. For this purpose, the REDFIT IDC no longer requires a mating connector to be soldered, but only the layout needs to be on the board. With its size, REDFIT IDC can be accommodated on almost all PCBs and at the same time is easy to operate. The plastic pins allow the REDFIT IDC to be firmly secured on the board during the programming process and ensures a reliable connection without having to fix the connector by hand.

In order to optimize the REDFIT IDC for debugging applications, a second board layout was developed, which increases the number of mating cycles to 50 and makes plugging and unplugging the REDFIT IDC even more convenient. Only the vibration stability of 10 g cannot be ensured with this layout. If this is needed for debugging applications, the permanent layout can still be used.

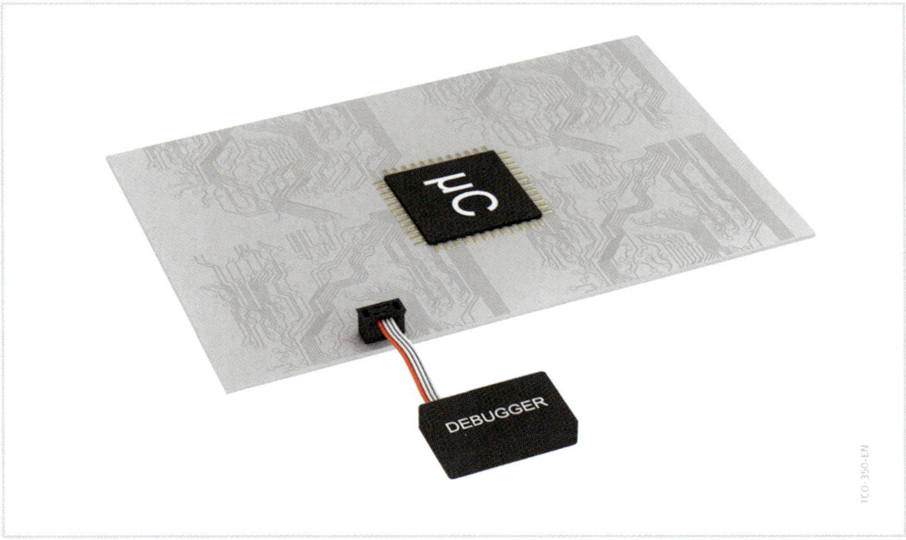

Fig. 1.404: SKEDD at debugging application

Adapters are required to simplify use of the REDFIT IDC as an interface for common debuggers. Layouts for four common types of adapter board are available on the Würth Elektronik homepage. You can find more debugging layout examples under the following link:
www.we-online.com/redfit.

Conclusion

A separate circuit board layout available for debugging applications. This reduces the REDFIT IDC insertion forces allowing a significant increase of possible mating cycles – from 10 up to 50 cycles. Since no connector base is required for debugging, this can be saved with SKEDD technology.

3.6.2 Optional system expansion

With the REDFIT IDC, systems can be prepared for expansions at almost no extra cost. While conventional connectors require a socket to be soldered in to expand the system in the future (e.g. through optional peripherals or similar), SKEDD technology allows sufficient space for the connector layout on the board. If required, a REDFIT IDC can be connected retrospectively (Figure 1.404).

This means that the costs for extensions are only incurred when they are actually needed.

Fig. 1.405: Easy and cost neutral system expansion while placing layouts

3.6.3 Solderless mounting

If the entire process of a soldered connection is considered, costs of at least 0.11 € per component are incurred for the soldering and component placement processes alone (data from REFA [German Association for Work Design, Business Organization and Business Development], collected by electronics manufacturers over 10 years) (Table. 1.40).

THT-placement		
Hourly rate	36	€
Handling time per component	2	sec.
Component placement	6	sec.
Manual optical inspection	3	sec.
Total	**11**	**sec.**
Costs	**0.11**	**€/piece**

Tab. 1.38: Average costs of soldering process (REFA data)

So SKEDD technology can save further costs compared to soldered connectors. There is no need for component placement or soldering a connector base. The SKEDD connector is mounted after soldering.

The use of SKEDD components, such as the REDFIT IDC, often allows boards to be produced completely without THT solder components, which results in further savings potential. This saves expensive THT soldering processes with manual component placement or selective soldering.

In the case of maintenance, it is usually not possible to comply with the necessary circuit board processing guidelines, e.g. drying before soldering. If the circuit board is reheated, then delamination or the popcorn effect may occur. Thanks to SKEDD technology, the connector can be replaced by hand without additional soldering or thermal stress.

Conclusion

Solderless mounting of the REDFIT IDC allows boards to be realized without the THT soldering process. This results in major cost reductions. At the same time, the REDFIT IDC offers the reliability of soldered connectors.

For expandable systems (e.g. access control extended to include retina scanning) the REDFIT IDC offers major advantages.

3.6.4 Tool-less Mounting

No additional tools are required for final REDFIT IDC mounting. Connectors with SKEDD technology can be plugged and unplugged by hand (Figure 1.405). Despite tool-less mounting, a stable and reliable fit is guaranteed up to an acceleration of 10 g.

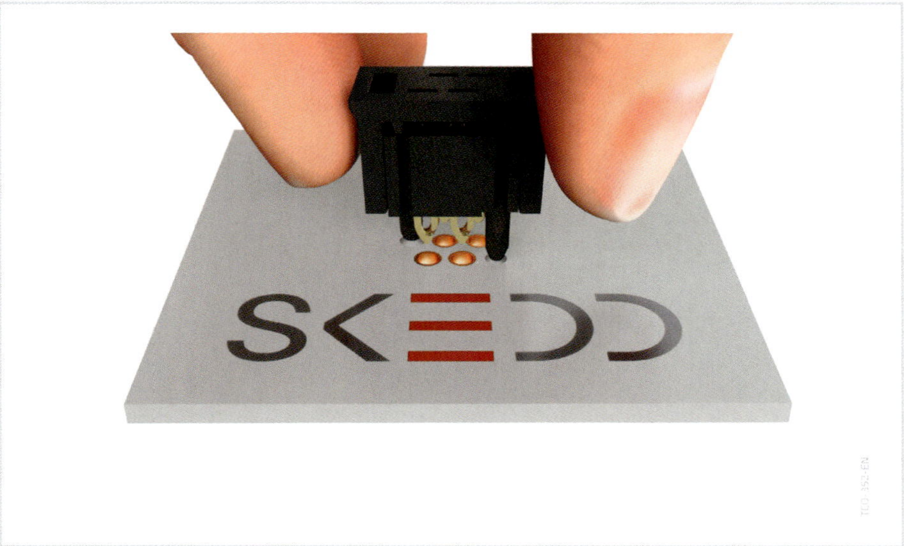

Fig. 1.406: Tool-less mounting

3.6.5 Double-sided component placement

Since the REDFIT IDC requires no soldering, it can be mounted from both sides, regardless of the direction of component placement on the PCB. For application on the underside of the PCB, the layout must be placed mirrored on the bottom layer (Figure 1.406). Miniaturization of all components on the PCB reduces the space available for plug connections. Mounting on the opposite side to component placement solves this problem.

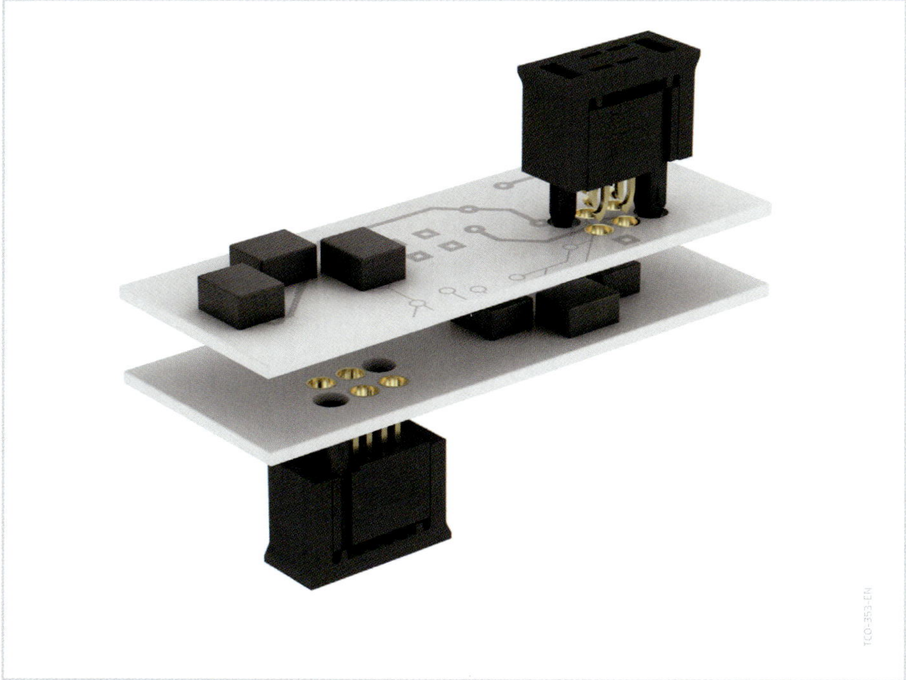

Fig. 1.407: Easy doubly sided placement

3.6.6 Daisy Chaining

In order to implement bus systems, the REDFIT IDC can be connected in series. With daisy chaining, several connectors can be pressed onto the same ribbon cable (Figure 1.407).

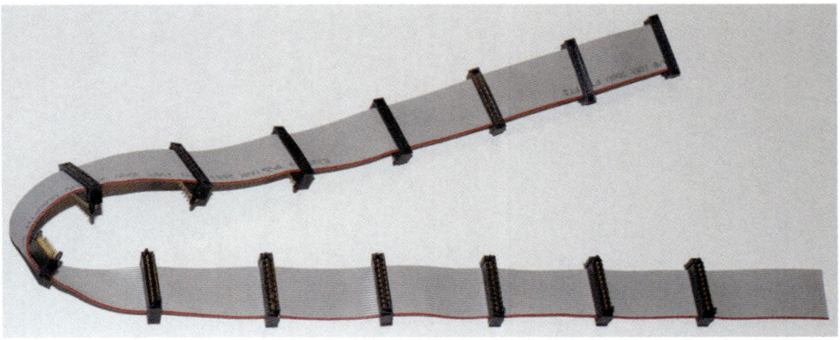

Fig. 1.408: Example for daisy chaining

Trilogy of Connectors

Glossary

Part 4: Glossary

4.1 Glossary

	Page
10/100 Base-T	301, 386
Active Locking	317, 386
Active pore	72
a-spot: The interface between two asperities in contact.	53, 55, 63
Alloy: A combination of two or more metals generally intended to improve mechanical performance.	12, 18, 25, 31, 38, 41, 44, 45, 48, 50, 52, 62, 63, 64, 66, 68, 70, 72, 85, 86, 106, 147, 155, 157, 175, 177, 211, 217, 224, 231, 258, 265, 271, 345
Anode: The positive terminal in electroplating.	38
Asperity: A high spot of surface roughness.	53, 55, 60
Attenuation	118, 315, 403
AWG (American Wire Gage): A measure of the diameter of round conductors. By definition 36 AWG corresponds to a diameter of 0.005 inches. As AWG numbers decrease, the conductor diameter increases. A change of 6 AWG corresponds to a doubling of the conductor diameter, e.g. 30 AWG has a diameter of 0.01 inches.	104, 179, 257, 385
Backplane: A printed wiring board designed to accept and interconnect other printed circuit boards or components. (Also called "Mother board")	125, 143
Balanced	118, 251, 399

4.1 Glossary

Term / Definition	Page
Ball Grid Array (BGA): An interconnection method using solder balls arranged in a grid pattern as contacts	
Barrel plating	43
Base metal: A metal susceptible to corrosion. Among connector materials, copper and copper alloys, nickel, silver and tin are base metals.	49, 56, 63, 70, 98, 205
Battery contacts	65, 84, 85
BTB	248, 256, 390
Bulk resistance	56, 220
Butt-Lead: A component lead configuration in surface mount technology where the lead abuts the soldering pad perpendicularly.	148
Cable: An assembly of wires or conductors.	13, 90, 107, 117, 142, 157, 176, 177, 244, 250, 252, 382, 387, 392, 395, 405, 412, 419
Cable, jacketed: An assembly of wires or conductors within a protective sheath.	177
Cable, ribbon: An assembly of wires or conductors in a planar configuration.	180, 252, 382, 391, 405, 419
Cable, transmission: An assembly of conductors and grounds in a planar configuration.	17, 119, 180, 258, 382, 395, 408, 412
Cantilever beam: A beam fixed at one end.	48, 86, 169, 205, 224

	Page
Card edge connector	145
Cathode:	
The negative terminal in electroplating, metal is deposited on the cathode.	38
Centerline Spacing:	
The distance between contacts, or receptacle positions, in a connector header or housing. (Also called "Pitch")	31
Chamfer:	
A bevel on edge of a printed circuit board, contact, crimp barrel or connector housing that permits easier insertion of components.	
Characteristic Impedance	133, 314, 382, 395
Circuit: A conductive path through which electrical current flows from voltage source to ground.	13, 17, 38, 56, 67, 78, 86, 117, 142, 150, 173, 211, 221, 252, 370, 415
Clearance	36, 153
Coaxial	13, 126, 180, 358, 399
Coding	261, 386
Compliant Press-in Pin:	
A solderless press-in termination between the specially formed tail of a contact and a plated through hole in a printed wiring board where deformation of the compliant pin section provides a residual force to create and maintain the integrity of the contact interface between the compliant pin and the plated through hole.	145
Conduction, electrical:	
The ability to carry a current.	

	Page
Conduction, thermal: The ability to dissipate heat from a higher temperature to a lower one.	
Conductivity, electrical: The ability to conduct current, the inverse of resistivity. Often expressed in terms of the percent conductivity of copper, percent IACS (International Annealed Copper Standard).	
Conductor: A component capable of carrying an electric current.	56, 66, 84, 85, 105, 118, 176, 177, 221, 266, 398
Conductor, solid: A single conductor	
Conductor, stranded: A bundle of individual (strands) conductors.	
Connector	12, 13, 15, 17, 18, 24, 31, 38, 53, 62, 70, 85, 86, 102, 118, 142, 146, 170, 177, 211, 214, 248, 249, 252, 370, 385, 393, 401, 415
Connector Mating/Unmating Force: The force required to mate/unmate the plug and receptacle halves of a connector couple.	56, 65, 75, 85, 148, 316, 147, 218
Connector Resistance	13, 57, 103, 214
Constriction Resistance	54, 63, 84, 85
Contact: The conducting part of a contact spring from the contact interface to the permanent connection.	12, 16, 17, 18, 25, 31, 38, 53, 62, 70, 85, 86, 102, 117, 142, 146, 168, 177, 211, 214, 249, 252, 370, 385

	Page
Contact Finish	12, 18, 33, 38, 56, 62, 70, 85, 93, 147, 172, 216, 249
Contact Finish Selection	12, 62
Contact Force	12, 18, 25, 31, 56, 64, 76, 86, 170, 179, 211, 215

Contact Mating Force:

The force required to mate a plug and receptacle contact pair.

Contact Normal Force:

The force created by deflection of the receptacle contact beam by the plug as the connector is mated. 12, 31, 50, 56, 74, 86, 221

Contact Lubricant:

An additive to improve the lubricity, wear resistance or corrosion resistance of a contact interface. 64, 82

Contact Interface Resistance 57, 73, 93, 221

Contact Resistance:

The resistance at the interface between the plug and receptacle contacts. 12, 33, 56, 66, 73, 93, 106, 122, 142, 173, 214, 266

Contact Resistance at Rated Current (CRRC) 218

Convection:

The transfer of heat by movement of air over the heated surfaces. Convection may be natural, based on temperature differentials, or forced, though the use of fans. 147

Corrosion:

Electrochemical deterioration of a contact spring due to the application environment. 13, 31, 41, 56, 62, 70, 85, 100, 136, 144, 178, 214, 251, 257

	Page
Corrosion Migration	64, 70, 231
Creepage	36
Crimp Barrel:	
An open or closed tubular barrel into which a solid or stranded conductor is inserted prior to crimping.	184
Crimp Height	189
Crimped Connections:	
A mechanical termination between conductor(s) and a specially designed crimp barrel to create and maintain contact interfaces between the conductors and the crimp barrel.	13, 92, 177
Crimping Force	189, 267
Crimping IDC	
Current:	
The flow of electrons through a circuit.	
Current Derating	
Current Rating:	
1. The process of determining the current which will produce a specified temperature rise under specified conditions. 2. The continuous current through a connector, under specified conditions, that produces a specified temperature rise at the hottest spot on the contacts.	173, 266
Dewetting:	
Withdrawal of solder from a previously solder coated surface during a soldering process.	147
Dielectric:	
An insulating material.	31, 180, 215, 266, 398

	Page
Dielectric Permittivity	398
Dielectric Strength:	
The maximum voltage a dielectric material can support, under specified conditions, without electrical breakdown.	31
Dielectric Withstanding Voltage:	
The maximum voltage applied under specified conditions without allowing a specified leakage current.	31, 215, 266
Dual beam contact:	
A dual spring receptacle contact system where mating occurs to two sides of the plug contact.	99
Duplex plating:	
Independent selective plating of a different plating on the mating area and permanent connection portion of a contact.	38
Durability:	
The number of mating/unmating, or fretting, cycles that a contact interface can support without exceeding specified variances in electrical and mechanical performance.	12, 33, 55, 64, 75, 85, 94, 172, 187, 214, 249, 258
Elastic Deformation:	
Deformation that is reversible when the load is removed.	99, 168
Elastic Modulus:	
The slope of a stress-strain curve obtained under tensile load prior to plastic deformation.	24, 48, 91, 205
Electrical Permittivity	397
Electrolyte:	
The conducting solution in an electroplating process	38

	Page
Electroplating (Electrodeposition):	
Deposition of a selected metal onto a substrate through electrochemical processes.	12, 38, 62, 71
Elongation:	
The increase in length of a sample under tensile loading.	48, 97, 209, 212
Engagement/Separation Force:	
The force required to engage/separate the plug and receptacle halves of a connector couple.	
Finish:	
1. Surface roughness. 2. A coating applied to contact surfaces to provide corrosion protection or improve contact interface performance	12, 18, 31, 38, 56, 62, 70, 85, 93, 147, 172, 216, 249, 291
Flange Mount Ability	388
Flash:	
A thin, electroplated layer.	29, 35, 42, 64, 72, 200, 231, 260
Force, contact:	
see Contact force	56
FPGA	411
Fretting:	
Small movements, of the order of microns, of a contact interface. Fretting motions may be due to mechanical drivers or differential thermal expansion effects.	12, 65, 70, 85, 94, 217, 251
Fretting Corrosion:	
The corrosion of wear debris produced by fretting motions.	65, 70, 85, 230, 251
Fretting Wear:	
The production of debris that occurs during fretting motions.	70, 251

	Page
Frequency Analysis	395
Friction:	
The force opposing the sliding motion of surfaces in contact due to an external force.	12, 55, 64, 74, 93, 171, 182, 257
Friction, coefficient of:	
The ratio of the friction force between two bodies and the force applied to the bodies.	
Gauge:	
A term describing the physical size of a wire. AWG and mm^2 are the two commonly used measurements.	177, 267
Gold Flash	42, 64, 72, 231
Gold Finish	64, 83
Ground:	
A common or reference point in an electrical circuit from which all voltages are measured.	13, 31, 118, 143, 180, 238, 272, 389, 393, 398, 410, 411
Gull-Wing:	
A lead configuration in surface mount components where the leads exit a component and are parallel to the printed wiring board to which the component will be soldered	148
Header:	
The plug portion of a two piece post/receptacle connector couple.	34, 67, 88, 210, 252, 370, 390, 411
High Current Contacts	65
High Frequency Signals	121, 358
HMI	314
IACS	45

	Page
Immersion Plating	43
Impedance	34, 117, 180, 283, 382, 395, 405, 411
Impedance Matching	118, 367, 397, 410
Insulation Displacement Termination/Connection (IDT/IDC)	
A mechanical termination technique in which a wire is forced into a specially designed slot in a terminal to displace the wire insulation and make an electrical contact to the conductor.	
Injection Molding:	
A manufacturing process in which liquid thermoplastic polymer is injected into a heated mold and cooled into a desired part configuration.	20, 35
Integrated Magnetics	305, 386
Interface	12, 13, 16, 17, 18, 29, 32, 38, 53, 62, 70, 85, 86, 118, 168, 187, 211, 215, 251, 257, 384
Intermetallic Compound:	
A compound of two metals, e.g. Cu3Sn4.	49, 66, 147, 216
I/O Interface – Ethernet PoE	387
Jacket:	
Outermost layer of insulating material of a cable.	177, 369
J-Lead:	
A lead configuration in surface mount components where the lead exits a component and is formed into a reverse "j" shape with the base of the "j" contacting the solder pad.	144, 148
Joule Heating	17, 35, 66, 99, 251
Keying	295, 389

	Page
Land:	
A metal coating on a printed circuit board to allow for mechanical or soldered contact.	142, 147
LCD	281, 395
LED	261, 268, 301, 393
Levels of Interconnection (LOI)	15, 248, 249
Line Impedance	283, 395
Loss	44, 56, 82, 94, 106, 133, 225, 251, 297, 392, 404
Low Level Contact Resistance (LLCR)	218
Mass Termination:	
Simultaneous mechanical termination of multiple conductors, usually by Insulation Displacement.	168
Mating Cycle	68, 70, 91, 346
Mating Force	56, 62, 82, 91, 215, 251, 364
Measurement Sequence	223
Mer:	
The repeating unit in a polymer which determines the characteristics of the polymer.	20
Micro USB 2.0	393
MiniUSB	393
Mixed Flowing Gas	73, 215
Modular Jack	281, 386

	Page
Nickel Finish	77
Nickel Underplate	12, 41, 63, 70, 148
Noble Metal: A metal not susceptible to corrosion, e.g. gold.	12, 62, 70, 215, 251
Ohm's Law: V = IR where V is voltage (volts), I is current (amperes) and R is resistance (ohms).	58, 223
Operating Temperature: The temperature at which a device is designed to operate at its rated current and voltage, usually specified as a maximum and minimum temperature range.	
Palladium Alloy	107, 214
Panel Mounting	264, 393
Passive Pore	72
PCB	13, 31, 81, 87, 103, 119, 142, 146, 168, 211, 223, 253, 370, 382, 385, 392, 402, 411, 415
Permanent Connection: A connection that is intended to be made only once.	13, 16, 31, 56, 87, 142, 148, 172, 177, 220, 250, 252
PET	29, 398
Pitch: The distance between contact positions in a connector, see centerline spacing.	68, 147, 159, 178, 253, 385, 395

4.1 Glossary

	Page
Plated-Through-Hole (PTH):	
A hole drilled through a printed wiring board that is subsequently plated with a metal to allow an electrical connection to be made to circuitry on, or in, the printed wiring board.	142, 146, 168
Plating:	
The application of a thin coating of metal onto a metallic substrate to improve the performance of the base metal.	18, 38, 63, 71, 85, 87, 144, 146, 172, 178, 211, 215, 257
Plug:	
The portion of a contact/connector that is inserted into the receptacle portion of the contact/connector.	15, 18, 33, 38, 53, 86, 128, 195, 220, 265, 415
PoE	305, 387
Polymer:	
A material consisting of many (poly) repeating units (mers).	12, 20, 31, 84, 152, 181, 219
Polymer, amorphous:	
A polymer where the arrangement of polymer chains is random.	
Polymer, crystalline:	
A polymer with regions of short range ordering of polymer chains in an amorphous polymer matrix.	
Polymer, Liquid Crystal (LCP):	
A polymer that contains short range ordered regions in the liquid state and maintains or increases that ordering when cooled.	
Polymer, thermoplastic:	
A polymer that can be heated into a liquid state repetitively.	
Polymer, thermosetting:	
A polymer that sets when heated.	
Pore Corrosion	64, 70, 231

	Page
Printed Circuit Board (PCB):	
An insulating board, usually fiberglass-epoxy, with selective circuitry on one or both sides of the board. Most PCB are multi-layered laminates.	87, 150, 370
Pull Strength	187
Rated Temperature:	
The maximum application temperature at which a connector can operate for an extended period of time and maintain its performance specifications.	
Rated Voltage:	
The maximum voltage at which a connector can operate for an extended period and maintain its performance specifications.	
Receptacle:	
The portion of a contact/connector that receives the plug portion of the contact/connector.	15, 18, 32, 38, 53, 86, 171, 220, 253
Reflection	133, 136, 314, 358, 396, 405
Reflectometry	395
Reflow Soldering:	
The process of taking a printed wiring board that has been prepared with solder paste and loaded with components and heating it to a temperature sufficient to reflow the solder thus soldering all the components to the board. All components being heated at a high temperature.	13, 87, 146, 348, 379
Residual Force:	
The force remaining in a deflected beam after elastic recovery.	168, 190
Resistance:	
Opposition to the flow of current through a circuit or component.	

	Page
Resistor	12, 13, 17, 29, 31, 45, 54, 63, 73, 85, 93, 102, 122, 142, 170, 180, 212, 214, 258, 407, 413
Ribbon Cable:	
A cable where wires/conductors are constrained into a planar configuration.	180, 252, 382, 391, 407, 411, 419
Rigid pin	145, 168
Robustness	251, 258, 388
Safety	28, 35, 110, 174, 214, 277
Selective Plating:	
The process of depositing a plating only on a particular functional area of a contact, e.g. at the mating interface or solder tail.	38, 259
Separable Connection:	
A connection that is intended to be made and unmade multiple times.	12, 16, 17, 31, 38, 86, 142, 171, 207, 250
Serrations	186
Shielded FFC	399
Shielding	13, 31, 51, 118, 265, 399
Signal Losses	297, 395
Silver Finish	68, 84, 85
Single Beam Contact:	
A single spring receptacle contact which mates to only one side of the plug contact.	226

	Page
SMT	29, 34, 88, 122, 142, 147, 168, 253, 370,391, 393, 395, 411
Solder:	
In connectors a tin alloy that melts at a relatively low temperature and bonds two metal components together by forming intermetallic compounds at the solder/component interfaces.	44, 87, 144, 146, 173, 220, 264, 379, 412, 417
Soldering	13, 29, 35, 38, 87, 106, 142, 146, 168, 211, 253, 379, 401, 416
Solder Paste:	
A mixture of solder particles, flux and binders that is screen printed onto a printed circuit board prior to reflow soldering.	150
Split Beam Contact:	
A spring contact that has been split lengthwise to provide two independent spring contacts.	
Strain Relief:	
Specially designed features on a contact or connector housing to minimize the transmission of mechanical stresses from a wire or cable to a contact or connector.	36, 186, 251, 253
Strand:	
A single conductor.	23, 130, 177, 252
Stress Relaxation:	
In a connector, a time dependent loss in contact force as a function of time and temperature.	25, 49, 99, 215, 251
Strip Length:	
The length of insulation that must be removed from a wire prior to crimping.	186
Strip Plating	12, 40

	Page
Surface Mount Technology (SMT):	
The electrical/mechanical connection of components to conductive patterns on the surface of a printed wiring board.	29, 142, 147, 272
Termination:	
1. A permanent connection. 2. The portion of a contact to which a permanent connection will be made.	38, 135, 148, 168, 178, 251, 269, 401
Termination Resistance:	
The resistance of a permanent connection.	
Time Domain	395
Tin finish	38, 56, 65, 70, 85, 251
Through-Hole Mounting:	
Connection of components to a printed wiring board using holes drilled through the board.	
Transmission	17, 119, 180, 258, 382, 395, 408, 412
TVS Diode	122, 393
Unbalanced	118, 398
USB	121, 248, 253, 374, 382, 392, 405, 412
USB 2.0	253, 392, 414
USB 2.0 Type A	393
USB 3.0	338, 392, 412
USB 3.0 Type A	393
USB 3.0 Type B	393

	Page
USB Type B	393
Vias	129, 391, 412
Voltage Breakdown	35, 182
Wave Soldering: A soldering process where the components to be soldered pass through a wave of liquid solder. The solder wets the solderable portion of the component or component leads to form the soldered connection.	13, 142, 146, 253
Wear: Loss of material on one, or both, components which are in contact under relative motion.	12, 54, 62, 70, 91, 172, 187, 215, 251, 351
Wear, adhesive: Wear processes resulting in significant wear debris.	55, 94
Wear, burnishing: Wear processes resulting in surface smoothing with little wear debris.	56
Wetting: In soldering, the ability of liquid solder to dissolve some of the solderable surfaces and form intermetallic compounds on cooling to create the desired electrical and mechanical bonds.	69, 146, 222
Whiskers	66
Wipe: The relative motion that occurs between two surfaces as the slide against one another	32, 187
Wire: A conductor plus insulation, if any.	14, 18, 57, 88, 104, 118, 142, 177, 222, 250, 252, 382, 385, 405

	Page
Yield Strength:	
The load applied to produce a specified amount of permanent deformation under a tensile load. In connectors a 0.2 percent offset is generally specified.	24, 48, 53, 91, 190
Young's modulus:	
See Elastic Modulus.	190
ZIF Connector	32, 317, 395

4.1 Glossary